Vitellogenin in Fishes- Diversification, Biological Properties, and Future Perspectives

Vaseeharan Baskaralingam
Rapeepun Vanichviriyakit
Editors

Vitellogenin in Fishes- Diversification, Biological Properties, and Future Perspectives

Springer

Editors
Vaseeharan Baskaralingam
Department of Animal Health
and Management
Alagappa University
Karaikudi, Tamil Nadu, India

Rapeepun Vanichviriyakit
Anatomy Department
Mahidol University
Bangkok, Thailand

ISBN 978-981-99-5339-4 ISBN 978-981-99-5340-0 (eBook)
https://doi.org/10.1007/978-981-99-5340-0

This Springer imprint is published by the registered company Springer Nature Singapore Pte Ltd.
The registered company address is: 152 Beach Road, #21-01/04 Gateway East, Singapore 189721, Singapore

This book is dedicated, with great reluctance, to everyone who stood beside me during the writing of this book, holding my hand as I went along; in front of me in case I just wanted it over too quick; behind me, pushing me forward with good ideas; and sometimes on top of me to pound something into my head; under me when I needed a cushion to lay upon.

—Vaseeharan Baskaralingam

Foreword I

The opportunity to write the foreword for this book fills me with joy and honor. The primary goal of publishing *Vitellogenin in Fishes: Diversification, Biological Properties, and Future Perspectives* is to disseminate the undersigned's expertise to those who are interested. The goal of writing this book on this issue is multifaceted. First and foremost, I would like to extend my sincere congratulations to the Publishers and Editors, Prof. B. Vaseeharan and Dr. Rapeepun Vanichviriyakit, for taking the initiative to compile for the first time within these pages all baseline to advanced knowledge about vitellogenin in fishes and expertise gathered from across the world on the depth of the history of vitellogenin, Previtellogenesis, their diversification and classification in fishes, and multiple vitellogenin systems in fish.

This work on *Vitellogenin in Fishes: Diversification, Biological Properties, and Future Perspectives* has been authored by top experts in the subject and provides a complete account of the most current findings. The evolution of vitellogenesis and its ecological significance, as well as the molecular and cellular mechanisms that regulate vitellogenin synthesis and transport, are only a few of the many topics covered in the book. The contributors of this book highlight the important issues and concepts that continue to be addressed and also provide insight into the most recent study conclusions.

The goal of this book is to inspire upcoming researchers and spread knowledge and findings to a wider audience. Vitellogenin (Vtg), an estrogen-inducible precursor protein of the egg yolk, serves as a sign of exposure to substances in the environment that have estrogenic properties. It is anticipated that the main egg yolk precursor protein, vitellogenin (Vtg), will provide growing embryos and larvae with a protein- and lipid-rich diet. However, the roles played by Vtg and its progeny, the yolk proteins lipovitellin (Lv) and phosvitin (Pv), go beyond merely nutritive ones.

The Vtg and its involvement in the host's innate immune defence are illustrated in this book through a number of roles. This book goes into greater detail regarding the several vitellogenin genes, each having a unique promoter region and a variable susceptibility to estradiol induction, as well as the numerous Vtg proteins themselves, each with different levels of posttranslational modification. This might provide the authors with new opportunities, skill sets, and, in some cases, a second career. This book also includes all the most recent research investigations on fish vitellogenin as well as cutting-edge technologies and research areas that still need to be investigated to further our understanding of this substance.

Former Vice Chancellor, Alagappa University P. Ramasamy
Karaikudi, Tamil Nadu, India

Foreword II

Dr. Baskaralingam Vaseeharan is a microbiologist who has devoted his research to gaining a deeper understanding of methods to control pathogens in aquaculture. He has contributed to the embryology field through his research of vitellogenin. He has complied his findings to advance research in the areas of fish reproduction and reproductive biology.

The word "vitellogenin" comes from the Latin word vitellus, which is also known as yolk nutrient protein. Vitellogenin is a glycolipophosphoprotein, which is considered to be a significant contributor to egg yolk protein. It is primarily found in all oviparous species, including fish, amphibians, reptiles, birds, invertebrates, and select mammals. Oviparous animals are classified by their ability to lay eggs from which the embryo develops and receives nutrients. The vitellogenins are synthesized and secreted by the liver, transported in the blood to the ovary, and taken up by growing oocytes. Vitellogenin is a significant source of nutrients during the early stages of development for egg-laying vertebrates and invertebrates. Recent studies have shown that vitellogenin has a broader array of functions in the body and is a crucial embryonic nutrient that all fish species produce. Dr. V. Baskaralingam's book is a compilation of his vast research experience in this area and is the first to discuss vitellogenin in fish, covering a variety of topics, including its history, functions, and more.

Vitellogenins are proteins ranging from 350 to 600 kDa. Alanine is one of the most abundant amino acids comprising vitellogenin's polypeptides. Some fish species contain polyalanine regions suggesting that alanine might play an essential role as an intermediate compound in carbohydrate metabolism. The common abbreviations for vitellogenin are Vg, VG, and Vtg, which are members of the large lipid transfer protein superfamily. Fish vitellogenins have complex structures and functions. Fishes, amphibians, and avian species have two forms of vitellogenins, VtgAB and VtgCD. Vitellogenins in the blood can serve as markers for the onset of puberty and the progression of gonad maturation in female fishes. Vitellogenins are involved in carrying ions, such as calcium, magnesium, iron, zinc, copper, and various minerals and vitamins, such as retinoids and carotenoids. They transport regulatory compounds in fish egg yolk, including lipid-soluble steroids and thyroid hormones. Vitellogenins play a role in depositing phospholipids in freshwater and marine fish, but the form of phospholipids differs. In teleost fish, it deposits neutral lipids such as triacylglycerides and wax or steryl esters. Other functions of vitellogenins in fish include hemagglutination and acting as antimicrobial agents, working to protect the host from certain bacterial and viral pathogens.

The chapters of this book have been organized systematically and comprehensively to advance the readers knowledge about vitellogenin in fish. The aspects of vitellogenin that the book will explore include vitellogenesis, its classifications, multiple vitellogenin systems, tools for identification and characterization, the functional and regulatory mechanisms, its immunocompetent activity, different concentration levels and the potential disruption to the endocrine system, as well as the application of vitellogenin as a biomarker. A complete understanding of the mechanism and function of vitellogenin in fish will open new avenues in the field of biotechnology and developmental biology to deliver targeted nutrients, minerals, or drugs to developing embryos. In this aspect, the editor and the authors of the first edition of this book have done an incredible job of providing valuable information to the readers.

United States Department of Agriculture (USDA-ARS) Palmy Jesudhasan
Fayetteville, AR, USA

Contents

Editors and Contributors

About the Editors

Vaseeharan Baskaralingam is a Professor and the Head of the Department of Animal Health and Management at Alagappa University, Karaikudi, Tamil Nadu, India, since 2008. He previously served as a postdoctoral researcher at National Tsing Hua University and National Taiwan Ocean University in Keelung, Taiwan, funded by National Science Council, Taipei, Taiwan. Upon successfully completing his postdoctoral research (2003–2007), he joined as a Senior Lecturer at the School of Biotechnology at VIT University, Vellore (2007–2008), Tamil Nadu, India. He is also a Visiting Professor at Mahidol University, Thailand (2022). His research revolves around Aquatic Animal Health Biotechnology. He has been conferred with various prestigious awards, notably "YOUNG SCIENTIST AWARD" in 2009 by the Department of Science and Technology, the Dr. M. Swaminathan Best Fisheries Scientist Award 2019 by PFGF, ICAR-CIFRI, Barrackpore, the Best Scientist Award—*Journal of Fisheries and Life Sciences* by the College of Fisheries, Mangaluru, Karnataka, India, and the TANSA-2019 Award by Tamil Nadu State Council for Science and Technology, Tamil Nadu, India. He has more than 22 years of teaching and research experience. He has published more than 250 research articles in peer-reviewed national and international journals. He is a member of various international scientific bodies such as the Zoological Society, Kolkata, Bactivac network—Institute of Immunology and Immunotherapy, University of Birmingham, UK, *the Journal of Fisheries and Life Sciences*, the Indian Science Congress Association, and the Asian Fisheries Society.

Rapeepun Vanichviriyakit is the Director of the Center of Excellence for Shrimp Molecular Biology and Biotechnology (Centex SHRIMP) and an Associate Professor at the Department of Anatomy, Faculty of Science, Mahidol University, Thailand (2017–present). She graduated with her Ph.D. in Anatomy from Mahidol University, Thailand, in 2007. She gained research experience in reproductive biology for about 2 years (2006–2007) in Prof. Nongnuch Tanpaichitr's lab at Ottawa Health Research Institute (OHRI), Canada. After that, she began working in the fields of aquatic animal reproductive biology and disease-associated histopathology research. She has more than 15 years of teaching experience in anatomy, cell biology, and comparative histology. She became an Associate Professor at Mahidol University in 2019. She has served as a referee for several international journals, including *Aquaculture*, *Fish and Shellfish Immunology*, *Fish Diseases*, *Science Asia*, *Journal of Invertebrate Pathology*, *Disease of Aquatic animals*, etc. She is also on the Editorial Advisory Board of the *Journal of Fish Disease* (©John Wiley & Sons Ltd) (2021–present). She has published more than 50 research articles in peer-reviewed international journals.

Contributors

Muthukumar Abinaya Crustacean Molecular Biology and Genomics Division, Biomaterials and Biotechnology in Animal Health Lab, Department of Animal Health and Management, Alagappa University, Karaikudi, Tamil Nadu, India

Chinnah Amutha Department of Animal Behaviour and Physiology, School of Biological Sciences, Madurai Kamaraj University, Madurai, Tamil Nadu, India

Chellathangam Anitha Department of Zoology, Holy Cross College (Autonomous) (Affiliated to Manonmaniam Sundaranar University, Tirunelveli), Nagercoil, Tamil Nadu, India

Maharajan Athisuyambulingam PG & Research Department of Zoology, Khadir Mohideen College (Affiliated to Bharathidasan University, Tiruchirappalli), Adirampattinam, Tamil Nadu, India

Vaseeharan Baskaralingam Biomaterials and Biotechnology in Animal Health Lab, Department of Animal Health and Management, Alagappa University, Karaikudi, Tamil Nadu, India

M. Chellapackialakshmi Department of Zoology, Thiagarajar College, Madurai, Tamil Nadu, India

Josephine Priyatharshini Chellappa Department of Zoology, Holy Cross College (Autonomous) (Affiliated to Manonmaniam Sundaranar University, Tirunelveli), Nagercoil, Tamil Nadu, India

M. Devaprakash Department of Molecular Microbiology, School of Biotechnology, Madurai Kamaraj University, Madurai, Tamil Nadu, India

Dharshana Dhinesh Department of Animal Behaviour and Physiology, School of Biological Sciences, Madurai Kamaraj University, Madurai, Tamil Nadu, India

Brisca Renuga Ferdinand Department of Zoology, Holy Cross College (Autonomous) (Affiliated to Manonmaniam Sundaranar University, Tirunelveli), Nagercoil, Tamil Nadu, India

Periyasamy Gnanaprakasam Department of Chemistry, Saveetha School of Engineering, Saveetha Institute of Medical and Technical Sciences, Chennai, Tamil Nadu, India

Ardhra Gopan Department of Animal Behaviour and Physiology, School of Biological Sciences, Madurai Kamaraj University, Madurai, Tamil Nadu, India

Ramachandran Ishwarya Mandapam Regional Centre, ICAR—Central Marine Fisheries Research Institute, Mandapam, Tamil Nadu, India

Rengarajan Jayakumar ICAR—Central Institute of Brackishwater Aquaculture (CIBA), Chennai, Tamil Nadu, India

Jeyaraj Jeyavani Biomaterials and Biotechnology in Animal Health Lab, Department of Animal Health and Management, Alagappa University, Karaikudi, Tamil Nadu, India

Anjugam Mahalingam Department of Marine Science, School of Marine Sciences, Bharathidasan University, Thiruchirapalli, Tamil Nadu, India

Johnson Vinoliya Josephine Mary Department of Zoology, Holy Cross College (Autonomous) (Affiliated to Manonmaniam Sundaranar University, Tirunelveli), Nagercoil, Tamil Nadu, India

Basil Rose Michael Rajam Department of Zoology, Holy Cross College (Autonomous) (Affiliated to Manonmaniam Sundaranar University, Tirunelveli), Nagercoil, Tamil Nadu, India

C. Nija Department of Zoology, Women's Christian College, Nagercoil, Tamil Nadu, India

Jeni Chandar Padua Department of Zoology, Holy Cross College (Autonomous) (Affiliated to Manonmaniam Sundaranar University, Tirunelveli), Nagercoil, Tamil Nadu, India

Santhanam Perumal Department of Marine Science, School of Marine Sciences, Bharathidasan University, Thiruchirapalli, Tamil Nadu, India

V. Brindha Priyadarisini Clinical Biotechnology Lab, Department of Microbial Biotechnology, Bharathiar University, Coimbatore, Tamil Nadu, India

Amirtha Mani Punitha Department of Zoology, Holy Cross College (Autonomous) (Affiliated to Manonmaniam Sundaranar University, Tirunelveli), Nagercoil, Tamil Nadu, India

Arokya Glory Pushpa Thiraviam Department of Zoology, Holy Cross College (Autonomous) (Affiliated to Manonmaniam Sundaranar University, Tirunelveli), Nagercoil, Tamil Nadu, India

C. Ragunath Unit of Aquatic Biotechnology, Department of Zoology, Bharathiar University, Coimbatore, Tamil Nadu, India

V. Ramasubramanian Unit of Aquatic Biotechnology, Department of Zoology, Bharathiar University, Coimbatore, Tamil Nadu, India

C. Ravi Department of Zoology, Thiagarajar College, Madurai, Tamil Nadu, India

Perumal Santhanam Department of Marine Science, School of Marine Sciences, Bharathidasan University, Thiruchirapalli, Tamil Nadu, India

M.S. Shabana Unit of Aquatic Biotechnology, Department of Zoology, Bharathiar University, Coimbatore, Tamil Nadu, India

SavariyarAdimy Prakash Shoba Department of Zoology, Holy Cross College (Autonomous) (Affiliated to Manonmaniam Sundaranar University, Tirunelveli), Nagercoil, Tamil Nadu, India

Ashokkumar Sibiya Biomaterials and Biotechnology in Animal Health Lab, Department of Animal Health and Management, Alagappa University, Karaikudi, Tamil Nadu, India

Mary Mettilda Bai Silvester Department of Zoology, Holy Cross College (Autonomous) (Affiliated to Manonmaniam Sundaranar University, Tirunelveli), Nagercoil, Tamil Nadu, India

N. Sivakumar Department of Molecular Microbiology, School of Biotechnology, Madurai Kamaraj University, Madurai, Tamil Nadu, India

Arunthathi Shyla Suganthi Department of Zoology, Holy Cross College (Autonomous) (Affiliated to Manonmaniam Sundaranar University, Tirunelveli), Nagercoil, Tamil Nadu, India

Govindan Tamilmani Mandapam Regional Centre, ICAR—Central Marine Fisheries Research Institute, Mandapam, Tamil Nadu, India

R. Thirumalaivasn Department of Molecular Microbiology, School of Biotechnology, Madurai Kamaraj University, Madurai, Tamil Nadu, India

X. Venci Candida Department of Zoology, Holy Cross College (Autonomous) (Affiliated to Manonmaniam Sundaranar University, Tirunelveli), Nagercoil, Tamil Nadu, India

Viswanathan Vinotha Biomaterials and Biotechnology in Animal Health Lab, Department of Animal Health and Management, Alagappa University, Karaikudi, Tamil Nadu, India

Ganapiriya Viswambaran PG & Research Department of Zoology, Khadir Mohideen College (Affiliated to Bharathidasan University, Tiruchirappalli), Adirampattinam, Tamil Nadu, India

Vitellogenin: Outline and History

1

C. Nija

Abstract

Protein precursors called vitellogenins are a crucial part of the process that leads to the development of yolk. There is a tendency for organisms that relate to egg development both exogenously and endogenously to create vitellogenin in their bodies. In the research that we are working on right now, we talk about the vitellogenins that are found in fish. The proportion of vitellogenins that is more prevalent in female fish compared to that which is seen in male and juvenile fish. They are generated in the liver as a response to estrogen exposure and are then delivered to the oocytes for production of yolk. In addition to the estrogens, they are able to be synthesized when they are subjected to a contact with molecules that behave similarly to estrogen. Additionally, by ingesting the antigens that they come into contact with, they boost the immunogenicity. In older days, researchers looked at their responses to pollutants and industrial effluents in huge aquatic ecosystems to see how organisms reacted to those factors. They have a chemical composition that consists of amino acids, sugars, phosphates, and lipids, among other things, and the response of individual compounds to other resources, as well as their antimicrobial activity and antioxidant activity, set them apart from other organisms in the environment in which they are able to survive.

Keywords

Vitellogenin · Estrogen · Oocytes · Immunogenicity · Yolk

C. Nija (✉)
Department of Zoology, Women's Christian College, Nagercoil, Tamil Nadu, India

V. Baskaralingam, R. Vanichviriyakit (eds.), *Vitellogenin in Fishes-Diversification, Biological Properties, and Future Perspectives*,
https://doi.org/10.1007/978-981-99-5340-0_1

1.1 Introduction

1.1.1 History and Evolution

Earlier in 1954 and 1965, the composition of proteins in the blood samples of mature female insects varied with the protein fractions of immature females and male insects. This was steadily connected with the proteins involved in maturity and vitellogenins. Observance of similar pattern of protein differentiation in amphibians and avian species proved that the proteins correspond with the vitellogenesis are the reason behind this. Since the unknown protein involved in the vitellogenesis had a major breakthrough, it was named as vitellogenins. Also, the site of synthesis of vitellogenins was intractable in spite of the studies done in insects and birds with ovary ablation and low involvement of fat bodies from tissues that was believed to be in charge of the formation of yolk. Later by series of experiments in few insects, the vitellogenin was precipitated and separated (Pan et al. 1969).

A decade later, the involvement of endocrine gland and juvenile hormones in silkworm was examined to know further about the vitellogenesis and inferred that they were not much needed for the formation of egg yolk (Pan 1977). This coincides with the experiment undertook earlier in the same species at two different stages (late pupal stage and yolk formation stage) that showed an increase and decrease of vitellogenin accordingly (Pan 1971). To understand the proteins of one organism, it is noteworthy to study another control organism that shares structural or functional characteristics. In the case of *Xenopus* and *Cecropia*, although they do not have anything in like manner, the latter was utilized to study the protein differentiation of the former. Vitellogenin of both species have similar chemical and physical properties no matter what their phenotype is (Pan and Wallace 1974). In fishes, the vitellogenins was identified and isolated in different timeline and are tabulated in Table 1.1.

More than 34,800 species of fishes exist in the world at present, and they impart a prime contribution in the ecosystem. Bionomics of fish helps in importing benefit to human and their diet. Nutriments of fish are widely spoken as they have essential

Table 1.1 Timeline and isolated vitellogenins from fishes

Year		
1961	Vanstone WE, Ho CW	Coho salmon
1962	Ridgeway	Catfish
1967	Thurston	Eel
1971	Plack P. A., Pritchard D. J. and Fraser N. W	Cat shark
1976	Emmerson and Peterson	Sockeye salmon
1977	Le Menn and Lamy	Female catfish
1978	Hara A, Hirai H	Rainbow trout, *S. gardneri*
1979	Korsgaard, B. and Peterson, I.	Viviparous eelpout
1980	Hara	Japanese eel
1980	Campbell C M, Idler D R	Juvenile rainbow trout

constituents that leads to good health and well-being. Fish eggs also known as fish roe are rich in fat-soluble vitamin D that goes along with omega 3 fatty acids so that the vitamins are readily absorbed by the body with the influence of the latter. Simultaneously vitamin B12 and other minerals in fish roe assist in metabolizing the food to provide energy and maintain good bone and dental health, balancing the transport of body fluids in and out of the cells and body. They have a special role in synthesizing collagen. Egg yolk, being the hefty proportion, attributes utmost sustenance and adds value to the whole. Formation of egg yolk, called vitellogenesis, is a crucial process through which researchers came to know about the protein vitellogenin.

The origination of the term "vitellogenin" began earlier in the twentieth century in the name of antigens. Later on, several experiments were achieved; thereby a gradual understanding of vitellogenins got off the ground. Vitellogenins are hormones that are produced by the liver as a result of exposure to environmental estrogens. It is then moved to the developing oocytes and becomes a part of the egg yolk (Engelmann 1979). Most of the oviparous animals produce vitellogenin. Fishes produce vitellogenins, and they have a major role in imparting essential nutrients such as protein and lipids to the egg and larvae that leads to its sustainability. They transport plasma lipids and also cumulate neutral lipids during the growth stage. Different types of vitellogenins are present in different fishes so as their structure and functions. The presence of vitellogenins were observed throughout the life cycle of fishes. In the very first egg stage, they are seen as a precursor of egg yolk. Eggs, the final product of oocyte growth and differentiation, develop into viable embryos after fertilization.

1.2 Vitellogenins in Male Fish

Generally, females produce vitellogenins in numerous species, while in contrast, male species of fishes produce vitellogenins in comparably low quantities. When male fishes are injected with estrogen intraperitoneally or through food or water, they produce more proteins with the same count of blood cells as that of the females. The estimation of vitellogenins in male fish can be confused with the greater number of antigenic sites as they were tested for enzyme linked immunosorbent assay (ELISA). Vitellogenins are easily breakable compound that can lead to rapid proteolytic distribution and may have increased number of binding sites. This causes the radioisotope-labelled protein to get bound with the inaccurate results. Also, the heterologous ELISA determines vitellogenins low when compared to the homologous ELISA.

The dosage of estrogens and xenoestrogens is not proportional to the amount of vitellogenins produced from the body of male fish. It was investigated with aquatic systems with profound supply of estrogens, continuous supply of chemicals that mimics estrogens, and also by the exposure of male fish toward the estrogens that are secreted and liberated in the water bodies. Fish food that are rich in estrogens are being imported in the water tanks. Commercially available fish foods originated from the fish viscera and steroids are major sources of estrogens and are added to the fish tanks that contains male fish to denote the reaction. Plant-based estrogens such as genistein and daidzein are included in making fish food specifically. Physical interactions of the proteins were analyzed by in vitro yeast 2 hybridization technique that proved the interconnection between the phytoestrogen and consequent activity of the hormones. Two growth conditions were set one with only male Fathead minnows and the other with male and female fishes. They were allowed to grow in optimal conditions with recirculating aqua pump system. Steroid estrogens and charcoal were revived every 14 days interval. Regular sampling was done in a period of 6 and 12 weeks to check the vitellogenin secretion. It was observed that male fishes fed commercial food and exhibited high levels of vitellogenin concentration, thereby maintaining it in the ecosystem. Fishes that are unexposed to the chemicals and food with estrogen had a decline in the generation of vitellogenins after particular extent. Also, there were immense quantities of vitellogenins in tanks that accommodate both sexes of fish (Beresford et al. 2011).

The molecules correspond to the vitellogenesis are deposited in the oocyte during its development. It is a complex process in which varied hormones interact with the principal component to effectuate the biochemical activity to occur (Peter and Crim 1979).

Vitellogenin gene duplication occurs that makes a great difference between the vertebrate and invertebrate. Although they share a common ancestor, there were lamprey and gnathostome ancestors. A loss of vitellogenin gene can be seen in mammals, platypus and frogs. Alternatively, the complete gene pattern enters in to the invertebral species especially cartilagenous fishes after duplication. This reoccurs to give rise to teleost population comprising of zebra fish, eel, acanthomorpha (teleost with spiny rays), and Atlantic salmon. It is clearly understood from the phylogenetic analysis that the vitellogenins from vertebrate and non-vertebrate share common ancestral genes. They underwent a series of duplication in a way that homologous genes are placed in the chromosomes of different species which imparts different structure and more or less similar functions to the lineages (Biscotti et al. 2018).

They belong to the large lipid transfer protein superfamily. It is a large apolipoprotein with molecular weight that varies from 280 to 550 kD. In 1971, coho salmon and cod were discovered with 390 and 400 kD; 550 kD of protein was obtained from Flounder and Cat fish in 1976 and 1981, respectively. Protein with molecular weight of 470 kD was identified in trout; approximately 280 kD of vitellogenin was derived from gold fish in 1980. Each individual subunit varies in molecular weight within the species. Eighty-five kD subunit of Japanese eel excludes from other teleost with molecular weight of subunits ranging from 10 to 220 kD. Vitellogenin has a heavy chain lipovitellin with four subdomains N sheet, C sheet, alpha helix and A sheet, phosvitin, lipovitellin, a von Willebrand Factor type D domain (vWFD) and C terminal coding region (Carducci et al. 2019). In the heavy chain lipovitellin, the hydrophobic constituents form the secondary and tertiary structures. Each subdomain embraces distinct receptors in which each binding particles can attach. N sheet has a receptor that binds with the oocytes, a binding site for zinc ions is located in the alpha helix, and an alanine-rich sequence resides on the A sheets which helps in embryo gluconeogenesis. Another functionality of the vitellogenin is the presence of serine-rich phosvitin domain that is capable of binding phosphates; in turn it attracts multivalent cations such as calcium, iron, titanium, magnesium, and others. Also, it possesses binding sites for carbohydrates that aids in glycosylation. The glycosylation sites after binding the carbohydrates and the metal ions together enhance the solubility of vitellogenins in marine environment. The light chain lipovitellin contains glycosylation sites to carry lipids. Cysteine residues generally play a significant role in the formation of tertiary structure of proteins. Here, vWFD region have conserved sequence of cysteine that supports the formation of dimers by folding. Fresh water marine organisms especially fishes mostly lack the availability of these ions in their habitat. Homologous genes from apolipophorins undergone certain changes such as transport of hydrophobic molecules to transform a new vitellogenin gene. Vitellogenins possess strong hydrophobic nature, and the fatty acid chain plays a crucial role in maintain the structure of lipoproteins.

Gene isoforms of vitellogenins vary with species so as their rapidity in yolk deposition. Molecular composition and functions vary from species to species and within the species. They possess different subunits, molecular weight, and the duration of protein synthesis and lipid formation. Vertebral and non-vertebral genes corresponding to vitellogenin had five conserved motifs (Chen et al. 1997), and it was further assured (Baker 1988). Ray-finned fish exhibits multiple genes that code for vitellogenin (Buisine et al. 2002). Vitellogenin A and B are the two paralogous genes seen in salmonid species. Two to 30 copy numbers were established when the genes were hybridized. Teleost fish from the taxon Acanthomorpha possess vitellogenin A and B that aid in egg buoyancy so as to maximize their population despite other factors (Matsubara et al. 2003). Similar set of seven multigenes—vitellogenin 1–7, were explored in zebra fish. Vitellogenins 1 and 3 came up with full set of sequences and also without phosvitin. It was concluded that phosvitin less vitellogenin was transitional between invertebrate and vertebrate vitellogenins (Wang et al. 2000).

Although there were no corroborations stating that the juvenile fish produces vitellogenins naturally, estrogen or substances that have the similar functions and receptors as estrogens were induced to stimulate the productivity of vitellogenins. They stick to the same mechanism as the other proteins does. Experiments says that vitellogenins obtained from the male cat fish was impregnated into the juvenile fish that leads to the gain in body weight, and also it sets off gametogenesis regardless of its gender. Immature trouts and carps were used for effluent studies which proved that there was presence of vitellogenins in the trials. Based on the evolutionary history, there were vitellogenins type A, B, C, and D, whereas two of these types are extinct and others are seen in the descendants of some fishes such as *Veraspermoseri*, *Ichthyomyzon unicuspis*, and teleosts. With regard to the existence of domains and subdomains, the vitellogenins are classified as complete and incomplete vitellogenins. Complete vitellogenins comprises all domains and chains, whereas the incomplete has nonexistence of phosvitin. Depending on the molecular weight, they are classified as vitellogenin 1, 2, and 3. Of these vitellogenin 1 and 2 possess relatively high molecular weight rather than vitellogenin 3 which is smaller.

1.3 Chemical Composition of Vitellogenin

As the name suggests, it is a protein; its major percentage of constituents is protein. This is common for all vertebrates and invertebrates. Lipids, phospholipids, and carbohydrates contribute less rather than the proteins. The chemical composition of trout vitellogenin is tabulated in Table 1.2.

Phospholipids contributes 13.3% of its total composition through which the individual class of phospholipids were estimated. The phospholipids composition is tabulated in Table 1.3.

Vitellogenins occurs both in vivo and in vitro claims a sequence of functions to be thoroughly understood.

1. Isolation of vitellogenins
2. Quantification of vitellogenins
3. Transport and uptake of proteins
4. Localization of proteins

1.3.1 Isolation of Vitellogenins

Different isolation techniques were adopted in different species except the basic steps. Fishes were reinjected with orthophosphate in one group and leucine in another group. The blood was then collected followed by the addition of EDTA with calcium or magnesium ions. It was then centrifuged and the supernatants were collected to proceed with chromatography that separates plasma proteins—phosvitin and lipovitellin. In the case of ovaries, they were extracted and treated with inorganic salt solution. It varies for different species just as sodium chloride is used for major species (trout, salmon, and cod); with some exceptions as in herring, it is magnesium sulfate. Precipitating it with ammonium phosphate which made the lipovitellins to dissolve. Resolving this by adding Tris-Cl finally eluted the proteins in different gradients which was taken for electrophoresis that showed bands of increasing mobility (De Vlaming et al. 1980). The basic steps involved in the protein isolation of amphibian is ineffectual in the case of gold fish, wherefore

Table 1.2 The chemical composition of trout vitellogenin

Component	Weight in percentage
Protein	79
Lipid	19
Phospholipid	13.3
Triglyceride	4.2
Cholesterol	1.5
Carbohydrate	0.3
Phosphoprotein	0.7
Calcium	0.7

Table 1.3 The phospholipids composition

Phospholipids	Weight in percentage
Phosphatidylcholine	83.2
Lysophosphatidylcholine	6.2
Phosphatidylethanolamine	3.8
Phosphatidylserine	3.2
Phosphatidylinositol	0.5
Sphingomyelin	3.1

chromatography was attempted to obtain the protein from Atlantic salmon, Asian sea bass, rainbow trout, and Medaka. Besides the isolation of protein, there were evidences showing that the instability of the same which was overcome by altering the presence and amount of certain protein and enzymes in due course (Silversand et al. 1993). The pooled serum, liver, and gonadal homogenate from E2 stimulated male zebrafish displayed a single peak on the chromatographic elution profile obtained using the Mono-Q column. This peak eluted at the right chloride ion concentration. The peak that was created was not present in the extracts that were obtained from unstimulated male and female fish; nevertheless, it was present in the samples that were collected from female fish that had been stimulated with estradiol.

The elution profile of extracts of the E2 stimulated female gonad, on the other hand, showed two peaks: one peak that eluted at the same chloride ion concentration as the major peak of serum and liver extract and an additional peak that eluted at a lesser chloride ion concentration. Both of these peaks were similar to the major peak of serum and liver extract. There was only a single peak seen on the chromatographic profile of homogenates that were generated from the gonads of untreated female fish. This peak eluted at a low chloride ion concentration, and there was no indication that there was another peak at a higher concentration.

1.3.2 Quantification of Vitellogenins

In earlier days, the comparison of serum proteins and phosphorus of teleost were performed to estimate and quantify vitellogenins that had its disadvantages as the organisms without protein phosphorus were lacking. Later in 1971, indirect semi-quantitative immunodiffusion was done to compute the percentage of vitellogenin. It was followed by the emergence of quantitative electrophoresis of *Esox lucius* in 1974. The drawbacks were mastered by another technique called densitometric polyacrylamide gel electrophoresis in rainbow trout and gold fish in 1979 and 1981. Rainbow trout proteins were quantitatively estimated in 1990 by single radial immunodiffusion method also. Radio immunoassays with radioisotope-labelled protein experienced instability, and so the procedure was altered, and a novel approach, ELISA, was introduced that masks the demerits caused by other previous techniques.

1.3.2.1 Synthesis and Expression of Vitellogenins in Fishes

In early years it was believed that the vitellogenins were synthesized only in female liver, and male fishes produced vitellogenins when they expose to estrogens. Expressions of vitellogenin genes in both males and female *Pimephales promelas*, *Fundulus heteroclitus*, and *Gambusia holbrooki* in different body fluids like plasma and mucus were separated using sodium dodecyl sulfate-polyacrylamide gel electrophoresis (Van Veld et al. 2005). The vitellogenins, when detected at the beginning, was below the normal percentage. After the treatment of diethylstilboestrol, the amount of vitellogenins raised for about 3 days and return to the normal level in the next 6 days (Williams et al. 1979). Female and male zebra fish were selected for the synthesis study of vitellogenins. When estradiol-β17 is induced to male and female zebra fish, it produced more vitellogenins. Recurrent dosages of the same hormone led to the production of more proteins of which serum proteins are more constituting about 68%, hepatoproteins 40%, and gonadotrophins 46%. The proteins produced in the liver has taken long duration as compared with the others.

Steroid-induced protein synthesis is common, and it occurs in the case of vitellogenin with not an exception. The size of oocytes favors an additional criterion to make the experiment in maximum success rate. 17cr-hydroxy-20p-dihydro-4-pregnene-3-one is a potent steroid which was used to promote the synthesis of vitellogenin. Fishes with GonadoSomatic index of 16% and more has the caliber to produce more oocytes.

1.3.2.2 Lipidation

Lipidation is a mandatory process by which the targeted proteins get modulated to increase their binding affinities. Female species of cod, turbot, and wolfish have greater relative fatty acid percentage (Silversand and Haux 1995). Saturated, monounsaturated, and polyunsaturated fatty acids of vitellogenins are indistinguishable in proportion regardless of their functions, structure, and nature. This enables the vitellogenin to go through lipidation that alters the affinities of proteins and similar compounds on the biological membrane of surrounding cells.

1.3.2.3 Glycosylation

Plecoglossus altivelis, *Tribolodon hakonensis*, and *Oryzias latipes* were taken for studying the glycosylation in vitellogenins. Unfertilized eggs possess free oligosaccharides at first and later on elucidating the three species; it incorporates glycophosphoproteins in elevated levels with N-linked glycan units. This proves that the glycosylation happened in the vitellogenins (Iwasaki et al. 1992).

1.3.2.4 Transfer and Uptake of Proteins to Oocytes

In the beginning, rainbow trout were used as a model organism for conducting in-depth investigations into the mechanism of vitellogenin recognition and the selectivity of its uptake in fish oocytes. These studies demonstrated that developing oocytes took up labelled vitellogenin in vitro at a rate that was slower than that of *Xenopus*. Furthermore, there was no evidence of selective uptake over serum albumin, which was observed in *Xenopus* under conditions that were comparable to those used in these studies.

Vitellogenins are transported via bloodstream in both exogenous and endogenous vitellogenesis. Fluorescein isothiocyanate (FITC)-conjugated vitellogenin protein was injected in the female gold fish to check the journey and after effects of the same. FITC conjugated protein was taken by the blood to the ovary where it gets involved in the synthesis of yolk and yolk proteins. This confirms that the proteins along with the carbohydrates are transferred to the embryo through the body fluids by means of trophotaeniae (Iida et al. 2019). Generally, the vitellogenins are taken up by the oocytes from the bloodstream readily. If the ovary is absent, the vitellogenins accumulate in the bloodstream which is taken by the liver and degraded by it. Phosphate groups in the vitellogenins have a wide role in permeating itself in to the cell membrane of oocytes.

In trout, a series of studies were performed which states that receptor-mediated endocytosis leads to the uptake of vitellogenins by oocytes. Compared with the bovine serum albumin and other proteins, the time taken for engulfing vitellogenins by oocytes is 60% more. The temperature, the size of oocytes, and the growth phase all together are responsible for the intake of vitellogenins. Higher temperature paves way for the generation of more vitellogenins. Intravenous administration of tagged vitellogenin or macromolecular tracers into a number of oviparous species has proven that such materials freely penetrate throughout the follicle, predominantly through an intercellular route. The ovarian matrix is enclosed by a squamous and largely impermeable ovarian epithelium.

Plasma proteins that are typically present in the blood stream and heterologous macromolecules are taken up by the oocytes, but vitellogenin is taken up by the oocytes in a selective manner. The phosphate content of vitellogenin is only about half of what it is in other vertebrates in the fish species that have been studied up to this point, which raises questions about the involvement of phosphate groups in vitellogenin uptake in fish and encourages further comparative studies, which could lead to interesting insights into recognition and receptor mechanism in general.

1.3.2.5 Changes in Liver Cells and Its Organelles

The liver has numerous estrogen receptors that is varied for different species including trout, hag fish, and salmon. Estrogen receptors of invertebrates are highly specific when compared with the other vertebral receptors that makes the invertebral receptors a template of analysis for the vertebral population. Protein synthesis of mature and immature fishes show a great deal with respect to time and duration. Fishes that secrete vitellogenins have a faster rate of protein synthesis, and the structural differences are clearly observed. The cellular organelles of red grouper fish vary in their morphology, say the nuclear envelope of liver cells is enlarged than the other. It was believed that the water and lipid content of liver cells were more to accommodate hence the size increased. Likewise, mitochondria, cisternae, endoplasmic reticulum, and the Golgi bodies too possess a difference in their physiological appearance. For trout, when it undergone endogenous vitellogenesis, there was not much differences except the glycogen granules in cytoplasm, whereas in exogenous vitellogenesis, major transformation occurs. The glycogen granules were reduced in the cytoplasm, endoplasmic reticulum is very well expanded, the Golgi complex are widened, and the mitochondria was coarsely packed. Salmon experienced a vast change such as an increase in the size of endoplasmic reticulum, an elevation in the protein, total mRNA, and nuclear level.

1.3.3 Localization of Vitellogenins

With the help of immunoperoxidase method, proteins coupled with vitellogenins were identified from ovarian tissue. Female fishes that are sexually matured and treated with estradiol showed immunoreactivity. Peripheral yolk spheres showed low immunoreactivity; on the other hand, follicular layer of oocytes exhibited high immunoreactivity. This is made visible by staining the cytoplasm of vitellogenin once they have anti-vitellogenin antiserum. Estradiol-injected male and female fishes unveiled antigenic substance when the liver tissues was being sectioned. This was completely absent in the non-injected fishes, and hence it is clear from the control group which has a lack of antigenic substance that the vitellogenins are localized in the liver tissue.

In the species *Brachydanio rerio*, a yolk protein precursor molecule is synthesized and secreted by the liver as a particular response to estrogen. This molecule is then released into the blood for delivery to the oocytes, where it is broken into yolk proteins by a proteolytic process.

Amino acid composition of vitellogenins derived from gold fish, rainbow trout, carp, Japanese eel, and striped bass is tabulated below in Table 1.4. Alanine is highly seen in all the species, followed by glutamic acid and leucine.

Unlike the vertebral and mammalian yolk formation, fishes follow modified mitochondrial evolution of organelles. Under the stimulation of estrogen, vitellogenins are synthesized in the liver which is then transported to the ovary by blood, and it was taken up by the oocytes. It was observed and confirmed from the studies undergone half century ago. Radioisotope-labelled amino acids of hepatoproteins were injected in the blood stream of zebra fish which was later seen in the ovary and in the liver. Similar study with phosphoproteins of liver was conducted in cat fish which exhibited vitellogenins with radioisotopes after 12 h of injection.

1.3.3.1 Response of Vitellogenin to Hormones

Balancing hormones in the piscine body remarkably regulates vitellogenins in the serum. Gonadotrophins are involved in the stimulation of pituitary gland to secrete follicle stimulating hormones and in turn leads to the synthesis of estrogen. It is responsible for producing vitellogenins, and the rest of the cascade flows automatically. Two sets of studies were undergoing: one with hypophysectomized and the other with ovariectomized catfish. In the former, gonadotrophins from salmon fish were injected into the catfish whose pituitary gland, responsible for generating estrogens, was ablated. But there was an increase in the vitellogenin level in serum. The second set of cat fish was injected with gonadotrophins from salmon. The

Table 1.4 Amino acid composition of vitellogenins derived from gold fish, rainbow trout, carp, Japanese eel, and striped bass

Amino acid	Gold fish	Rainbow trout	Carp	Japanese eel	Striped bass
Aspartic acid	6.5	8.4	6.7	7.3	7.6
Threonine	5.5	5.0	5.4	5.3	5.2
Serine	6.9	7.5	7.6	5.8	7.2
Proline	5.5	5.2	5.9	4.7	4.3
Glutamic acid	11.9	11.5	11.8	11.8	8.3
Glycine	4.6	4.2	5.1	5.6	4.2
Alanine	12.8	11.7	12.6	18.0	11.9
Valine	6.9	7.1	6.3	6.1	7.8
Methionine	2.0	2.6	1.9	2.8	2.8
Isoleucine	6.6	5.5	5.4	4.9	7.0
Leucine	10.8	9.5	10.5	7.8	10.8
Tyrosine	2.6	3.0	2.8	2.8	3.4
Phenylalanine	2.9	4.0	2.8	3.8	3.4
Lysine	2.3	7.1	6.3	5.9	7.3
Histidine	7.0	2.1	3.4	2.0	3.1
Arginine	4.9	4.5	5.0	4.8	4.9

removal of the ovary drops down the levels of vitellogenin in the serum which was recaptured by adding estradiol 17β, and this proves that albeit the absence of pituitary gland and ovary, the gonadotrophins and potent form of estrogen pave way for the production of vitellogenins. Gonadotrophin-induced plasma protein synthesis was proved in white-spotted char and gold fish. Estrone, a metabolite of estrogen and androgens, has a similar role in the production of lipids and phosphates during vitellogenesis. The effect of estradiol in individual component synthesis of vitellogenin was experimented in fishes like medaka, tilapia, gold fish, and rainbow trout. Other than the abovementioned hormones, luteinizing hormone, prolactin, and thyroid-stimulating hormones as well have association with the vitellogenin synthesis.

Teleost fishes are the major percentage of the total fish population. Subsistence of few species of fishes like tuna, herring, halibut, salmon, etc. makes teleost fishes of great value. The anatomy and color of some teleost fishes hold aesthetic beauty to be maintained in aquariums. They acclimatize themselves in tough conditions, i.e., from Arctic and Antarctic circles to hot springs with 100 °F. Vitellogenins exist in diverse forms in different species of teleosts. An approximate composition of components in teleost fishes vitellogenins is established earlier. It has 80% of protein and 20% of lipids comprising of phospholipids as the main class of lipid. Transmission electron microscopy was used to investigate oocyte differentiation and vitellogenesis in the oocytes of female demersal fish, *Kareius bicoloratus*. In the early vitellogenic phase, the Golgi complex in the cytoplasm is involved in the formation of yolk vesicles that contain yolk carbohydrates in the yolk vesicle of oocytes. Many pinocytotic vesicles formed by pinocytosis contain yolk precursors during this phase. The carbohydrates and these yolk precursors together get involved in the exogenous heterosynthetic vitellogenesis. Studies were conducted in analyzing the amino acid variation in the embryo and larval stage of another demersal fish, *Veraspermoseri*. There were changes in the proportion of proteins and free amino acid in the early egg phase and larval phase of the fish. Larva of the *Pleuronectes americanus* has the protein components of vitellogenin, whereas the lipid components benefit the embryo.

1.4 Immunogenicity and Antimicrobial Activity of Vitellogenins

Vitellogenins were obtained from *Branchiostoma belcheri tsingtaunese* which was then used for the hemagglutination study that proved the glomerulation of erythrocytes in grass carp fish and other few vertebrates. Apart from providing the nourishments, it brings forth the immunogenicity. Vitellogenins from *Hexagrammos otakii* has multifaceted functions: they play a crucial role in fostering the macrophage phagocytosis; they act as pattern recognition receptor in several pathogens by binding with lipopolysaccharides, lipoteichoic acid, peptidoglycan, laminarin, and glucan (Shi et al. 2006). Vitellogenins bound with peptidoglycan showed notable differences in the growth of *Staphylococcus aureus*. Carbohydrates on the other hand when linked with vitellogenins exhibit a remarkable inhibitory activity of *Escherichia coli* and *Staphylococcus aureus*. The peptide chain and carbohydrate residues of vitellogenins are responsible for the antibacterial activity of pathogens (Liu et al. 2009). Trypsin and carbohydrate-degrading compounds like trypsin and sodium periodate were treated with the cells and found that vitellogenins with low molecular mass are retained in the sample, thereby promoting the cell lysis, whereas the trypsin- and sodium periodate-treated cells become inactive.

Vitellogenins acquired from *Puntius conchonius* were tested for different assays in different time spans of which it showed a greater growth inhibition of *Escherichia coli* at an average concentration and elevated more at maximum concentration. The same vitellogenins were used for hemagglutination activity in which the activity was measured in a higher rate in male rosy barb than the female. Also, it was proven that the male rosy barb with respect to infection by *Escherichia coli* tends to produce vitellogenins. Similar activity was observed when *Danio rerio* was infected with both gram-negative and gram-positive bacterium *Citrobacter freundii* and *Staphylococcus homogenes*, respectively. *Danio rerio* has another evidence of inhibiting the growth of *Escherichia coli* and *Staphylococcus aureus* when lipopolysaccharides and lipoteichoic acid were injected in various doses. Atlantic salmon was studied for antiviral activity against infectious pancreatic necrosis virus. Infection of the virus decreased eventually in limited time duration by disclosing that the vitellogenins were responsible for the reduction in lysis of cells by virus (Munro et al. 2005).

1.5 Antioxidant Activity of Vitellogenins

A ground-breaking research was done at the end of twentieth century proving that the vitellogenins impede the oxidation of lipoproteins which is the causative agent for atherosclerosis. The existence of alpha tocopherols in Japanese eel to a greater extent makes the oxidation stress to fall. The reason behind the prevention of oxidation in Japanese eel is the presence of vitellogenins. Other fishes like rainbow trout, common carp, and sardines too had the same antioxidant property when compared with the antioxidants derived from plants and other sources. Transition metals have autoxidation property that declines the growth phase of oocytes. Vitellogenins terminate the reactions happens in the oxidation of plasma proteins as well as hinder the transition metals; in doing so it encourages the growth of oocytes. Its vitality can be seen in depleting the reactive oxygen species generated in cells. Although the presence of industrial effluent varies in the surrounding ecosystem, the CAT activity and TBARS activity of vitellogenins in rainbow fish does not drop to a lower level which clearly indicates that the protein has strong antioxidant property (Miranda et al. 2020). A salient factor that degrades the antioxidant activity is temperature. When tested with high temperatures, phosvitin withstands pasteurization, justifying their role in defeating the underlying reasons for oxidation. Intrusion of zinc and antioxidant activity was estimated in four groups of cat fish. The group with normal diet and zinc supplements were given to one group which predominantly supports the antioxidant activity. The group with accumulation of 50 mg of zinc showed higher antioxidant status rather than the other groups which deliberately shows that the limited pile of zinc enhances the antioxidant property of fish (Gupta et al. 2022).

1.6 Response of Vitellogenins to Contaminants and Endocrine-Disrupting Chemicals

With exposure to the chemicals in industrialized zones, the amount of vitellogenins got mitigated, and they were shoot up in the fresh water ecosystem (Pereira et al. 1992). Contradictorily, there were no such fluctuations in vitellogenin levels of starry flounder and English sole both in contaminated and non-contaminated water bodies (Spies et al. 1990; Johnson et al. 1988). Similarly, β naphthoflavone showed inverse effect of activity in stimulating vitellogenins in rainbow trout, i.e., low concentration of β naphthoflavone gave out high stimulation and vice versa (Anderson et al. 1996). Organic contaminants in the marine ecosystem reduce the production of proteins in piscine species (Chen et al. 1986). Anthropogenic effects in aquatic ecosystem seriously affects the marine environment. Endocrine-disrupting chemicals get in the way of normal body functions that engenders interruption in metabolism and endocrine system. Meanwhile, the xenoestrogens identical to the regular natural estrogens accumulate as a result of shedding untreated sewage water into the fresh water bodies. Consequently, fishes respond to these compounds tend to produce vitellogenins which act as a biomarker (Hansen et al. 1998). ELISA was used to get a wide knowledge about the response of vitellogenins to the estrogens and other chemicals in the aquatic system (Nilsen et al. 2004).

The presence of industrial surfactant, perfluorooctanoic acid, has a low impact in the production of vitellogenins. Compared to other effluents, the effect of perfluorooctanoic acid has reduced a mere percentage of proteins in the aquatic system (Miranda et al. 2020). Rainbow trout fish of two groups—one with parasite infected and the other with noninfected—were exposed to 17β-estradiol. Infected fish, when encounters the hormone, synthesized lower hepatic vitellogenins compared to the other group (Burki et al. 2012). In order to evaluate, in a natural population of cyprinid fish, the biological consequences that are associated with exposure to estrogenic chemicals, a field study was carried out. To evaluate the reproductive health of common carp (*Cyprinus carpio*) from three rivers receiving sewage treatment plant (STP) effluents (Guadarrama, Henares, and Jarama rivers, Spain) and from a reference site, the gonadosomatic index (GSI), plasma vitellogenin (VTG) level, and histological observations of gonads were performed (Lozoya river). It was shown that there were no significant differences in male GSI between the sites. Only 18% of the fish sampled from the Guadarrama river had abnormally high plasma VTG levels. In among the guys that were investigated, no oocytes were discovered to be present within the testis. According to the findings, the estrogenic chemicals that are released into the environment as a result of the operation of STPs are not capable of exerting sufficient influence over the gonadal development of the fish that were investigated.

Due to the presence of serine moieties, which make up a substantial portion of the vitellogenin structure, it is common knowledge that vitellogenin is an abundant source of phosphate in fish, just as it is in amphibians. The protein phosphate level of fish, on the other hand, is only around 50% of that of its equivalents in avian and amphibian organisms, which may perhaps indicate a lower serine content. Phosvitin, which is often contained in a low-molecular-weight form, is the primary source of protein phosphorous in ovarian yolk.

There is a wide range of variation in the amount of phosphorous found in each species, and there have been reports of phosvitin being nearly or entirely absent from preparations of yolk. In addition to phosphate, teleost vitellogenin readily binds ions like calcium, magnesium, or iron, and as a result, it serves as an essential source of nutrients for the developing oocytes. In point of fact, the ability of vitellogenin to function as a chelating material is the rationale behind the application of EDTA in the process of isolating vitellogenin from the other plasma proteins.

In fish vitellogenin, the lipid concentration is twice as high as that seen in other groups of vertebrates, in contrast to the comparatively low phosphate level of the substance. According to research done on goldfish, rainbow trout, sea trout, and dogfish, the lipid content of fish vitellogenin typically falls somewhere around 20% mark in terms of weight. The polar lipovitellin component of yolk is composed of the vast majority of this lipid substance. Gel electrophoresis or chromatographic methods have both been utilized in order to provide approximations of the molecular weight of the vitellogenin molecule. Their dependability, on the other hand, is questionable because of the unusual approach that was utilized and the quantity of proteolytic breakdown that occurred.

1.6.1 Retinoids and Carotenoids

Animal-based retinal, derived from retinoid, is a principal component that abides in eggs of fish. They play a vital role in regulating and signaling molecules during the development of the embryo, and obviously they are transported to the eggs that are hatched outside the female fish. The vitellogenins are bound by retinals and stored in the biosystem when there is no lipid globules (Irie and Seki 2002). Transcripts of retinol-binding protein in the ovary and oocytes propound a balancing method of synthesizing follicles (Lubzens et al. 2003). Astaxanthin, a carotenoid, is carried by the vitellogenin to the ovary. Carotenoid levels in the muscle decreased significantly during spawning migration, while serum carotenoid levels increased significantly. At the beginning of spawning, vitellogenin were seen in the blood of salmon which gradually decreased at the time of upstream migration and has more astaxanthin. The presence and absence of the carotenoid during the different growth phase insists that they were transported through the bloodstream to the ovaries (Ando et al. 1986).

1.6.2 Vitellogenins in Aquaculture

Factors that affect the growth of fishes and the hormonal activity in aquaculture are increase in water temperature, climatic change, and others. It causes deleterious effects in the function of endocrine, gametogenesis, reproduction, and maturation. A distraction in gonadal steroid synthesis and vitellogenins production from liver was affected in Atlantic salmon when they were exposed to elevated temperatures. This in turn led to the low estrogen receptors survival and loss of egg production. Rainbow trout experienced the same effect when they were fed in water systems with high temperatures. In such cases, luteinizing hormone facilitates the entire synthesis process and maintains the estrogen receptor dynamics. Temperature rise is not much unfavorable for male fishes except its effect in spermiation of both Atlantic salmon and rainbow trout (King and Pankhurst 2010).

The Patagonian toothfish or Chilean seabass is a demersal fish with a lifespan of 50 years and is well known for its international demand. Commercialization of the fish is too risky as they fail to reproduce in the growth conditions provided at the time of aquaculture. With minimal ideas of complete knowledge, it is strenuous to maintain the fish in a closed environment. This was over cede by isolating, identifying, and incubating the fish protein. Polyclonal antibody discrete for vitellogenins as antigens was created so as to perform ELISA combined with SDS-PAGE. Analysis of naturally grown Chilean sea bass with environmentally controlled species showed the increase in growth of oocytes and the entire piscine population (Amthauer et al. 2021).

Temperature difference leads to the increase and decrease of vitellogenin, estradiol, and testosterone in Atlantic salmon. During austral autumn spawning, the mean body mass of fishes increased remarkably, gonadosomatic index level shoot up to a higher extent, estradiol and testosterone ascend multiple times, and vitellogenins scaled up from below average level to extraordinary level. In contrast to the aforementioned levels, the fishes grown in mid-summer showed neither an increase nor decrease in hormone levels (King and Pankhurst 2003).

1.7 The Role of Ascorbic Acid in Vitellogenesis

Studies says that apart from the ascorbic acid synthesized naturally in our body systems, the dietary ascorbic acid impacts vitellogenesis. Natural along with dietary ascorbic acid when accumulates in the body of fish especially post vitellogenesis phase leads to the hydroxylation of amino acids that promotes collagen integrity and also improved quality of fish eggs (Waagbu et al. 1988). A decrease in ascorbic acid level directly deteriorates the level of hormones that are responsible for the synthesis of vitellogenins. Moreover, the fishes became anemic with no increased mortality rate.

1.8 Sex Identification in Fishes by Vitellogenins

The Hapuku fish cultured in New Zealand is sexually monomorphic. To identify the genders, three methods were followed, namely, (a) ultrasound imaging technique, (b) plasma vitellogenin enzyme-linked immunosorbent assay (ELISA), and (c) plasma sex steroid (17β-estradiol (E2) and 11-ketotestosterone (11-KT)) radioimmunoassay (RIA). These techniques were applied 2 months before spawning season or during spawning. One hundred percent vigorous result were attained from radioimmunoassay; 95% of accurate results were obtained from ultrasound imaging. Ninety-two percent of precise results were acquired from vitellogenin- enzyme-linked immunosorbent assay (Kohn et al. 2013).

1.9 Vitellogenins as Carrier Compound

For the purpose of conducting research on developmental biology, vitellogenins have been suggested as a device that might be used to selectively transfer ingredients into the egg yolk of goldfish, zebrafish, and carps. Additionally, recombinant vitellogenins have been suggested for use as larval feeds. The nutrients, which are collectively transported from the liver to the ovary in the form of circulating yolk precursors called vitellogenins, consist of substances that are derived from the mother and include proteins, carbohydrates, lipids, vitamins, minerals, and ions. These nutrients are essential for the development of the embryo.

1.10 Application of Vitellogenins

In female fishes, especially those raised in captivity, circulating vitellogenins have been used as markers to track the process of gonad maturation and the onset of puberty. It has also been possible to determine the gender of fishes that do not display sexual dimorphism by analyzing the presence of vitellogenins in their blood, mucus, and muscle. Evaluation of the appropriate maturational proteolysis of yolk proteins produced from vitellogenin, or of their expression. A model of the maturation of the ovary illustrates the mechanisms of maturational proteolysis of yolk proteins and ion transport during oocyte hydration. Ooplasm clarification and oocyte hydration are both related to the proteolysis of vitellogenin-derived yolk proteins. This process requires acidification of the yolk vesicles in order to activate cathepsin proenzymes, which then produces free amino acids (FAAs) that act as osmotic effectors and influence oocyte hydration. Ooplasm clarification and oocyte hydration are both necessary steps in the fertilization process. In addition, the RNAs and proteins produced by vitellogenesis in fish have been applied in the role of egg quality indicators.

Additionally, vitellogenins are widely used to assess the exposure of aquatic animals to endocrine-disrupting chemicals, particularly those that mimic the action of estrogens. This practice is common in habitats where animals are exposed to water. Because vitellogenins are formed in response to endogenous, they are potentially suitable candidates for use as markers in the evaluation of estrogenic chemicals. However, because different forms of vitellogenins have varying degrees of sensitivity to the induction caused by estrogen, it is important to take into account the particular type of vitellogenin that is being evaluated in order to identify endocrine disruptive substances. For the purpose of conducting research on developmental biology, the vitellogenins have been proposed as a biotechnology that could be used to specifically deliver materials into the egg yolk of goldfish, zebrafish, and carps. Additionally, recombinant vitellogenin have been proposed for use as larval feeds.

References

Amthauer R, Cárdenas F, Reyes A, Valenzuela A, Dantagnan P, Vidal R, Vargas-Chacoff L (2021) Vitellogenesis in the Patagonian toothfish (Dissostichus eleginoides) conditioned to a recirculating aquaculture system. Gen Comp Endocrinol 307:113768

Anderson MJ, Olsen H, Matsumura F, Hinton DE (1996) In vivo modulation of 17b-estradiol-induced vitellogenin synthesis and estrogen receptor in rainbow trout (Oncorhynchus mykiss) liver cells by b-naphthoflavone. Toxicol Appl Pharmacol 137:210–218

Ando S, Takeyama T, MutsuoHatano. (1986) Transport associated with serum vitellogenin of carotenoid in chum salmon (*Oncorhynchus keta*). Agric Biol Chem 50:557–563

Baker ME (1988) Is vitellogenin an ancestor of apolipoprotein B-100 of human low-density lipoprotein and human lipoprotein lipase? Biochem J 255(3):1057–1060

Beresford N, Brian JV, Runnalls TJ, Sumpter JP, Jobling S (2011) Oestrogenic activity of tropical fish food can alter baseline vitellogenin levels in male fathead minnow (*Pimephales promelas*)

Biscotti MA, Barucca M, Carducci F, Canapa A (2018) New perspectives on the evolutionary history of vitellogenin gene family in vertebrates. Genome Biol Evol 10(10):2709–2715

Buisine N, Trichet V, Wolff J (2002) Complex evolution of vitellogenin genes in salmonid fishes. Mol Gen Genomics 268:535–542

Burki R, Krasnov A, Bettge K, Rexroad CE, Afanasyev S, Antikainen M, Burkhardt-Holm P, Wahli T, Segner H (2012) Pathogenic infection confounds induction of the estrogenic biomarker vitellogenin in rainbow trout. Environ Toxicol Chem 31:2318–2323

Carducci F, Biscotti MA, Canapa A (2019) Vitellogenin gene family in vertebrates: evolution and functions. Eur Zool J 86:233

Chen TT, Reid PC, Van Beneden R, Sonstegaard RA (1986) Effect of Aroclor 1254 and Mirex on estradiol-induces vitellogenin production in juvenile rainbow trout *Salmo gairdneri*. J Fish Aquat Sci 43:169–173

Chen J-S, Sappington TW, Raikhel AS (1997) Extensive sequence conservation among insect, nematode, and vertebrate vitellogenins reveals ancient common ancestry. J Mol Evol 44:440–451

De Vlaming VL, Wiley HS, Delahunty G, Wallace RA (1980) Goldfish (*Carassius Auratus*) vitellogenin: induction, isolation, properties and relationship to yolk proteins. Comp Biochem Physiol B Comp Biochem 67(4):613–623

Engelmann F (1979) Insect vitellogenin: identification, biosynthesis, and role in vitellogenesis. Adv Insect Physiol 14:49–108

Gupta G, Srivastava PP, Gangwar M, Varghese T, Chanu TI, Gupta S, Ande MP, Krishna G, Jana P (2022) Extra-fortification of zinc upsets vitellogenin gene expression and antioxidant status in female of *Clarias magur* brooders. Biol Trace Elem Res 200:1861–1871

Hansen PD, Dizer H, Hock B, Marx A, Sherry J, McMaster M, Blaise C (1998) Vitellogenin—a biomarker for endocrine disruptors. Trends Anal Chem 17(7):448–451

Iida A, Arai HN, Someya Y, Sano K (2019) Mother-to-embryo vitellogenin transport in a viviparous teleost *Xenotocaeiseni*. Proc Natl Acad Sci U S A 116(44):22359–22365

Irie T, Seki T (2002) Retinoid composition and retinal localization in the eggs of teleost fishes. Comp Biochem Physiol B Biochem Mol Biol 131(2):209–219

Iwasaki M, Seko A, Kitajima K, Inoue Y, Inoue S (1992) Fish egg glycophosphoproteins have species-specific N-linked glycan units previously found in a storage pool of free glycan chains. J Biol Chem 267(34):24287–24296

Johnson LL, Casillas E, Collier TK, McCain BB, Varanasi U (1988) Contaminant effects on ovarian development in English sole (Parophrys vetulus) from Puget Sound. Washington. Can. J. Fish. Aquat. Sci. 45:2133–2146

King HR, Pankhurst NW (2003) Ovarian growth and plasma sex steroid and vitellogenin profiles during vitellogenesis in Tasmanian female Atlantic salmon (*Salmo salar*). Aquaculture 219:797–813

King HR, Pankhurst NW (2010) Temperature and salmonid reproduction: implications for aquaculture. J Fish Biol 76:69–85

Kohn YY, Lokman PM, Kilimnik A, Symonds JE (2013) Sex identification in captive hapuku (Polyprion oxygeneios) using ultrasound imagery and plasma levels of vitellogenin and sex steroids. Aquaculture 384–387:87–93

Liu Q-H, Zhang S-C, Li Z-J, Gao C-R (2009) Characterization of a pattern recognition molecule vitellogenin from carp (*Cyprinus carpio*). Immunobiology 214(4):257–267

Lubzens E, Lissauera L, Levavi-Sivan B, Avarrea J-C, Sammar M (2003) Carotenoid and retinoid transport to fish oocytes and eggs: what is the role of retinol binding protein? Mol Asp Med 24(6):441–457

Matsubara T, Nagae M, Ohkubo N, Andoh T, Sawaguchi S, Hiramatsu N, Sullivan CV, Hara A (2003) Multiple vitellogenins and their unique roles in marine teleosts. Fish Physiol Biochem 28:295–299

Miranda AF, Trestrail C, Lekamge S, Nugegoda D (2020) Effects of perfluorooctanoic acid (PFOA) on the thyroid status, vitellogenin, and oxidant–antioxidant balance in the Murray River rainbowfish. Ecotoxicology 29(2):163–174

Munro ES, Gahlawat SK, Acosta FA, Ellis AE (2005) In infectious pancreatic necrosis virus carrier Atlantic salmon, *Salmo salar L.*, post-smolts, almost all kidney macrophages ex vivo contain a low level of non-replicating virus. J Fish Dis 29(1):43–48

Nilsen BM, Berg K, Eidem JK, Kristiansen S-I, Brion F, Porcher J-M, Goksoyr A (2004) Development of quantitative vitellogenin-ELISAs for fish test species used in endocrine disruptor screening. Anal Bioanal Chem 378:621–633

Pan ML (1971) The synthesis of vitellogenin in the Cecropia silkworm. J Insect Physiol 17:677–689

Pan ML (1977) Juvenile hormone and vitellogenin synthesis in the Cecropia silkworm. Biol Bull 153:336

Pan ML, Wallace RA (1974) Cecropia vitellogenin: isolation and characterization. Am Zool 14:1239–1242

Pan ML, Bell WJ, Telfer WH (1969) Vitellogenic blood protein synthesis by insect fat body. Science 165(3891):393–394

Pereira JJ, Ziskowski J, Mercaldo-Allen R, Kuropat C, Luedke D, Gould E (1992) Vitellogenin in flounder (*Pleuronectes americanus*) from Long Island Sound and Boston Harbor. Estuaries 15(3):289–297

Peter RE, Crim LW (1979) Reproductive endocrinology of fishes: gonadal cycles and gonadotropin in teleosts. Annu Rev Physiol 41:323–335

Shi X, Zhang S, Pang Q (2006) Vitellogenin is a novel player in defense reactions. Fish Shellfish Immunol 20(5):769–772

Silversand C, Haux C (1995) Fatty acid composition of vitellogenin from four teleost species. J Comp Physiol B 164:593–599

Silversand C, Hyllner SJ, Haux C (1993) Isolation, immunochemical detection, and observations of the instability of vitellogenin from four teleosts. J Exp Zool 267(6):587–597

Spies T, Bresnahan M, Bahrain S, Arnold D, Blanck G, Mellins E, Pious D, DeMars R (1990) A gene in the human major histocompatibility complex class II region controlling the class I antigen presentation pathway. Nature 348(6303):744–747

Van Veld PA, Rutan BJ, Sullivan CA, Johnston D, Rice CD, Fisher DF, Yonkos LT (2005) A universal assay for vitellogenin in fish mucus and plasma. Environ Toxicol Chem 24(12):3048–3052

Waagbu R, Thorsen T, Sandnes K (1988) Role of dietary ascorbic acid in vitellogenesis in rainbow trout (*Salmo gairdneri*). Aquaculture 80(1989):301–314

Wang H, Yan T, Tan JT, Gong Z (2000) A zebrafish vitellogenin gene (vg3) encodes a novel vitellogenin without a phosvitin domain and may represent a primitive vertebrate vitellogenin gene. Gene 256(1–2):303–310

Williams DL, Wang S-Y, Capony F (1979) Multiple response patterns to oestrogenic stimulation in the avian liver. In: Hormonal steroids proceedings of the fifth international congress on hormonal steroids, pp 231–236

Previtellogenesis and Vitellogenesis

2

Ashokkumar Sibiya and Vaseeharan Baskaralingam

Abstract

Early in the twentieth century, immunological techniques were used to pinpoint a specific antigen in the blood of gravid female fish during investigations on sex discrimination. The principal precursor of egg yolk protein, which is currently known as vitellogenin and is generated in the female liver before being secreted into the circulation and integrated into the egg, is currently identified as this particular antigen. The majority of the protein in yolk is made up of phosphoproteins and lipoprotein precursors. Vitellogenin serves as the wonderful provider and a component of the yolk, the key food to the developing embryos of egg layers (oviparous), including vertebrates and invertebrates. It transfers proteins and certain lipids from the liver through the blood to the maturing oocytes. Oocytes arrest in prophase I of the meiotic cell cycle, during which they expand as the building blocks for early embryonic development (vitellogenesis). In many vertebrates, fully grown postvitellogenic oocytes under prophase I arrest are unable to be fertilized until they mature. This current chapter focusses on a brief overview of previtellogenesis and vitellogenesis.

Keywords

Vitellogenin · Oocytes · Oviparous · Vertebrates · Prophase I

A. Sibiya · V. Baskaralingam (✉)
Biomaterials and Biotechnology in Animal Health Lab, Department of Animal Health and Management, Alagappa University, Karaikudi, Tamil Nadu, India
e-mail: vaseeharanb@alagappauniversity.ac.in

V. Baskaralingam, R. Vanichviriyakit (eds.), *Vitellogenin in Fishes-Diversification, Biological Properties, and Future Perspectives*,
https://doi.org/10.1007/978-981-99-5340-0_2

Abbreviations

E2	17-estradiol
EDCs	Endocrine-disrupting chemicals
ERs	Estrogen receptors
FSH	Follicle-stimulating hormone
GnRH	Gonadotropin-releasing hormone
JH	Juvenile hormone
VLDL	Very-low-density lipoprotein
Vtg	Vitellogenin

2.1 Introduction

Vitellogenesis is the oocyte developmental stage defined by the synthesis of egg yolk proteins and the buildup of those proteins ooplasm (Latin *vitellus*, meaning "yolk," and Greek *genesis*, meaning "creation"). Previtellogenic (primary growth) and vitellogenic (secondary growth) stages of ovarian follicle development can be distinguished during which the principal lipid and protein resources necessary for embryonic and larval development are stored within the egg (Reading et al. 2017). Vitellogenin (Vtg) protein are present at high concentrations in mature females during the reproductive season, because of its activity as an egg-yolk precursor protein (Hiramatsu et al. 2015). Hara et al. (2016) says that, in vertebrates, it is synthetized during vitellogenesis in the liver in response to estradiol-17β from the ovary, which is under the control of gonadotropins from the pituitary gland. After secretion into the bloodstream, vitellogenin is taken up by developing oocytes, where it is cleaved into smaller yolk proteins, including phosvitin, lipovitellin, and β-component. Vtg is a complex lipoglycophosphoprotein with a high molecular mass that binds to other elements like calcium, iron, zinc, and others. In oviparous vertebrates, Vtg is often a lipoglycophosphoprotein that acts as a key precursor of the proteins found in egg yolks, which are then stored as vital nutrients for upcoming embryogenesis. All oviparous species, including invertebrates, fish, amphibians, reptiles, birds, and monotremes, create Vtgs in their females. The fish Vtg protein has been employed as a sensitive biomarker for determining estrogenic activity in aquatic environments (Tran et al. 2019). As a biomarker for assessing the impacts of endocrine system-disrupting substances prevalent in various waters, Vtg has gained interest. During vitellogenesis an sequential process of the following events occur which include the release into circulation, bloodstream transport of vitellogenin to the target tissue, uptake of vitellogenin by developing oocytes, and conversion of vitellogenin into storage forms (Ho 1987). Vitellogenin serves as the wonderful provider and a component of the yolk, the key food to the developing embryos of egg layers (oviparous), including vertebrates and invertebrates. It transfers proteins and certain lipids from the liver through the blood to the maturing oocytes (Fig. 2.1). Vtgs are converted into yolk proteins in the oocyte and are primarily stored as yolk granules, which are also known as globules or platelets, and occasionally in an amorphous

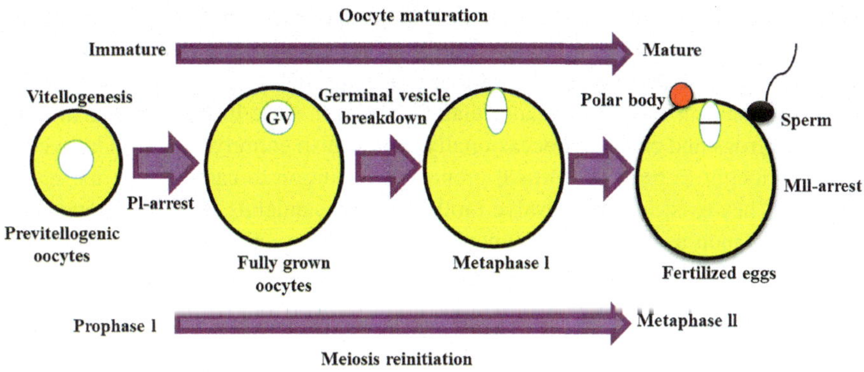

Fig. 2.1 Oocyte maturation and meiotic cell cycle

compartment (fluid yolk). Numerous vitellogenin variations have been found, according to recent protein and gene investigations. Fish Vtgs display a complicated evolutionary history that manifests a sizable discrepancy in structure and function (Hiramatsu et al. 2015). A substrate Vtg rich in nutrients, proteins, and lipids is provided for larval growth and embryonic development by the large amount of yolk mass contained in teleost eggs. Egg yolk reserves provide the whole nutritional requirements for oviparous animals that are growing young. Before the onset of external feeding, the ovulated egg must contain adequate nutrients to support progeny growth. In order for maturing oocytes to grow into eggs, the process of vitellogenesis entails feeding them with the necessary yolk nutrients. The maternally produced components that make up these nutrients include proteins, carbohydrates, lipids, vitamins, minerals, and ions; they are all transported from the liver to the ovary in the form of circulating vitellogenins, which are precursors to the yolk (Reading et al. 2017). Vitellogenin is the major substance found in the yolk of vertebrate eggs. All fishes produce vitellogenins, which are crucial nutrients for growing embryos. Therefore, the current chapter concentrated on providing a succinct summary of previtellogenesis and vitellogenesis.

2.2 Previtellogenesis

Previtellogenesis is the event that trigger vitellogenesis. During this period no synthesis or buildup of the food reserve material, the yolk occurs during this phase although the volume of the primary oocyte's nucleus and cytoplasm significantly increases. The amount of cytoplasm has increased in both quality and quantity. The number of mitochondria increases, the endoplasmic reticulum with ribosomes network becomes more intricate, and the Golgi bodies produce cortical granules in addition to carrying out their regular duties (Kumari n.d.).The developing oocyte's nucleus enlarges during this period as a result of the abundant nuclear sap production. The yolk nucleus of the Balbiani is a black body that appears in one location outside the nucleus, generally close to the Golgi complex. The term "germinal vesicle" now refers to this enormous, fluid-inflated oocyte. The centrosome is surrounded by the Golgi bodies in immature oocytes. They can grow into a sizable spherical mass in some mammals' mature oocytes, settle in the subcortical cytoplasm of frogs and chicks, or occasionally even vanish entirely. Oocyt's Golgi complex is thought to produce cortical granules in addition to carrying out its regular activity. They exist in some bivalve mollusks, some annelids, fish, frogs, and some mammals (such as rabbits and humans), but not all insects, gastropod uropodeles, all birds, and all mammals (rat and guinea pig). These granules are created in the oocyte's core by the cisternae of the Golgi complex, and they eventually travel to the periphery where they are organized in a layer very near the plasma membrane.

2.3 Vitellogenesis

The process through which maturing oocytes in the ovary accumulate yolk is known as vitellogenesis. This maturation phase is seasonal in most fish species, and spawning only happens once every annual reproductive cycle. Therefore, any factor affecting the vitellogenic cycle can significantly lower an individual's reproductive success (number of eggs, hatching rate, and embryo viability). The entire population of the species could then be impacted by this. In fish, gonadotropins and estrogens work together to start and control the largely exogenous synthesis of vitellogenin. Hepatocytes produce the species-specific protein vitellogenin, which is then actively sequestered by mature oocytes after being released into the bloodstream (Nicolas 1999). The liver's production of the yolk precursor protein vitellogenin and the oocytes' uptake of it are the key events in oogenesis (Johnson et al. 1991). The gonadotropins are released into the bloodstream once oogenesis is begun, where they travel to the ovaries to stimulate oocyte growth and ultimately ovulation. Additionally, they encourage the follicle cells to produce estrogen (Nagler and Idler 1992). This activates the transcription of vitellogenin, which has the desired effect. Following release into the bloodstream, vitellogenin travels to the ovaries. Through a capillary network in the thecal layer of the follicle, it enters the follicle. Vitellogenin exits the capillaries and travels to the oocyte surface via channels between the follicle cells. Vitellogenin is then absorbed into yolk platelets under gonadotropin stimulation via receptor-mediated endocytosis (Tyler et al. 1991).

2.4 Vitellogenesis and Previtellogenesis in Invertebrate

In order to ensure the continuation of their genes, female insects often lay enormous quantities of eggs and devote a lot of resources to this goal. A haploid set of chromosomes, enough nutrients to keep the developing embryo well-fed until the larva or nymph encloses and starts feeding, and a set of determinants to direct the organization and progression of embryogenesis, including the differentiation of a new cluster of germ cells, are the essential components of an insect egg. In Drosophila, the female enters adulthood with an immature ovary, but juvenile hormone (JH) causes it to start producing yolk protein (vitellogenesis) and follicle maturation on the first day of adulthood. The differentiation of the previtellogenic follicle is complete, and the constructed follicle is discharged into the vitellarium (oocyte-nurse cell complex surrounded by follicular epithelial cell monolayer). From previtellogenesis through vitellogenesis, followed by the shift to choriogenesis and ovulation occurs. The oocyte obtains material from the nurse and follicle cells as well as the hemolymph during vitellogenesis. While some of these elements are nutrients required for the completion of embryogenesis, others are differently distributed factors that establish pattern formation throughout early embryonic development (Swevers et al. 2005). The follicle enters the choriogenic stage after the provisioning of the oocyte is finished, but before it is ejected into the oviduct. During this time, the oocyte is covered by the chorion after being first sealed inside a vitelline membrane. It is suggested that DILPs at the ovarian level will speed up follicular cell proliferation and encourage the change from previtellogenesis to vitellogenesis. Furthermore, the existence of a parallel autocrine/paracrine system that controls follicle growth and maturation is suggested by the expression of DILP5 in follicular cells of vitellogenic follicles (Ikeya et al. 2002).

2.5 Vitellogenesis and Previtellogenesis in Vertebrates

Since the early 1900s, when the formation of egg yolk proteins and low-molecular-weight compounds like amino acids were assumed to be responsible for oocyte growth, chicken oogenesis has been investigated. It was first demonstrated in 1974 that Vtg was the precursor of egg yolk proteins based on a biochemical study using African clawed frogs, *Xenopus laevis*. This new theory proposed that egg yolk proteins are first synthesized in the maternal liver and then transferred to growing oocytes via the blood. Vtg is a protein that is exclusively expressed in the female blood serum during vitellogenesis, and some of the general properties of Vtg in oviparous animal include when estrogen is given to male or young fish, plasma Vtg is produced. Vtg is a complex protein with a high molecular mass that is composed of sugar, lipid, and phosphorus and binds other elements like calcium, iron, and zinc. It is also a precursor to the egg yolk proteins that are reactive with antibodies made against egg extracts. The pituitary gland release follicle-stimulating hormone (FSH) into the bloodstream that causes the synthesis of the sex steroid hormone estrogen (estradiol-17), which triggers vitellogenesis, in the follicle cells surrounding growing oocytes (Hara et al. 2016). The Vtg gene in the nucleus is affected by estrogen via estrogen receptors after it binds to sex steroid hormone-binding globulin in the blood and is transported to hepatocytes (Fig. 2.2). Hepatocytes' combination of estrogen and the estrogen receptor attach to the Vtg gene's promoter region, activating the gene to start and speed up transcription. Following the start of translation of the Vtg transcript products, the Vtg proteins go through lipidation, phosphorylation, and glycosylation steps before being released into the circulation.

Fig. 2.2 Vitellogenin induction in female fish

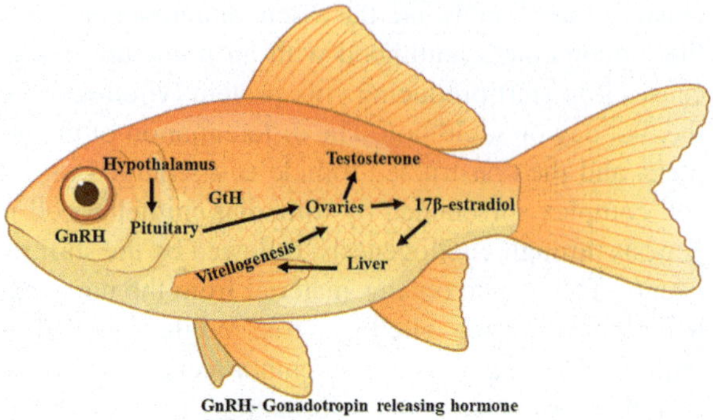

Blood Vtg enters the cell after binding to the Vtg receptor on the oocyte plasma membrane. When Vtg reaches an oocyte, a cathepsin D-like enzyme specifically breaks it down and causes molecular cleavage to create Lv, Pv, and ′-c, which are then stored in the cell. This is known as the "single Vtg model," which was initially put forth when it was thought that Vtg was made up of just one molecular type of protein. However, the "many Vtg model" with more complex processes is currently recognized as a more accurate explanation of vitellogenesis as a result of the recent discovery of multiple species of Vtg.

2.6 Synthesis of Vitellogenin

Vitellogenins are proteins that are the principal food source in the yolk of oviparous vertebrate eggs. Some species' ovulated eggs include 80–90% of the entire dry mass of yolk that comes from Vtg. The generation of Vtg is a gonadotropin-dependent process that is regulated by a number of factors, including photoperiod, nutritional condition, and water temperature. The hypothalamus can be stimulated by a variety of environmental stimuli to release the gonadotropin-releasing hormone (GnRH). In response to GnRH, the pituitary gland releases GTHs, which boosts oocyte vitellogenin absorption and E2 production in follicle cells. They are created in the liver under the strict supervision of E2, released into the bloodstream, and then taken up by the developing oocytes in the ovary under the direction of GTH. Specific receptors then cleave them into lipovitellins and phosvitins (Li and Zhang 2017). Estradiol will cause the transcription of Vtgs when it binds to the hepatocyte-based Vtg receptor. Initial production of the protein backbone for Vtg occurs on membrane-bound ribosomes, and posttranslational alterations such as lipidation, glycosylation, and phosphorylation are then made (Tramunt et al. 2021) (Fig. 2.3).

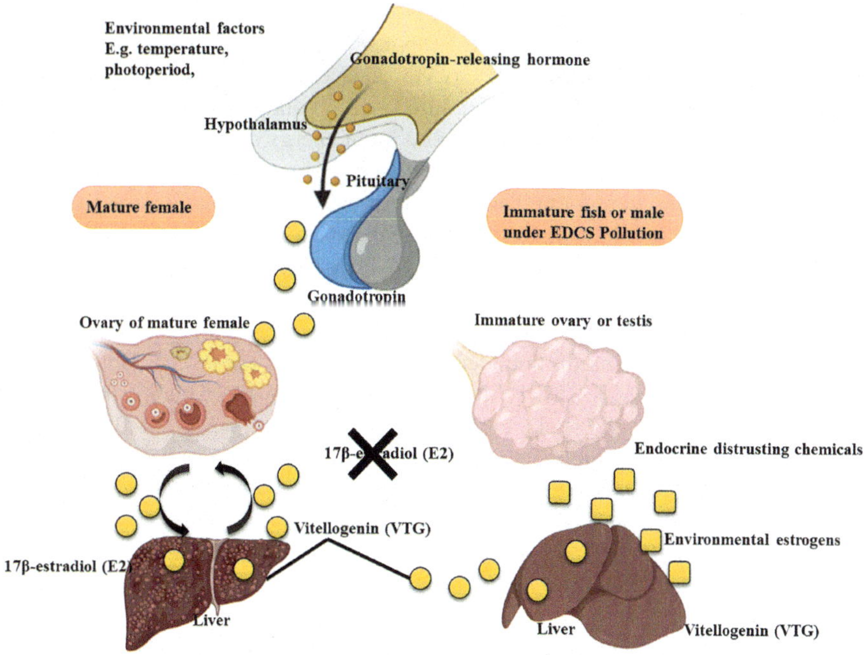

Fig. 2.3 Mechanism of vitellogenin synthesis

2.7 Mode of Action

Gonadotropins regulate the cyclical or seasonal nature of vitellogenesis. Numerous internal and external factors affect the brain's (hypothalamus) ability to produce gonadotropin-releasing hormone, including inherent biorhythms, nutritional status, and seasonal changes in day length and water temperature (GnRH). Follicle-stimulating hormone (FSH), which theca and granulose cells of the ovarian follicle secrete in response to GnRH, is released by pituitary gonadotrophs (E2). The ovarian steroid hormone 17-estradiol (E2), which is produced under the control of the hypothalamic-pituitary-gonad axis, is the most significant inducer of vitellogenin (Vtg) expression (Nelson and Habibi 2013). The majority of fish species experience vitellogenesis after being exposed to estradiol-17 (E2), and it can even occur in places where it is not typically formed, like the hepatocytes in the male liver. Estradiol-17, also known as E2, is created by the ovarian follicular cells, transported through the blood, and binds to SHBGs before entering the liver cells either through diffusion or receptor-mediated absorption (Hara et al. 2016). Growth hormone, thyroid hormone, cortisol, androgens, and other estrogenic steroids like estrone may also help to induce hepatic vitellogenesis in some species. Hagfish (*Eptatretus stoutii*) vitellogenesis is frequently less responsive to E2 and may be impacted by other factors, such as the period of time following feeding. In the liver cells of the Mozambique tilapia, *Oreochromis mossambicus*, circulating E2 interacts to one of the three estrogen receptors (ERs) (Davis et al. 2007). The liver produces Vtg throughout the reproductive season in response to estrogen stimulation, which is then released into the bloodstream. Oocyte surface Vtg receptors take up plasma Vtg, which cathepsin then breaks down into smaller yolk proteins. The first stage of oocyte production, vitellogenesis, requires a lot of energy and resources. Poor vitellogenin oocyte absorption leads to poor larval development and greater egg mortality. Egg fertility and young-child survival drastically decrease when the Vtg genes are deleted in female zebrafish. Fish reproduction success therefore depends on high Vtg production. Vtg A, Vtg B, and Vtg C, which are encoded by vtgaa, vtgab, and vtgc, respectively, are the three groups of piscine Vtg (Cui et al. 2017). Grey mullet (*Mugil cephalus*), an Acanthomorpha species that lays pelagic eggs, has an oocyte ratio of 4:13.3:1, while barfin flounder has an oocyte ratio of 9:15:1. Both of these species have Vtg C that is B5% of the total Vtg-derived yolk, which is the same as what has been seen in the sticky egg-laying species white perch. As illustrated by the fact that the Vtg C component of the yolk can reach as high as 25% in mosquito fish and striped bass, two species that spawn neutrally buoyant eggs, the final yolk composition of eggs can vary greatly between fish species (Williams et al. 2017). Studies done on salmonids, as well as several other marine and freshwater species, in vivo and in vitro have revealed that maternal VLDL is a significant source of ooplasm lipids and that the lipase-dependent, non-endocytotic pathway is the main mechanism for absorbing VLDL-associated lipids. Although the lipoprotein receptor-mediated pathway does not appear to be involved in the formation of ooplasm lipid droplets, there is still a possibility that Ldlr is involved in this process in some species, including anguillid eelVtg has been widely used as a biomarker to measure estrogenic pollution in aquatic habitats as a foundation.

2.8 Conclusion

Particularly in aquaculture, circulating Vtgs have functioned as markers for the start of puberty and the course of gonad maturation in female animals. In aquaculture, animals that don't exhibit sexual dimorphism, the presence of Vtg in the blood, mucus, and muscle has also been used to determine a fish's gender. Additionally, Vtgs are commonly used to evaluate animal exposure to endocrine-disrupting chemicals (EDCs), particularly EDCs that mimic the effects of estrogens, in aquatic habitats. Since Vtgs are produced in response to endogenous E2, they are potentially ideal markers for assessment of estrogenic EDCs. However, given the differences in the different types of Vtgs' (or Vtg transcripts') sensitivity to being induced by estrogen(s), consideration should be given to the particular type of Vtg being evaluated to detect EDCs. Only a few research have looked at the interactions between vitellogenesis and previtellogenesis. Future research should concentrate on physiological examination of the complete Vtg process within a species, from Vtg synthesis in the liver to the migration of proteins from the circulation into the egg yolk.

Acknowledgments The authors would like to thank and acknowledge the Ministry of Human Resource Development, Government of India, and Alagappa University, Karaikudi, for providing support in RUSA Phase 2.0 grant sanctioned No. F.24-51/2014-U, Policy (TNMulti-Gen), Department of Education, Government of India.

References

Cui XF, Zhao Y, Chen HP, Deng SP, Jiang DN, Wu TL, Zhu CH, Li GL (2017) Cloning, expression and functional characterization on vitellogenesis of estrogen receptors in *Scatophagus argus*. Gen Comp Endocrinol 246:37–45

Davis LK, Hiramatsu N, Hiramatsu K (2007) Induction of three vitellogenins by 17 beta-estradiol with concurrent inhibition of the growth hormone-insulin like growth factor 1 axis in a euryhaline teleost, the tilapia (Oreochromis mossambicus). Biol Reprod 77:614–625

Hara A, Hiramatsu N, Fujita T (2016) Vitellogenesis and choriogenesis in fishes. Fish Sci 82(2):187–202

Hiramatsu N, Todo T, Sullivan CV, Schilling J, Reading BJ, Matsubara T, Ryu YW, Mizuta H, Luo W, Nishimiya O, Wu M (2015) Ovarian yolk formation in fishes: molecular mechanisms underlying formation of lipid droplets and vitellogenin-derived yolk proteins. Gen Comp Endocrinol 221:9–15

Ho SM (1987) Endocrinology of vitellogenesis. In: Hormones and reproduction in fishes, amphibians, and reptiles. Springer, Boston, MA, pp 145–169

Ikeya T, Galic M, Belawat P, Nairz K, Hafen E (2002) Nutrient-dependent expression of insulin-like peptides from neuroendocrine cells in the CNS contributes to growth regulation in Drosophila. Curr Biol 12(15):1293–1300

Johnson LL, Casillas E, Myers MS, Rhodes LD, Olson OP (1991) Patterns of oocyte development and related changes in plasma 17-β estradiol, vitellogenin, and plasma chemistry in English sole *Parophrys vetulus* Girard. J Exp Mar Biol Ecol 152(2):161–185

Kumari A (n.d.) Previtellogenesis and vitellogenesis. Dept. of Zoology, L.S. College, Muzaffarpur. https:\\www.lscollege.ac.in/sites/default/files/e-content/Previtellogenesis%20and%20vitellogenesis_1.pdf

Li H, Zhang S (2017) Functions of vitellogenin in eggs. Oocytes 63:389–401

Nagler JJ, Idler DR (1992) In vitro ovarian estradiol-17β and testosterone responses to pituitary extract and corresponding serum levels during the prespawning to vitellogenic phases of the reproductive cycle in winter flounder (*Pseudopleuronectes americanus*). Comp Biochem Physiol A Physiol 101(1):69–75

Nelson ER, Habibi HR (2013) Estrogen receptor function and regulation in fish and other vertebrates. Gen Comp Endocrinol 192:15–24

Nicolas JM (1999) Vitellogenesis in fish and the effects of polycyclic aromatic hydrocarbon contaminants. Aquat Toxicol 45(2–3):77–90

Reading BJ, Sullivan CV, Schilling J. Vitellogenesis in fishes. Reference module in life sciences. Elsevier BV. Amsterdam. https://doi.org/10.2017;1016:03076-4

Swevers L, Raikhel A, Sappington TW, Shirk P, Iatrou K (2005) Vitellogenesis and post-vitellogenic maturation of the insect ovarian follicle

Tramunt B, Montagner A, Tan NS, Gourdy P, Rémignon H, Wahli W (2021) Roles of estrogens in the healthy and diseased oviparous vertebrate liver. Metabolites 11(8):502

Tran TKA, Yu RMK, Islam R, Nguyen THT, Bui TLH, Kong RYC et al (2019) The utility of vitellogenin as a biomarker of estrogenic endocrine disrupting chemicals in molluscs. Environ Pollut 248:1067–1078

Tyler CR, Sumpter JP, Campbell PM (1991) Uptake of vitellogenin into oocytes during early vitellogenic development in the rainbow trout, *Oncorhynchus mykiss* (Walbaum). J Fish Biol 38(5):681–689

Williams VN, Reading BJ, Amano H (2017) Proportional accumulation of yolk proteins derived from multiple vitellogenesis precisely regulated during vitellogenesis in striped bass (Morone saxatilis). J Exp Zool A 321:301–315

Diversification and Classification of Vitellogenin in Fishes

3

Mary Mettilda Bai Silvester,
Arokya Glory Pushpa Thiraviam, Josephine Priyatharshini Chellappa, and Basil Rose Michael Rajam

Abstract

Vitellogenin (Vtg) is a protein synthesized by the liver in response to estrogen expressed in the female bloodstream during vitellogenesis. It is a high-molecular-mass complex protein consisting of sugar, lipid, and phosphorus with other binding elements such as calcium, iron, and zinc. It is the precursor of the lipoproteins and phosphoproteins that makes up most of the protein content of yolk. Vitellogenin transports proteins and some lipids from the liver through the blood to the growing oocytes and functions as the incredible provider and part of the yolk, the vital nutrient to the developing embryos of egg layers (oviparous), both vertebrates and invertebrates. In the oocyte, Vtgs are processed into yolk proteins, stored mostly as yolk granules also termed as globules or platelets and sometimes in an amorphous compartment (fluid yolk). Recent protein and gene analyses have revealed the presence of several vitellogenin variants. Fish Vtgs exhibit complex evolutionary history expressing significant disparity in structure and function. This chapter deals with the diversification and classification of piscine vitellogenin.

Keywords

Vitellogenin · Egg yolk protein · Lipoglycophosphoprotein · Lipovitellin · Phosvitin · Piscine vitellogenin

M. M. B. Silvester (✉) · A. G. Pushpa Thiraviam · J. P. Chellappa · B. R. Michael Rajam
Department of Zoology, Holy Cross College (Autonomous) (Affiliated to Manonmaniam Sundaranar University, Tirunelveli), Nagercoil, Tamil Nadu, India

V. Baskaralingam, R. Vanichviriyakit (eds.), *Vitellogenin in Fishes-Diversification, Biological Properties, and Future Perspectives*,
https://doi.org/10.1007/978-981-99-5340-0_3

Abbreviations

β′-c	β′-component
E2	17β-estradiol
EDTA	Ethylenediaminetetraacetic acid
FAA	Free amino acids
FSSP	Female-specific serum protein
HA	Hydroxyapatite
Lv	Lipovitellin
LvH	Lipovitellin heavy chain
LvL	Lipovitellin light chain
MS	Mass spectrometry
PAGE	Polyacrylamide gel electrophoresis
Pv	Phosvitin
SDS-PAGE	Sodium dodecyl sulfate polyacrylamide gel electrophoresis
Vtg	Vitellogenin
VtgAa	Acanthomorph A-type vitellogenin
VtgC	C-type vitellogenin

3.1 Introduction

Vitellogenin (Vg) is the precursor of vitellin (egg yolk protein). It is a female-specific protein expressed in the blood or body fluid during oogenesis of gravid females and can react with antibodies raised against egg extracts. This vitellogenin is synthesized in the female liver by the induction of hepatocytes with estrogen and is secreted into the blood to be incorporated into the egg (Hara et al. 2016). Vtg is a high-molecular-mass complex lipoglycophosphoprotein consisting of sugar, lipid, and phosphorus and binding with other elements such as calcium, iron, zinc, etc. Vtgs are produced in the females of all oviparous animals including invertebrates, fish, amphibians, reptiles, birds, and monotremes. Vitellogenins (Vtgs) are expressed in somatic tissues like vertebrate liver (Romano et al. 2004), blood and insect fat body (Tufail and Takeda 2008), and body fluid. The term "vitellogenin" was first used by Pan et al. (1969) to describe a female-specific protein expressed in the hemolymph of cecropia moth (vitellin + gene) meaning the source of egg yolk. Since then, the same term has been used for the proteins specifically expressed in the blood of maturing females of oviparous vertebrates. Though it is a matrotrophic protein, Vtg can be detected in the male fishes upon induction with 10^{-8} M estradiol-17β (Kanetoshi et al. 2004), estrogenic environment.

While vitellogenin regulation is diverse, they are usually regulated in a sex- and stage-specific manner being expressed in specific somatic tissues of adult females (Raikhel and Dhadialla 1992). Analyses in teleosts (e.g., salmonids) indicate that following increased production of follicle-stimulating hormone (FSH), the follicles are induced to produce estrogen, which then triggers the hepatic synthesis of Vtg (Specker and Sullivan 1994). The Vtgs recruit lipids and other nutrients before being secreted into the circulations. Graving fish oocytes selectively accumulate the circulating Vtg via receptor-mediated endocytosis (Stifani et al. 1990). These endocytosed vesicles fuse with lysosomes in the peripheral ooplasm forming multivesicular bodies where Vtg undergoes limited proteolysis giving rise to a number of yolk proteins (Bergink and Wallace 1974; Christmann et al. 1977; Hara and Hirai 1978; Carnevali et al. 1999; Hiramatsu et al. 2002; Matsubara et al. 2003; Polzonetti-Magni et al. 2004; Romano et al. 2004; Babin et al. 2007; Finn 2007a) such as lipovitellin (Lv), phosvitin (Pv), β′component (β′c), and C-terminal peptide (Matsubara et al. 1999, 2003; Hiramatsu et al. 2002). It's basic primary structure from the N-terminus is expressed, as NH_2-LvH-Pv-LvL-β'c-C terminal peptide COOH (Plack et al. 1971; Matsubara et al. 1999; Hiramatsu et al. 2002, 2006). For the derived proteins, lipovitellin (Lv) is a larger hydrophobic subunit that carry lipids, while phosvitin (Pv) is a smaller subunit with a high degree of phosphorylation (Raikhel and Dhadialla 1992). Fish Vtgs share a complex evaluating history, and they exhibit considerable variation in structure and function. Vitellogenin display a high degree of structural conservation, though they are diverged in regulation and copy number (Tufail et al. 2014).

3.2 General Characteristic of Vitellogenins

All vitellogenins (Vtgs) possess certain important common characteristics.

1. All Vtgs are phospholipoglycoprotein with a protein backbone carrying lipids, carbohydrates, and phosphate functional groups.
2. Vtgs are female-specific blood/body fluid proteins.
3. Vtgs are precursors synthesized in the liver/fat body in response to estrogen and induce the synthesis of yolk proteins.
4. Vtgs are bulky and complex proteins with molecular mass ranging from 300 to 600 kDa.
5. The ion-binding properties of Vtg (e.g., calcium, zinc, cadmium, iron, etc.) enable Vtg to supply minerals to the oocytes.

3.3 Diversification and Classification of Vitellogenin

Vitellogenins are massive proteins, comprising homodimers of two identical polypeptides each consisting of a linear array of yolk protein domains bearing phosphate, carbohydrate, and lipid components. Multiple alignment of Vtg sequences revealed five conserved regions (Lim et al. 2001). Complete forms of Vtg are made up of five yolk protein domains in a linear series as depicted in the Fig. 3.1: amino terminus (N)-lipovitellin heavy chain (LvH), phosvitin (Pv), lipovitellin light chain (LvL), *β′*-component, and C-terminal peptide (Ct)-carboxy terminus-(C).

Fig. 3.1 Yolk protein domains of a complete vitellogenin

Using immunological techniques, a number of female-specific serum proteins, i.e., vitellogenin, had been discovered in a number of species of teleosts (Plack et al. 1971; Le Menn 1979; Hara et al. 1983, 1986). Until the 1990s only a single type of Vtg was reported in teleosts. The presence of dual Vtg transcripts in the mummichog *Fundulus heteroclitus* as a result of cDNA cloning (LaFleur Jr et al. 1995a, b), and more than two Vtg transcripts or translated products in a large number of teleost fishes (Hiramatsu et al. 2002, 2005; Matsubara et al. 2003) insist on the need for classification of vitellogenin.

3.4 Classification of Vitellogenin Based on Degradation

Two models have been proposed to explain the types of vitellogenin based on their composition of substances or molecules found in it.

3.4.1 Single Vtg Model

According to this model, Vtg was considered to be made up of a single molecular species of protein, i.e., on degradation vitellogenin produces only amino acids.

3.4.2 Multiple Vtg Model

According to this model, Vtg was considered to be made up of many components, i.e., on degradation vitellogenin produces substances like lipovitellin, phosvitin, etc.

In order to understand the components of vitellogenin, it is better to look into the formation and cleaving of vitellogenin.

3.5 Formation of Vitellogenin

Estrogen secreted from the ovarian follicle triggers the synthesis of Vtg in the liver, which is then secreted into the blood, incorporated into oocytes, and cleaved to generate multiple egg yolk proteins (Reading and Sullivan 2011). Estrogen binds with estrogen receptors in hepatocytes and forms the complex estrogen-estrogen receptor complex. This complex in hepatocytes binds to the promoter region of the Vtg gene, which activates the gene to initiate and accelerate transcription. Then, translation of the Vtg transcript products begins, and the Vtg proteins undergo modification processes such as lipidation, phosphorylation, and glycosylation within the hepatocytes and then secreted into the blood.

3.6 Cleaving or Degradation of Vitellogenin

Blood Vtg binds to the Vtg receptor on the oocyte plasma membrane and is taken into the egg cell. Vtg that enters an oocyte is degraded by a cathepsin D-like enzyme specifically and undergoes molecular cleavage. This cleaving or proteolysis of vitellogenin during vitellogenesis produce more types of egg yolk proteins such as Lv (lipovitellin), Pv (phosvitin), and β'-c, which are stored in eggs (ooplasm).

The initial degradation of plasma vitellogenin during vitellogenesis in the ooplasm is termed the first proteolysis, and later degradations during final maturation of egg and embryogenesis are called the second and third proteolysis, respectively (Hiramatsu et al. 2002).

3.6.1 The First Proteolysis

The initial degradation of plasma vitellogenin during vitellogenesis in the ooplasm produces egg yolk proteins such as Lv (lipovitellin), Pv (phosvitin), and β'-c.

3.6.2 The Second Proteolysis

Further degradation of the egg yolk proteins during the final maturation of the oocytes is referred to as the second proteolysis. In the second degradation step, the majority of Pv and β'-c egg yolk proteins are degraded into free amino acids (FAA). This phenomenon was first reported in mummichogs (Wallace and Begovac 1985; Wallace and Selman 1985). The second proteolysis is an adaptation to maintain buoyancy of the egg and is observed in marine and brackish water fishes that lay their eggs in the marine water. In these fishes, the oocytes absorb large amounts of water during the final stages of maturation (Greeley Jr et al. 1986; Matsubara et al. 1995). This second proteolysis is not observed in salmonids, which lay their eggs in freshwater (Hiramatsu et al. 2002).

3.6.3 The Third Proteolysis

The third proteolysis of egg yolk proteins occurs during the embryogenesis of a few species. The three types of egg yolk proteins (Lv, Pv and β'-c) were subjected to separate degradations after fertilization, and that the degradation of Lv to smaller products and dephosphorylation of Pv occurred after the eyed embryo stage, while β'-c didn't undergo any degradation throughout embryonic development (Hiramatsu et al. 2002).

Protein and gene analyses have revealed the presence of a number of vitellogenin variants. Various Lv-Pv conjugations such as LvH-Pv, Pv-Lvl, and LvH-Pv-Lvl have been found in the yolk of many species of fishes. The Lv variants derived from each of these Vtgs were degraded differently during oocyte maturation (Matsubara et al. 2003).

3.7 Classification of Vtg Based on Phosvitin

Based on the presence or absence of phosvitin, fish or piscine Vtgs are divided into two main types, namely, complete and incomplete (Fig. 3.2). The complete Vtg has a deduced primary amino acid sequence composed of the five egg yolk protein regions (LvH, Pv, LvL, β'-c, and C-terminal coding domain). The second major group, the incomplete Vtg, is named as "Vtg type C" (VtgC), mainly consisting of LvH and LvL, which are the main components of lipoprotein in fish eggs and devoid of phosvitin. In other words the Pv-less Vtg is considered to be an incomplete Vtg (Hiramatsu et al. 2006).

Based on the homology analysis, the complete Vtg was further divided into two subgroups, type A (VtgA) and type B (VtgB). VtgA and VtgB have highly homologous primary structures and share similar characteristics such as molecular weight. Hence the purification of individual proteins VtgA and VtgB are difficult. In contrast, VtgC has a lower molecular weight than the other two types (VtgA and VtgB), and it is relatively easy to purify.

3.7.1 Classification Based on Molecular Evolution of Vtgs

This classification is based on the analyses of deduced amino acid sequences of Vtg gene transcripts (Finn and Kristoffersen 2007). The structure and function of fish Vtgs has diversified during the evolution. Fish Vtg genes are considered to have diversified through whole-genome duplication (WGD) events and also via lineage-specific tandem gene duplication (TGD), followed by neofunctionalization (Fig. 3.3).

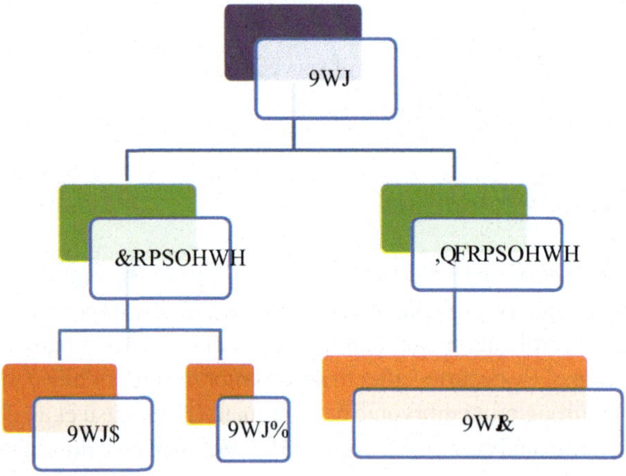

Fig. 3.2 Classification of vitellogenin

Fig. 3.3 Pathway of diversification of Vtgs

The chordate Vtg named as VtgABCD observed in silver lamprey *Ichthyomyzon unicuspis*is considered to be the ancestor of vertebrate Vtg appeared during the first round of whole-genome duplication (1R-WGD). In the subsequent 2R-WGD, Vtgs were differentiated into VtgAB (chondrostean vitellogenin; found in chondrostean fish, amphibians, and birds) and VtgCD. During the 3R-WGD, VtgAB allegedly differentiated to yield VtgA (found in all teleosts) and VtgB, and VtgCD was further divided into VtgC (fish VtgC or Pv-less Vtg) and VtgD (Fig. 3.4). Thus, the ancestral chordate VtgABCD, present in silver lamprey, arose after the *fi*rst round of WGD gave rise to VtgAB and VtgCD after the second round. The VtgCD subsequently gave rise to VtgC, present in most major *fi*sh taxa, and to VtgD, which is extinct.

The Vtgs formerly named VtgA and VtgB become VtgAa and VtgAb paralogs, respectively. VtgA experienced further lineage-specific gene duplication within teleosts, resulting in the formation of various paralogous and orthologous Vtg subtypes (e.g., VtgAa and VtgAb). The highly advanced group of fishes belonging to Paracanthopterygii and Acanthopterygii possess all three Vtg orthologous (VtgAa, VtgAb, and VtgC). In contrast, paralogous Vtg variants found in some other fish cannot be categorized as VtgAa, VtgAb, or VtgC. Salmonidae family in Protacanthopterygii, which was phylogenetically differentiated little earlier, contains two types of Vtgs: (1) complete form of VtgA, named as salmonid-type A-type Vtg: VtgAs, and (2) VtgC, incomplete form. Ostariophysi fishes have Ostariophysian-type A-type Vtg in two forms such as VtgAo1 and VtgAo2. Elopomorpha has Elopomorpha-type A-type Vtgs in three forms like VtgAe1, VtgAe2, and VtgAe3. In addition to these paralogous VtgA, both Ostariophysi and Elopomorpha have the orthologous VtgC (Reading and Sullivan 2011; Yamane et al. 2013; Williams et al. 2014; Mushirobira et al. 2013; Wu et al. 2014). Analyses of Vtg gene synteny (Babin 2008; Finn et al. 2009) indicated that the A-type and C-type Vtgs have a much longer evolutionary history (BP ~425 million). Comparative genomics analysis suggested the occurrence of two whole-genome duplication events, one before and the other after the divergence of ray-finned and lobe-finned fishes before the evolution of teleost fishes. Synteny analysis arranges these three types of pre-teleostVtg (Vtg A, Vtg B, and Vtg C) in the conserved Vtg gene cluster (VGC) which existed for over B450 million years (Andersen et al. 2017). VtgC, the incomplete vitellogenin also known as phosvitinless Vtg, is present in teleost fishes.

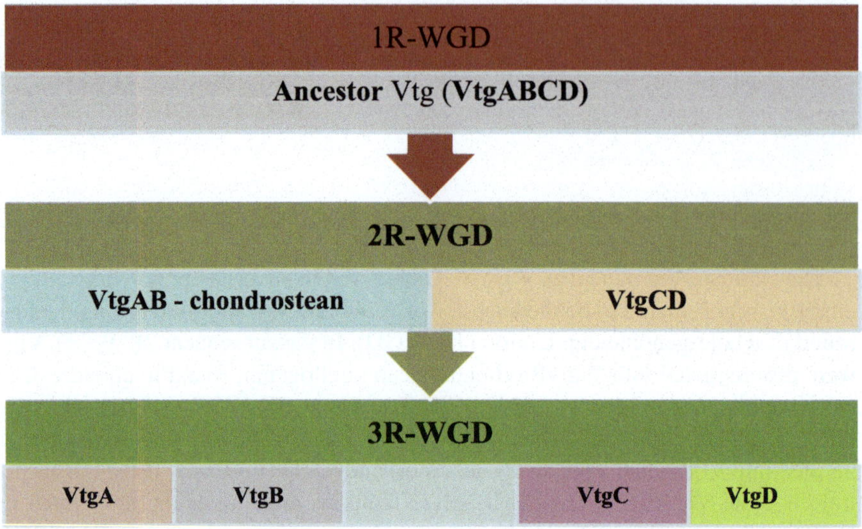

Fig. 3.4 Evolution of Vtg variants via three rounds of whole genome duplication (WGD) and some lineage-specific gene duplications (Finn and Kristoffersen 2007)

Proto-VtgCs is a subtype of vitellogenin present in Holocephali—elephant shark, chimera, West Indian coelacanth. It has shortened Pv domains. Pre-teleostean C type vitellogenin with complete Pv domains as the tandem VtgC1 and Vtg C2 is found in the spotted gar, a Holostean. Also partial Vtg gene deletions are observed among teleosts. The Ostariophysan VtgAo1 found in zebrafish (Cypriniformes) has Pv but doesn't have its β'-c and C-terminal coding domains. In the Atlantic salmon, the VtgAsb is reduced to its LvH domain. In the rainbow trout, *Oncorhynchus mykiss*, the locus of the gene encoding VtgAsa is duplicated in tandem 20 times, and the gene encoding VtgAsb is totally deleted.

The complete Vtg (VtgA and/or VtgB) and incomplete Vtg (VtgC) have been successfully isolated and purified as two different proteins with distinct molecular weights and antigenicity in various fish species (Ding et al. 1989; Kishida and Specker 1993; Shimizu et al. 2002; Ohkubo et al. 2003; Amano et al. 2010) including tilapia (*O. aureus* and *O. mossambicus*), medaka (*Oryzias latipes*), Japanese goby (*Acanthogobius flavimanus*), Sakhalin taimen (*Parahucho perryi*). In addition, three distinct types of Vtg have been purified or detected in mosquitofish *Gambusia affinis* (Sawaguchi et al. 2005), red sea bream *Pagrus major* (Sawaguchi et al. 2006), and grey mullet *Mugil cephalus* (Amano et al. 2007). Two or three types of Vtgs have been detected as proteins with distinct antigenicity in various existing fish species (Hiramatsu et al. 2006). Occurrence of diversified vitellogenins in different groups of fishes and the degradability of the major component, lipovitellin heavy chain is given in Table 3.1.

Table 3.1 Occurrence of diversified vitellogenins in diverse groups of fishes and the degradability of LvH (Hara et al. 2016)

Sl. no.	Type of Vtg	Category of the fish (superorder/order/sub-class)	Name of the fish	LvH processing during oocyte maturation
1.	VtgAa	Acanthopterygii	Atlantic halibut Barfin flounder Three-spined stickleback Red seabream	Fully degraded
			Mosquitofish Common mummichog Mangrove killifish Medaka Japanese common Gobi	Cleaved—partially degraded
		Paracanthopterygii	Haddock	Fully degraded
2.	VtgAb	Acanthopterygii	Atlantic halibut Barfin flounder Red seabream	Not degraded
			Mosquitofish Common mummichog Medaka Three-spined stickleback	Not degraded
		Paracanthopterygii	Haddock	Not degraded
3.	VtgAsa	Acanthopterygii	Chinook salmon Coho salmon Rainbow trout Arctic chart Brook trout Brown trout Grayling White fish	Not degraded
4.	VtgAsb	Paracanthopterygii	Atlantic salmon White fish Grayling	Not degraded
5.	VtgAo	Ostariophysi	Common carp Fathead minnow Zebrafish	May be nicked, but not degraded
6.	VtgAe	Elopomorpha	Japanese eel Conger eel	Degraded
7.	VtgAB	Chondrostei	White sturgeon	Not degraded
8.	VtgC	Acanthopterygii	Red seabream Torafugu Three-spined stickleback Mosquitofish Medaka Japanese common goby	Not degraded
		Ostariophysi	Zebrafish	Not degraded
9.	VtgABCD	Hyperoartia	Silver lamprey	Not degraded

3.8 Distribution of Diverse Vitellogenin in Different Fishes

In oviparous vertebrates including fishes, Vtg serves as a major precursor of egg protein carrying essential nutrients for future embryogenesis. Fish vitellogenins not only act as precursor for yolk protein, but they provide proper egg buoyancy, metabolic water for embryonic development. and provide a planned supply of carbohydrate, protein, and lipid nourishment to growing embryos during their development (Reading et al. 2018). An iron-binding female-specific serum protein (FSSP) was reported in teleost fishes, chum salmon (*Oncorhynchus keta*), and rainbow trout (*O. mykiss*) by Hara and Hirai (1976), which was isolated and identified as Vtg. Urist and Schjeide (1961) reported the occurrence of identical yolk precursor proteins, vitellogenin, in the blood of most oviparous vertebrate species like teleost fishes, amphibians, and reptiles.

Nath and Sundararaj (1981) semi-purified Vtg (550 kDa) from plasma of E2-treated male and female Singhi (*Heteropneustes fossilis*) by gel filtration. Vitellogenins, Vtg1/VtgA (Hamazaki et al. 1987) and Vtg2/VtgB (Shimizu et al. 2002) were purified from the ascites of Medaka, *Oryzias latipes*. Ding et al. (1989) identified two forms of vitellogenin from the plasma and gonads of male *Oreochromis aureus*. Nath et al. (1992) purified two molecular forms of Vtg, Vtg1/VtgA (430 kDa) and Vtg2/VtgB (240 kDa), from the plasma of E2-treated rohu, *Labeo rohita*, by gel chromatography method. Vitellogenin Vtg-200 and Vtg-130 were identified in tilapia, *Oreochromis mossambicus* by Johanning and Specker (1995).

Two different vitellogenins, Vtg1 and Vtg2, were identified by LaFleur Jr et al. (1995a) in mummichog, *Fundulus heteroclitus*. Wang et al. (2000) demonstrated the presence of Vtg cDNA (Vtg3/VtgC) encoded an unusual (incomplete) Vtg lacking a phosvitin (polyserine) domain and having low sequence similarity to other fish Vtg in zebrafish, *Danio rerio*. Two vitellogenin types, Had1 and Had2, were identified in haddock, *Melanogrammus aeglefinus* by Reith et al. (2001). Vitellogenin was isolated from the plasma of E2-treated murrel through gel filtration followed by ion-exchange chromatography (Sehgal and Goswami 2001).

Two major vitellogenins, VtgA and VtgB, were identified in the serum of barfin flounder, *Verasper moseri* (Matsubara et al. 1999), and haddock, *Melanogrammus aeglefinus* (Reith et al. 2001), using biochemical analysis and cDNA cloning. Hiramatsu et al. (2002) purified two complete Vtg, VtgA and VtgB, along with Pv-poor Vtg (VtgC) from the plasma of E2-treated white perch, *Morone americana*. A vitellogenin with low phosphorus content was reported in the serum or ascites fluid of estradiol-treated tilapia, *Oreochromis mossambicus* (Kishida and Specker 1993), and medaka, *Oryzias latipes* (Shimizu et al. 2002).

Matsubara et al. (2003) detected all three forms of Vtg in red sea bream (*Pagrus major*), white-edged rockfish (*Sebastes taczanowskii*), mummichog (*Fundulus heteroclitus*), striped mullet (*Mugil cephalus*), and mosquitofish (*Gambusia affinis*). They also identified VtgA and VtgB in barfin flounder and *Walleye pollock* and VtgA and Pv-less Vtg in Japanese common goby (*Acanthogobius flavimanus*). In addition, Vtg and Pv-less were detected in white-spotted char (*Salvelinus leucomaenis*), zebrafish (*Danio rerio*), and Japanese eel (*Anguilla japonica*). They also identified Vtg in Pacific herring (*Clupea pallasii*).

The vitellogenin VtgA and VtgB were present in Paracanthopterygii and Acanthopterygii, and Pv-less Vg is widely distributed among teleosts (Matsubara et al. 2003). Ohkubo et al. (2003) isolated VtgAa type (Vg-530) and VtgC type (Vg-320) from the serum of Japanese goby (*Acanthogobius flavimanus*). Two forms of Vtg, VtgA (600 kDa), VtgB (400 kDA), and C-type, were purified from the plasma of E2-treated mosquitofish, *Gambusia affinis* (Sawaguchi et al. 2005). Sehgal and Goswami (2005) identified three changed isomeric forms of Vtg in the blood of E2-treated murrel (*Channa striata*). The native Vtg (530 kDa) showed three protein bands on native polyacrylamide gel electrophoresis (PAGE) which resolved into a single peptide (175 kDa) on sodium dodecyl sulfate (SDS) PAGE.

From the red seabream (*Pagrus major*), RsbVtgA and RsbVtgB (complete forms of Vtg) and incomplete PvlVtg (RsbPvlVtg) were isolated by Sawaguchi et al. (2006). Ndiaye et al. (2006) identified vitellogenins Vtg1 and Vtg2 from the plasma of *Oreochromis niloticus* with a molecular mass of 130 kDa and 170 kDa, respectively. Amano et al. (2007) purified vitellogenins VtgA, VtgB, and VtgC with molecular masses of 570, 580, and 335 kDa, respectively, from grey mullet (*Mugil cephalus*). Using gel filtration on UltrogelAcA 34 followed by adsorption chromatography on hydroxyapatite (HA)-Ultrogel, two forms of vitellogenin, HAI and HAII, of 75 kDa and 85 kDa, respectively, were isolated from the estradiol E2-treated plasma of Indian major carp, *Cirrhinus mrigala* by Maitra et al. (2007).

Kolarevic et al. (2007) identified three forms of vitellogenins: VtgAa, VtgAb, and VtgC in goldsinny wrasse *Ctenolabrus rupestris*. Amano et al. (2010) identified the vitellogenins, VtgAs and VtgC, from Sakhalin taimen (*Parahucho perryi*). Reading et al. (2011) reported the presence of multiple Vtgs (VtgAa, VtgAb, and VtgC) subtypes in white perch (*Morone americana*).

Yamane et al. (2013) purified a vitellogenin with a molecular mass of 560 kDa in its intact state and which was found to be 210 kDa under reduced condition on SDS-PAGE from the fish cloudy catshark, *Scyliorhinus torazame*. A Vtg having a molecular weight of 482 kDa was purified from the plasma of E2-induced Asian catfish, *Clarias batrachus* by Garnayak et al. (2013). This study revealed that Vtg synthesis was induced in male *C. batrachus* following three intraperitoneal injections of 17β-estradiol (E2). It was purified from E2-induced plasma by precipitation with EDTA and $MgCl_2$ followed by gel filtration chromatography on Sephacryl S-300-HR. The purified protein on native gradient PAGE appeared as a single band with apparent molecular weight of ~482 kDa. Further, this protein was found to be positive for carbohydrate, lipid, and phosphorus and thus was confirmed as vitellogenin. In SDS-PAGE, two major polypeptides corresponding to ~97 and ~67 kDa were observed by them.

Multiple Vtgs named as salmonid A-type Vtgs (VtgAs) and C-type Vtg (VtgC) were identified in cutthroat trout (*Oncorhynchus clarkii*) by Mushirobira et al. (2015). Vitellogenins VtgAa, VtgAb, and VtgC were isolated from the liver, plasma, and ovary of striped bass, *Morone saxatilis* using label-free quantitative mass spectrometry (MS) by Williams et al. (2014).

Table 3.2 Distribution of different types of vitellogenin in fishes (Hara et al. 2016)

Sl. no.	Type of vitellogenin	Name of the fishes
1.	VtgAa	White perch, striped bass, red seabream, flathead mullet, Atlantic halibut, barfin flounder, mummichog, Western mosquito fish, haddock
2.	VtgAs	White spotted char, rainbow trout, cutthroat trout
3.	VtgAb	Mummichog, red tailed splitfin, Western mosquito fish, haddock, flathead mullet, Atlantic halibut, barfin flounder, red seabream, white perch, striped bass
4.	VtgAe1	Whitespotted conger, Japanese eel, Atlantic herring
5.	VtgAe2	Japanese eel, zebrafish
6.	VtgAe3	Japanese eel
7.	VtgAo1	Oriental weatherfish, zebrafish, Chinese minnow
8.	VtgAo2	Zebrafish, common carp
9.	VtgAB	White sturgeon
10.	VtgAB2	Zebra finch, herring gull
11.	VtgC	Zebrafish, oriental weatherfish, Japanese common goby, cutthroat trout, whitespotted char, Western mosquito fish, red tiled splitfin, flathead mullet, red seabream, white perch, striped bass
12.	VtgABCD	Catshark

Yilmaz et al. (2016) identified VtgAa, VtgAb, and VtgC in European sea bass, *Dicentrarchus labrax.* Reading et al. (2017) reported the production of VtgAa, VtgAb, and VtgC in the ovoviviparous mosquitofish (*Gambusia affinis*) and the viviparous redtail (*Xenotoca eiseni).* Different levels of transcripts encoding the three forms of Vtg (VtgAa, VtgAb, and VtgC) were assessed in wild greater amberjack (*Seriola dumerili*) by Pousis et al. (2017).

Mahapatra et al. (2017) purified two forms of Vtg, Vtg1 and Vtg 2, from the plasma of estradiol- E2-treated Indian walking catfish (*Clarias batrachus*) by gel filtration and adsorption chromatography. The molecular masses of Vtg1 and Vtg2 were found to be 375 and 450 kDa, respectively. Yilmaz et al. (2014, 2018a, b) reported the presence of VtgI, VtgII, and VtgIII in zebrafish (*Danio rerio*). The distribution of different types of vitellogenin in fishes is shown in the Table 3.2.

3.8.1 Distribution of Vitellogenin Within the Fish Body

In vitellogenic females, vitellogenins were strongly expressed in the liver. Studies in zebrafish, *Danio rerio*, revealed that the liver is not the only organ which show the presence of vitellogenins but also in extrahepatic tissues like the heart, brain (Yin et al. 2009), skin (Jin et al. 2008), and gill (Islinger et al. 2003), white adipose tissue (Tingaud-Sequeira et al. 2012), intestine, ovary, muscle, and estrogen-treated testes of male fish (Wang et al. 2005). In white cloud mountain minnow, *Tanichthys albonube*, the expression of vitellogenin was observed in the gill, ovary, and testes (Wang et al. 2010). Vitellogenin was also detected in the testes and kidney of spotted ray, *Torpedo marmorata* (Del Giudice et al. 2011). In Chinese rare minnow,

Gobiocypris rarus, vitellogenin was expressed in the heart and brain (Ma et al. 2009). Juvenile salmon were induced to express vitellogenin in the skin when exposed to nonylphenol (Arukwe and Roe 2008). The estrogen-treated female zebrafish expressed the presence of vitellogenin in the heart, liver, spleen, kidney, skin, muscle, gill, eye, fin, blood, and brain (Zhong et al. 2014).

Vitellogenin mRNA was expressed in the testis of E2-treated males of white sturgeon *Acipenser transmontanus* (Bidwell and Carlson 1995); zebrafish, *Danio rerio* (Wang et al. 2005); and of medaka, *Oryzias latipes*, treated with E2 or nonylphenol, an alkylphenol with estrogenic activity (Kobayashi et al. 2005; Koger et al. 1999). The expression of vitellogenin in many extrahepatic tissues may be due to the wide distribution of white adipose tissue which has been identified to express the Vtg1 gene (Tingaud-Sequeira et al. 2012). Some viviparous fish species have maintained Vtg genes and a large amount of yolk nutrient in their eggs (Sawaguchi et al. 2005; Vega-Lopez et al. 2007).

3.8.2 Distribution of Vitellogenin Proteins in Intraovarian Embryo

In fishes, Vtg protein is one of the maternal nutrients supplied into the intraovarian embryo. Mother-to-embryo vitellogenin transport in a viviparous teleost fish *Xenotoca eiseni* was studied using fluorescent immunohistochemistry and immunoelectron microscopy (Iida et al. 2019). Vtg proteins are distributed in intracellular vesicles in the epithelial cells, and the mesenchymal cell surface of the trophotaeniae, the pseudo-placenta. These findings revealed that the maternal components including Vtg proteins dissolve in the ovarian fluid as secreted proteins and could be absorbed as macromolecules into the trophotaeniae through the epithelial cells, retaining their antigenicity. The study concluded that the maternal vitellogenins in interstitial fluid and blood serum could be secreted into the ovarian fluid from the ovarian tissues without any changes in their molecular weight, and then absorbed into the intraovarian embryo via the trophotaeniae.

3.9 Conclusion

Vitellogenin (Vtg), a female-specific plasma egg yolk protein precursor, is the rich source of nutrients for the developing embryo. It's synthesized in the liver under the induction of estrogen. There are diversified forms of vitellogenin in different groups of fishes. Vtgs are broadly classified into complete and incomplete vitellogenins and named as VtgA, VtgB, VtgC, etc.

Acknowledgments The authors with a deep sense of gratitude acknowledge the constant support and encouragement given by the College management.

Conflict of Interest The authors have no conflicts of interest to declare.

References

Amano H, Fujita T, Hiramatsu N, Sawaguchi S, Matsubara T, Sullivan CV, Hara A (2007) Purification of multiple vitellogenins in grey mullet (*Mugil cephalus*). Mar Biol 152:1215–1225

Amano H, Mochizuki M, Fujita T, Hiramatsu N, Todo T, Hara A (2010) Purification and characterization of a novel incomplete-type vitellogenin protein (VtgC) in Sakhalin taimen (*Hucho perryi*). Comp Biochem Physiol 157A:41–48

Andersen Ø, Xu C, Timmerhaus G, Kirste KH, Naeve I, Mommens M, Tveiten H (2017) Resolving the complexity of vitellogenins and their receptors in the tetraploid Atlantic salmon (*Salmo salar*): ancient origin of the phosvitin-less VtgC in chondrichthyan fishes. Mol Reprod Dev 84(11):1191–1202

Arukwe A, Roe K (2008) Molecular and cellular detection of expression of vitellogenin and zona radiata protein in liver and skin of juvenile salmon (*Salmo salar*) exposed to nonylphenol. Cell Tissue Res 331:701–712

Babin PJ (2008) Conservation of a vitellogenin gene cluster in oviparous vertebrates and identification of its traces in the platypus genome. Gene 413:76–82

Babin PJ, Carnevali O, Lubzens E, Schneider WJ (2007) Molecular aspects of oocyte vitellogenesis in fish. In: Babin PJ, Cerda J, Lubzens E (eds) The fish oocyte: from basic studies to biotechnological applications. Springer, Dordrecht, pp 39–76

Bergink EW, Wallace RW (1974) Precursor-product relationship between amphibian vitellogenin and yolk proteins, lipovitellin and phosvitin. J Biol Chem 249:2897–2903

Bidwell CA, Carlson DM (1995) Characterization of vitellogenin from white sturgeon, *Acipenser transmontanus*. J Mol Evol 41:104–112

Carnevali O, Carletta R, Cambi A, Vita A, Bromage N (1999) Yolk formation and degradation during oocyte maturation in seabream, *Sparus aurata*: involvement of two lysosomal proteinases. Biol Reprod 60:140–146

Christmann JL, Grayson MJ, Huang RCC (1977) Comparative study of hen yolk phosvitin and plasma vitellogenin. Biochemistry 16:3250–3256

Del Giudice G, Prisco M, Agnese M, Verderame M, Limatola E, Andreuccetti P (2011) Expression of vitellogenin in the testis and kidney of the spotted ray, *Torpedo marmorata* exposed to 17-beta-estradiol. Gen Comp Endocrinol 174:318–325

Ding JL, Hee PL, Lam TJ (1989) Two forms of vitellogenin in the plasma and gonads of male *Oreochromis aureus*. Comp Biochem Physiol 93B:363–370

Finn RN (2007a) Vertebrate yolk complexes and the functional implications of phosvitin and other subdomains in vitellogenins. Biol Reprod 76:926–935

Finn RN, Kristoffersen BA (2007) Vertebrate vitellogenin gene duplication in relation to the "3R hypothesis": correlation to the pelagic egg and the oceanic radiation of teleosts. PLoS One 2:e169

Finn RN, Kolarevic J, Kongshaug H, Nilsen F (2009) Evolution and differential expression of a vertebrate vitellogenin gene cluster. BMC Evol Biol 9:2

Garnayak SK, Mohanty J, Rao TV, Sahoo SK, Sahoo PK (2013) Vitellogenin in Asian catfish, *Clarias batrachus*: purification, partial characterization and quantification during the reproductive cycle by ELISA. Aquaculture 392–395:148–155

Greeley MS Jr, Calder DR, Wallace RA (1986) Changes in teleost yolk proteins during oocyte maturation: correlation of yolk proteolysis with oocyte hydration. Comp Biochem Physiol 84B:1–9

Hamazaki TS, Iuchi I, Yamagami K (1987) Purification and identification of vitellogenin and its immunohistochemical detection in growing oocytes of the teleost, *Oryzias latipes*. J Exp Zool 242(3):333–341

Hara A, Hirai H (1976) Iron-binding activity of female-specific serum proteins of rainbow trout (*Salmo gairdneri*) and chum salmon (*Oncorhynchus keta*). Biochim Biophys Acta 437:549–557

Hara A, Hirai H (1978) Comparative studies on immunochemical properties of female specific serum protein and egg yolk proteins in rainbow trout (*Salmo gairdneri*). Comp Biochem Physiol 59B:339–343

Hara A, Takano K, Hirai H (1983) Immunochemical identification of female-specific serum protein, vitellogenin, in the medaka, *Oryzias latipes* (teleosts). Comp Biochem Physiol 74A:135–141

Hara A, Takemura A, Matsubara T, Takano K (1986) Immunochemical identification of female-specific serum proteins in a viviparous fish, the white-edged rockfish (*Sebastes taczanowskii*), during vitellogenesis and pregnancy, and after estrogen treatment. Bull Fac Fisheries Hokkaido Univ 37:101–110

Hara A, Hiramatsu N, Fujita T (2016) Vitellogenesis and choriogenesis in fishes. Fish Sci 82:187–202

Hiramatsu N, Ichikawa N, Fukada H, Fujita T, Sullivan CV, Hara A (2002) Identification and characterization of proteases involved in specific proteolysis of vitellogenin and yolk proteins in salmonids. J Exp Zool 292:11–25

Hiramatsu N, Cheek AO, Sullivan CV, Matsubara T, Hara A (2005) Vitellogenesis and endocrine disruption. In: Mommsen TP, Moon TW (eds) Biochemistry and molecular biology of fishes: environmental toxicology, vol 6, pp 431–471

Hiramatsu N, Matsubara T, Fujita T, Sullivan CV, Hara A (2006) Multiple piscine vitellogenins: biomarkers of fish exposure to estrogenic endocrine disruptors in aquatic environments. Mar Biol 149:35–47

Iida A, Araib HN, Someyac Y, Inokuchic M, Onumad TA, Yokoie H, Suzukie T, Hondoa E, Sanof K (2019) Mother-to-embryo vitellogenin transport in a viviparous teleost *Xenotoca eiseni*. Proc Natl Acad Sci U S A 116(44):22359–22365

Islinger M, Willimski D, Volkl A, Braunbeck T (2003) Effects of ethinyl estradiol on the expression of three estrogen-responsive genes and cellular ultrastructure of liver and testes in male zebrafish. Aquat Toxicol 62:85–103

Jin YX, Wang WY, Sheng GD, Liu WP, Fu ZW (2008) Hepatic and extra hepatic expression of estrogen-responsive genes in male adult zebrafish (*Danio rerio*) as biomarkers of short-term exposure to 17beta-estradiol. Environ Monit Assess 146:105–111

Johanning KM, Specker JL (1995) Characterization of yolk proteins during oocyte development of tilapia, *Oreochromis mossambicus*. Com Biochem Physiol B Biochem Mol Biol 112(2):177–189

Kanetoshi A, Katsura E, Fujimoto T, Kojima H, Hori Y, Fukada H, Takahara S, Hara A (2004) Study on the screening test of endocrine disrupting chemicals using carp (*Cyprinus carpio*) hepatocyte culture: vitellogenin induction test by estrogens. Rep Hokkaido Inst Pub Health 54:1–6

Kishida M, Specker JL (1993) Vitellogenin in tilapia (*Oreochromis mossambicus*): induction of two forms by estradiol, quantification in plasma and characterization in oocyte extract. Fish Physiol Biochem 12:171–182

Kobayashi K, Tamotsu S, Yasuda K, Oishi T (2005) Vitellogeninimmuno histochemistry in the exposed to 17β-estradiol and 4-nonylphenol in the liver the testis of the medaka *Oryzias latipes*. Zool Sci 22:453–461

Koger CS, Teh SJ, Hinton DE (1999) Variations of light and temperature regimes and resulting effects on reproductive parameters in medaka (*Oryzias latipes*). Biol Reprod 61(5):1287–1293

Kolarevic J, Nerland A, Nilsen F, Finn RN (2007) Goldsinny wrasse (*Ctenolabrus rupestris*) is an extreme VtgAa-type pelagophil teleost. Mol Reprod Dev 75(6):1011–1020

LaFleur GJ Jr, Byrne BM, Kanungo J, Nelson LD, Greenberg RM, Wallace RA (1995a) *Fundulus heteroclitus* vitellogenin: the deduced primary structure of a piscine precursor to noncrystalline, liquid-phase yolk protein. J Mol Evol 41:505–521

LaFleur GJ Jr, Byrne BM, Haux C, Greenberg RM, Wallace RA (1995b) Liver-derived cDNAs: vitellogenins and vitelline envelope protein precursors (choriogenins). In: Goetz FW, Thomas P (eds) Proceedings of the fifth international symposium on the reproductive physiology of fish. University of Texas, Austin, pp 336–338

Le Menn F (1979) Some aspects of vitellogenesis in a teleostean fish: *Gobius niger* L. Comp Biochem Physiol 62A:495–500

Lim EH, Teo BY, Lam TJ, Ding JL (2001) Sequence analysis of a fish vitellogenin cDNA with a large phosvitin domain. Gene 277:175–186

Ma LW, Li DL, Wang JW, He JY, Yin Z (2009) Effects of adrenergic agonists on the extra hepatic expression of vitellogenin in heart and brain of the Chinese rare minnow (*Gobiocypris rarus*). Aquat Toxicol 91:19–25

Mahapatra S, Kabita SK, Bhattacharya D, Sarkar S, Juin SK, Maitra S, Nath P (2017) Purification and development of ELISAs for two forms of vitellogenin in Indian walking catfish, *Clarias batrachus* (L.). Fish Physiol Biochem 43(2):477

Maitra S, Sahu R, Trehan N, Garg SK, Nath P (2007) Circannual variation in plasma vitellogenin and gonadotropin II levels in relation to annual ovarian cycle in female mrigal, *Cirrhinus mrigala*. Aquaculture 265(1–4):370–384

Matsubara T, Adachi S, Ijiri S, Yamauchi K (1995) Changes of lipovitellin during *in vitro* oocyte maturation in Japanese flounder, *Paralichthys olivaceus*. Fish Sci 61:478–481

Matsubara T, Ohkubo N, Andoh T, Sullivan CV, Hara A (1999) Two forms of vitellogenin, yielding two distinct lipovitellins, play different roles during oocyte maturation and early development of Barfin flounder, *Veraspermoseri*, a marine teleost that spawns pelagic eggs. Dev Biol 213:18–32

Matsubara T, Nagae M, Ohkubo N, Andoh T, Sawaguchi S, Hiramatsu N, Sullivan CV, Hara A (2003) Multiple vitellogenins and their unique roles in marine teleosts. Fish Physiol Biochem 28:295–299

Mushirobira Y, Mizuta H, Luo W, Morita Y, Sawaguchi S, Matsubara T, Hiramatsu N, Todo T, Hara A (2013) Changes in levels of dual vitellogenin transcripts and proteins in cutthroat trout *Oncorhynchus clarkii* during ovarian development. Nippon Suisan Gakkaishi 79:175–189

Mushirobira Y, Mizuta H, Luo W, Todo T, Hara A, Reading BJ, Sullivan CV, Hiramatsu N (2015) Molecular cloning and partial characterization of a low-density lipoprotein receptor-related protein 13 (Lrp13) involved in vitellogenin uptake in the cutthroat trout (*Oncorhynchus clarkii*). Mol Reprod Dev 82(12):986–1000

Nath P, Sundararaj BI (1981) Isolation and identification of female-specific serum lipophosphoprotein (vitellogenin) in the catfish, *Heteropneustes fossilis*. Gen Comp Endocrinol 43:184–190

Nath P, Bhakta M, Mitra K (1992) Demonstration of two forms of vitellogenin in serum of estradiol-17 beta-treated Indian major carp, Labeo rohita. Indian J Exp Biol 30(6):464–469

Ndiaye P, Forgue J, Lamothe L, Cauty C, Tacon P, Lafon P, Davail B, Fostier A, Le Menn F, Núñez J (2006) Tilapia (*Oreochromis niloticus*) vitellogenins: development of homologous and heterologous ELISAs and analysis of vitellogenin pathway through the ovarian follicle. J Exp Zool A Comp Exp Biol 305:576–593

Ohkubo N, Mochida K, Adachi S, Hara A, Hotta K, Nakamura Y, Matsubara T (2003) Development of enzyme-linked immunosorbent assays for two forms of vitellogenin in Japanese common goby (*Acanthogobius flavimanus*). Gen Comp Endocrinol 131:353–364

Pan ML, Bell WJ, Telfer WH (1969) Vitellogenic blood protein synthesis by insect fat body. Science 165:393–394

Plack PA, Pritchard DJ, Fraser NW (1971) Egg proteins in cod serum: natural occurrence and induction by injections of oestradiol 3-benzoate. Biochem J 121:847–856

Polzonetti-Magni AM, Mosconi G, Soverchia L, Kikuyama S, Carnevali O (2004) Multihormonal control of vitellogenesis in lower vertebrates. Int Rev Cytol 239:1–46

Pousis C, Mylonas CC, DeVirgilio C, Gadaleta G, Santamaria N, Passantino L, Zupa R, Papadaki M, Fakriadis I, Ferreri R, Corriero A (2017) The observed oogenesis impairment in greater amberjack *Seriola dumerili* (Risso, 1810) reared in captivity is not related to an insufficient liver transcription or oocyte uptake of vitellogenin. Aquac Res 49(67):243

Raikhel AS, Dhadialla T (1992) Accumulation of yolk proteins in insect oocytes. Annu Rev Entomol 37:217–251

Reading BJ, Sullivan CV (2011) The reproductive organs and processes-vitellogenesis in fishes. In: Farrell AP (ed) Encyclopedia of fish physiology: from genome to environment. The reproductive organs and processes. Reference module in life sciences. Elsevier, Maryland Heights, MO, pp 635–646

Reading BJ, Hiramatsu N, Sullivan CV (2011) Disparate binding of three types of vitellogenin to multiple forms of vitellogenin receptor in white perch (*Morone americana*). Biol Reprod 84:392–399

Reading BJ, Sullivan CV, Schilling J (2017) Vitellogenesis in fishes. In: Reference module in life sciences. Elsevier, Amsterdam. 152(6):1215–1225

Reading BJ, Andersen LK, Ryu Y, Mushirobira Y, Todo T, Hiramatsu N (2018) Oogenesis and egg quality in finfish: yolk formation and other factors influencing female fertility. Aust Fish 3(4):45

Reith M, Munholland J, Kelly J, Finn RN, Fyhn HJ (2001) Lipovitellins derived from two forms of vitellogenin are differentially processed during oocyte maturation in haddock (*Melanogrammus aeglefinus*). J Exp Zool 291:58–67

Romano M, Rosanova P, Anteo C, Limatola E (2004) Vertebrate yolk proteins: a review. Mol Reprod Dev 69:109–118

Sawaguchi S, Koya Y, Yoshizaki N, Ohkubo N, Andoh T, Hiramatsu N, Sullivan CV, Hara A, Matsubara T (2005) Multiple vitellogenins (Vtgs) in mosquitofish (*Gambusia affinis*): identification and characterization of three functional Vtg genes and their circulating and yolk proteins products. Biol Reprod 72:1045–1060

Sawaguchi S, Kagawa H, Ohkubo N, Hiramatsu N, Sullivan CV, Matsubara T (2006) Molecular characterization of three forms of vitellogenin and their yolk protein products during oocyte growth and maturation in red sea bream (*Pagrus major*), a marine teleost spawning pelagic eggs. Mol Reprod Dev 73(6):719–736

Sehgal N, Goswami SV (2001) Biochemical changes in the liver of the Indian freshwater murrel, *Channa punctatus* (Bloch) during estradiol-induced vitellogenin synthesis. Fish Physiol Biochem 24:149–155

Sehgal N, Goswami SV (2005) Vitellogenin exists as charge isomers in the Indian freshwater murrel, *Channa punctatus* (Bloch). Gen Comp Endocrinol 141(1):12–21

Shimizu M, Fujiwara Y, Fukada H, Hara A (2002) Purification and identification of a second form of vitellogenin from ascites of Medaka (*Oryzias latipes*) treated with estrogen. J Exp Zool 293:726–735

Specker JL, Sullivan CV (1994) Vitellogenesis in fishes: status and perspectives. In: Davey KG, Petter RE, Tobe SS (eds) Perspectives in comparative endocrinology. National Research Council, Ottawa, pp 304–315

Stifani S, Le Menn F, Rodriguez JN, Schneider WJ (1990) Regulation of oogenesis: the piscine receptor for vitellogenin. Biochem Biophys Acta 1045:271–279

Tingaud-Sequeira A, Knoll-Gellida A, Andre M, Babin PJ (2012) Vitellogenin expression in white adipose tissue in female teleost fish. Biol Reprod 86:38

Tufail M, Takeda M (2008) Molecular characteristics of insect vitellogenins. J Insect Physiol 54:1447–1458

Tufail M, Nagaba Y, Elgendy AM, Takeda M (2014) Regulation of vitellogenin genes in insects. Entomol Sci 17:269–282

Urist MR, Schjeide OA (1961) The partition of calcium and protein in the blood of oviparous vertebrates during estrus. J Gen Physiol 44(4):743–756

Vega-Lopez A, Ortiz-Ordonez E, Uria-Galicia E, Mendoza-Santana EL, Hernandez-Cornejo R, Atondo-Mexia R, García-Gasca A, García-Latorre E, Dominguez-Lopez ML (2007) The role of vitellogenin during gestation of *Girardinichthys viviparus* and *Ameca splendens*; two goodeid fish with matrotrophic viviparity. Comp Biochem Physiol A Mol Integr Physiol 147(3):731–742

Wallace RA, Begovac PC (1985) Phosvitins in *Fundulus* oocytes and eggs: preliminary chromatographic and electrophoretic analyses together with biological considerations. J Biol Chem 260:11268–11274

Wallace RA, Selman K (1985) Major protein changes during vitellogenesis and maturation of *Fundulus* oocytes. Dev Biol 110:492–498

Wang H, Yan T, Tan JTT, Gong ZA (2000) Zebrafish vitellogenin gene (*vg3*) encodes a novel vitellogenin without a phosvitin domain and may represent a primitive vertebrate vitellogenin gene. Gene 256:303–310

Wang H, Tan JTT, Emelyanov A, Korzh V, Gong Z (2005) Hepatic and extrahepatic expression of vitellogenin genes in the zebrafish *Danio rerio*. Gene 356:91–100

Wang RL, Gao Y, Zhang LH, Zhang YK, Fang ZQ, He JG, Zhang WM, Ma GZ (2010) Cloning, expression, and induction by 17-beta estradiol (E-2) of a vitellogenin gene in the white cloud mountain minnow *Tanichthys albonubes*. Fish Physiol Biochem 36:157–164

Williams VN, Reading BJ, Hiramatsu N, Amano H, Glassbrook N, Hara A, Sullivan CV (2014) Multiple vitellogenins and product yolk proteins in striped bass, *Morone saxatilis*: molecular characterization and processing during oocyte growth and maturation. Fish Physiol Bichem 40:395–415

Wu M, Nishimiya O, Nakamori M, Soyano K, Todo T, Hara A, Hiramatsu N (2014) Molecular cloning and characterization of the expression profiles of vitellogenin transcripts in the dojo loach (*Misgurnus anguillicaudatus*) in response to 17α-ethinyl estradiol and 17β-estradiol administration. Zool Sci 31:202–212

Yamane K, Yagai T, Nishimiya O, Sugawara R, Amano H, Fujita T, Hiramatsu N, Todo T, Matsubara T, Hara A (2013) Characterization of vitellogenin and its derived yolk proteins in cloudy catshark (*Scyliorhinus torazame*). Fish Physiol Biochem 39:373–390

Yilmaz O, Patinote A, Nguyen T, Bobe J (2014) Multiple vitellogenins in zebrafish (*Danio rerio*): quantitative inventory of genes, transcripts and proteins, and relation to egg quality. Fish Physiol Biochem 44:1509–1525

Yilmaz O, Prat F, Ibanez AJ, Koksoy S, Amano H, Sullivan CV (2016) Multiple vitellogenins and product yolk proteins in European sea bass (*Dicentrarchus labrax*): molecular characterization, quantification in plasma, liver and ovary, and maturational proteolysis. Comp Biochem Physiol B Biochem Mol Biol 194–195:71–86

Yilmaz O, Patinote A, Nguyen T, Bobe J (2018a) Multiple vitellogenins in zebrafish (*Danio rerio*): quantitative inventory of genes, transcripts and proteins, and relation to egg quality. Fish Physiol Biochem 44:1509–1525

Yilmaz O, Patinote A, Nguyen T, Com E, Pineau C, Bobe J (2018b) Genome editing reveals reproductive and developmental dependencies on specific types of vitellogenin in zebrafish (*Danio rerio*). https://www.biorxiv.org/cgi/reprint/456053v1

Yin N, Jin X, He J, Yin Z (2009) Effects of adrenergic agents on the expression of zebrafish (*Danio rerio*) vitellogenin Ao1. Toxicol Appl Pharmacol 238:20–26

Zhong L, Yuan L, Rao Y, Li Z, Zhang Z, Liao T, Xu Y, Dai H (2014) Distribution of vitellogenin in zebrafish (*Danio rerio*) tissues for biomarker analysis. Aquat Toxicol 149:1–7

State of the Art of Multiple Vitellogenin System in Fishes

4

Jeyaraj Jeyavani and Vaseeharan Baskaralingam

Abstract

In most oviparous vertebrates, including fish, the establishment of the yolk and eggshell proteins is essential for oocyte formation (zona radiata proteins). Unlike vitellogenesis, which involves a series of processes in which the liver produces vitellogenin, which is then secreted into the circulation where it is confined into the developing oocytes before being broken by a proteolytic action to make yolk proteins and ultimately deposited in the oocytes. Many investigations have been made on various fish species to understand more about vitellogenin and yolk protein levels, their makeup, and their roles in fish reproduction. Generally, it has a linear chain with domains such as terminal amino acid-lipovitellin, phosvitin, β-carotene component-c-terminal carboxyl group. In fishes, there are two types of vitellogenin: complete (VgA and B) and incomplete (Vg C). There are numerous claims that the meroblastic or holoblastic cleavage pattern, the features of the egg, and the placental or nonplacental method of reproduction ultimately depend on the existence or lack of the vitellogenin (Vtg) gene family in vertebrates (pelagic or benthic). Using advanced techniques multiple vitellogenins was determined in fishes.

Keywords

Vitellogenin · Fishes · Multiplicity · Yolk protein

J. Jeyavani · V. Baskaralingam (✉)
Biomaterials and Biotechnology in Animal Health Lab, Department of Animal Health and Management, Alagappa University, Karaikudi, Tamil Nadu, India
e-mail: vaseeharanb@alagappauniversity.ac.in

V. Baskaralingam, R. Vanichviriyakit (eds.), *Vitellogenin in Fishes-Diversification, Biological Properties, and Future Perspectives*,
https://doi.org/10.1007/978-981-99-5340-0_4

4.1 Introduction

Vitellogenin is the unique yolk protein that is occurred in the gravid fish at the time of oocyte developmental phase. These proteins were synthesized in the liver of fishes, transported via blood, and reached into oocytes by internalization process (Receptor meditated endocytosis or proteolytic cleavage by capthasin D) (Opresko and Karpf 1987; Sire et al. 1994). And it aids to create multiple yolk proteins such as lipovitellin subunits, phosvitin subunits, and β-component (Matsubara et al. 1995), from which lipovitellin subunits and phosvitin subunits were stored in yolk globules and β-component are soluble in cytoplasmic inclusion and it was utilized by embryos (Arukwe and Goksøyr 2003; Finn and Fyhn 2010; Patiño et al. 2022; Reading et al. 2018). It also helps develop embryo development in oviparous aquatic organisms (Wiegand 1982). In recent years, it was found that the extent of vitellogenin multiplicity in fishes due to its proteolytic cleavage and their derived yolk protein has not been fully explored. These vitellogenin are estrogen-prompted protein in the maternal helps to develop the internal lining of the egg membrane. This protein has recently been used as a biomarker for evaluating the effects of endocrine disruptors (Sumpter and Jobling 1995; Arukwe and Goksøyr 2003). The general structure of vitellogenin is very similar in vertebrates (fishes) (Fig. 4.1) and invertebrates (Arthropoda). It has three domains: N-terminus vitellogenin domain, domain unknown function (DUF 1944, DUF 1943), and von Willebrand factor type D domain (vWD) (C-terminus end). In starting of N—the terminus end having single peptide includes heavy chain lipovitellin, (Lvh) phosphorylated serine-rich phosvitin (Pv), lipovitellin light chain (LVL), and c-terminal end having β-component (β-c) and c terminal coding region (CT) (Zhang et al. 2015). In fishes, there are three types of vitellogenin protein available in order of fishes, namely, Paracanthopterygii and Acanthopterygii; their expression is found in the transcription level (Hiramatsu et al. 2006; Rawat et al. 2013a, b), from which two vitellogenin proteins (VgAa and VgAb) have a complete structure with a yolk protein domain (Hiramatsu et al. 2006; Finn and Kristoffersen 2007; Rawat et al. 2013a, b). Generally, vitellogenin protein has some following features: (a) lipoglycophosphoprotein (M.W 300–600 KD), (b) precursor of yolk proteins, (c) female specific plasma protein, (d) produced by

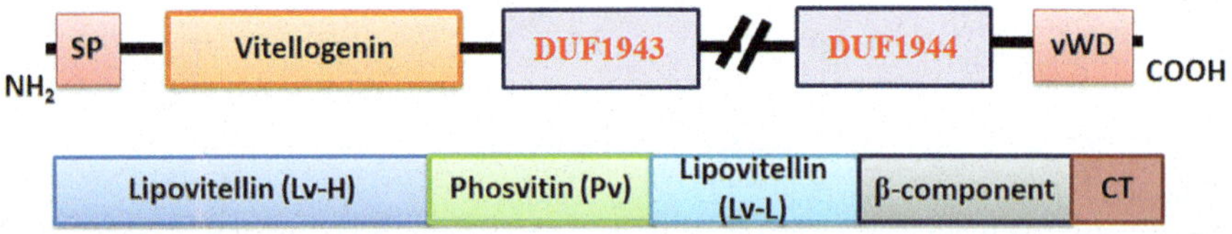

Fig. 4.1 Composition of vitellogenin (Teleost). It consists of non-Willebrand factor, signal peptide, N-terminal end, vitellogenin, and domain of unknown function (1943 and 1944). Lipovitellin light chain (LVL), heavy chain (Lvh), phosphorylated serine-rich phosvitin (Pv), and c-terminal end with -component (-c) and c terminal coding area are all included in a single peptide (CT)

induction of estrogen, and (e) they are carrier protein with ionic and lipid components (iron, calcium, zinc, etc.) (Hiramatsu et al. 2006).

4.2 General Process of Vitellogenesis

The hypothalamus-pituitary-gonad neuroendocrine axis's activation is the first step in the process of vitellogenesis (Fig. 4.2). This leads to an upsurge in follicle-stimulating hormone levels in the circulatory system, which in turn stimulates the growth of follicular cells as well as prompts the liver to begin producing vitellogenin (Speeker and Sullivan 1994; Eckelbarger and Davis 1996). This vitellogenin is released in the blood, and it was utilized by oocytes for their growth; some amounts are stored in the ooplasm of yolk granules (Hiramatsu et al. 2006). The oocytes' surface of fishes has a receptor with high affinity with vitellogenin called vitellogenin receptor which mediates the endocytosis process. Invagination of the vitellogenin receptor in clathrin-coated pits could occasionally result in the formation of vesicles. This invaginated vesicle that merges with lysosomes (ooplasm) having cathepsin led to the conversion of vitellogenin into yolk protein (Carnevali et al. 1999; Polzonetti-Magni et al. 2004; Romano et al. 2004; Hiramatsu et al. 2006). The egg's yolk protein serves as a source of nutrition and energy for the developing embryo (Lubzens et al. 2017).

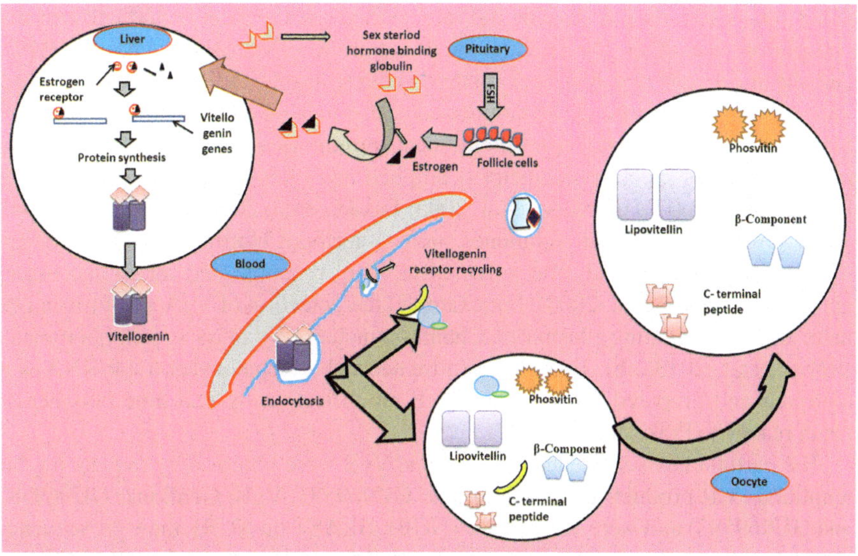

Fig. 4.2 Process of vitellogenesis

4.3 Vitellogenesis in Fish

Massive molecular mass protein of vitellogenin is a compressed by carbohydrates, lipids, phosphorous, and minerals as well as calcium, iron, and zinc. Whenever estrogen is administered, vitellogenin is synthesized in male or juvenile fish and in the blood serum of female fish during the process of vitellogenesis. Vitellogenin levels fluctuated in the serum of female fishes (*Sakhalin taimen*) nearly about 1–10 mg/mL during oogenesis (Hiramatsu et al. 1997). During the vitellogenesis process, degradation of vitellogenin proteins such as Lv, Pv, and β′-c takes place. Initially, the degradation of yolk protein is called first proteolysis, and further degradation of the protein tale is at the time of embryogenesis. The first and second degradations of yolk protein occurred in mummichogs (Hiramatsu et al. 2002), and finally the remaining yolk protein of seawater and brackish water fishes, which matured into final oocytes, is able to absorb a large amount of water (Greeley Jr et al. 1986; Matsubara et al. 1995). In contrast, freshwater salmonids do not exhibit secondary proteolysis (Wallace and Begovac 1985; Wallace and Selman 1985). Barfin flounder yolk proteins of Pv and β′-c were degraded into the amino acid at the time of second proteolysis (Matsubara and Sawano 1995).

4.4 Detection of Fish Vitellogenin

The presence of vitellogenin in fish serves as one of the most important indicators of estrogen exposure in aquatic habitats (Hiramatsu et al. 2005). Several methods are available to detect and quantify the vitellogenin protein in fish circulation (blood) and hepatocytes (production): there are two types of methods, namely, direct and indirect methods. The indirect method uses protein-bound calcium and phosphorus to estimate the vitellogenin concentration (Nath and Sundararaj 1981), and the measurements of direct method include single radial immunodiffusion, immunoelectrophoresis, radioimmunoassay, enzyme-linked immunosorbent assay, immunochromatography, chemiluminescent immunoassay, and liquid chromatography with mass spectrophotometry (Campbell and Idler 1980; Maitre et al. 1985; Prakash et al. 2007; Hiramatsu et al. 2005; Fukada et al. 2001; Zhang et al. 2004). Also, advanced molecular techniques such as northern blotting, reverse transcription and quantitative real-time polymerase chain reaction, and micro and macro gene array (Larkin et al. 2003). For example, the detection of VgA and VgB in isomers in the circulating plasma of Channa punctatus by mass spectrophotometer (Rawat et al. 2013a, b). Estradiol-17β-treated fish's hepatocytes, total RNA was collected and prepared for VgA and VgB amplification by real-time polymerase chain reaction (Prakash et al. 2007).

The partial cDNA sequences of VgA and VgB were obtained by sequencing the amplified PCR products (VgA: GenBank: GU969581, VgB: GenBank: GU969582 and GU969583) and were submitted to NCBI. BLAST on NCBI, the VgA sequence showed a strong similarity to the VgA of fishes from the orders Perciformes, Beloniformes, Cyprinodontiformes, and Gadiformes, while the VgB sequence showed a similarity to the VgB of fishes from the orders Perciformes, Pleuronectiformes, and Gadiformes. The amplified cDNA fragments' amino acid sequences may be as follows: VgB using primers FB1RB1 687–839 as compared to the amino acid sequence of VgB from Veraspermoseri ((GenBank: BAD93696) showing the highest homology; and primers FB2RB2 (687–839 as compared to the amino acid sequence of VgB from *Morone americana* GenBank: AAZ17416) showing highest homology) showed the highest homology. VgA 522–686 as compared to the amino acid sequence (Rawat et al. 2013a, b).

4.5 Multiplicity of Vitellogenin

Earlier vitellogenin in fishes was identified by various immunological methods. The middle 1990s identified a single type of vitellogenin in teleosts. After that, a dual type of vitellogenin was discovered in *Fundulus heteroclitus* by CDNA cloning (Hiramatsu et al. 2006). The reproductive physiology of fish has been significantly affected by the structural and functional multiplicity of vitellogenin. The nomenclature and categorization of dual or multiple vitellogenin proteins and their corresponding genes have grown complex, even for the finding of several vitellogenins within a single species. The following classification system for various teleost vitellogenin was recently developed to reduce this confusion.

Generally, there are two types of vitellogenin, namely, complete and incomplete, reported in Fig. 4.3. In complete vitellogenin A (VgA and VgB) have amino acid (N-terminal ends)—LvH–Pv–LvL-β-component—C-terminal peptide and carboxyl group and this vitellogenin elucidate during anion-exchange chromatography using a higher concentration of sodium chloride. VgA structure homologous with some fishes including barfin flounder, mummichog, *Melanogrammus aeglefinus*, etc. and final maturation of oocytes, LvH region, was degraded. In contrast, the LvH region was partially or fully degraded in complete vitellogenin B during oocyte maturation. In incomplete vitellogenin (Vg C), Pv domain was absent. It consists of amino acid (N-terminal ends)-LvH-LvL-β-component-C-terminal peptide and carboxyl groups, and this structure is more homologous to *Danio rerio*, *Acanthogobius flavimanus* fishes (Hiramatsu et al. 2006; Hara et al. 2016). Both complete (VgA, VgB) and incomplete (Vg C) vitellogenins have different molecular weight and was effectively isolated and purified from various fishes including *Oreochromis mossambicus* (Kishida and Specker 1993), *Oreochromis aureus* (Ding et al. 1989), *Oryzias latipes* (Shimizu et al. 2002), *Acanthogobius flavimanus* (Ohkubo et al. 2003), and *Sakhalin taimen* (Amano et al. 2010). Interestingly, it was noted that fishes such as *Gambusia affinis*, *Morone americana*, *Pagrus major*, and *Sebastes taczanowskii* showed all three of these vitellogenin types (VgA, B, and C) (Sawaguchi et al. 2005, 2006). The distribution of multiple vitellogenins among the teleost was investigated using phylogenetic relationship classification approaches.

Several vitellogenin gene transcripts were discovered in fish species such as *Fundulus heteroclitus* (mummichog) (LaFleur Jr et al. 1995), *Verasper moseri* (barfin flounder) (Matsubara et al. 1999), *Melanogrammus aeglefinus* (Reith et al.

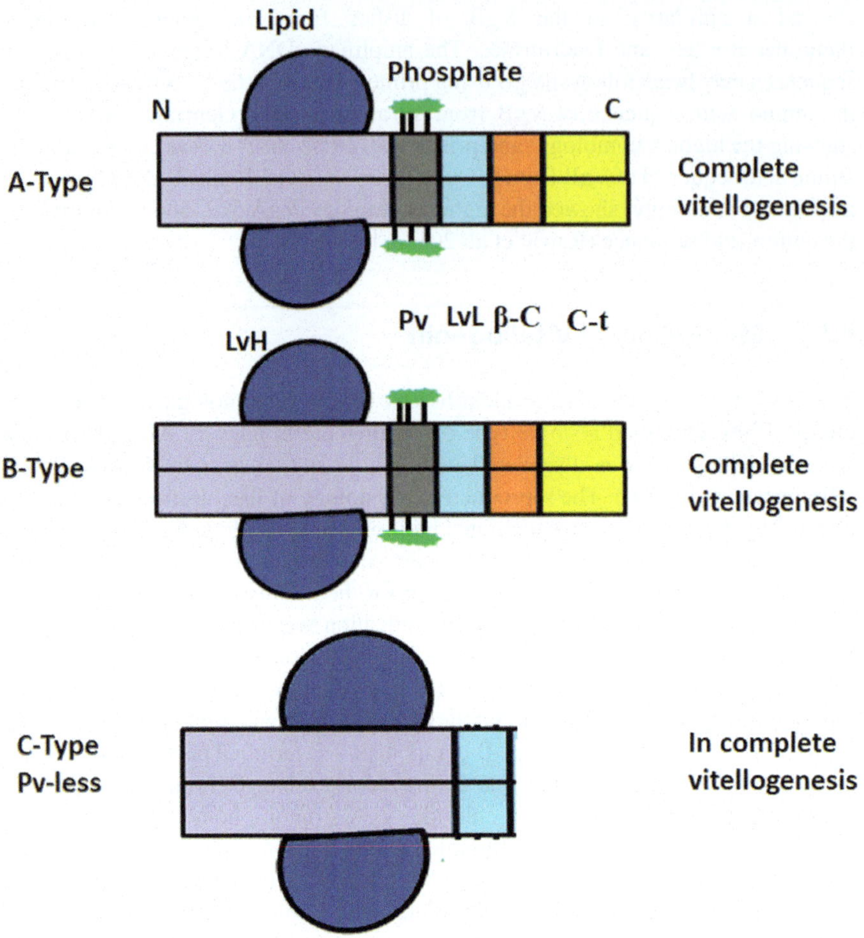

Fig. 4.3 Domain structure of fish vitellogenin

2001), *Oreochromis aureus* (Lee et al. 1994), and *Cyprinodon variegatus* (Bowman et al. 2000). In zebrafish, gene encoding Pv (phosvitin domain) region was absent in vitellogenin (Wang et al. 2000). Additionally, a genomic investigation of rainbow trout identified 10 pseudogenes and 20 active vitellogenin genes. With only ≤3% sequence variability, the vitellogenin genes in rainbow trout are predicted to encode substantially identical proteins. This species, which lays very big eggs, likely needs this gene redundancy in order to produce substantial levels of vitellogenin (Trichet et al. 2000).

Following particular tandem gene duplication and neofunctionalization, whole genome duplication was used to discover how vitellogenin structure and function were diversified during the duration of evolution. By employing transcription of the vitellogenin gene and inferred amino acid sequences, Finn and Kristoffersen (2007) demonstrated that chordates vitellogenin is the ancestor of the vitellogenin found in vertebrates and their molecular relation constructed by the phylogenetic tree (Finn and Kristoffersen 2007; Hara et al. 2016).

Although it is unknown whether these gene products differ in their immunological or functional characteristics, the haploid rainbow trout genome had 10 pseudogenes and 20 complete Vg genes (Trichet et al. 2000). There are at least three primary forms of piscine vitellogenin; it will take more sequencing data and examination of the evolutionary distribution of the various vitellogenin types to fully and accurately classify these proteins. The physiological importance of many vitellogenins in teleosts is still being verified, but it is clear that each vitellogenin serves a unique purpose.

4.6 Conclusion

It is important to purify and describe vitellogenins from various fish species and genera to determine their multifaceted nature. In order to isolate cDNAs and to determine whether distinct genes are responsible for various Vgs, it is important to identify and define the vitellogenin genes. The composition of vitellogenin-derived yolk proteins should be thoroughly researched because it may affect egg quality. Furthermore to establish their precise role in fish reproduction, different vitellogenin forms can also be investigated in vivo and in vitro in fish. The study will also be focused on identifying additional roles, such as hormone and ion transporter and availability of pure vitellogenin from various fish species.

References

Amano H, Mochizuki M, Fujita T, Hiramatsu N, Todo T, Hara A (2010) Purification and characterization of a novel incomplete type vitellogenin protein (VgC) in Sakhalin taimen (Hucho perryi). Comp Biochem Physiol 157A:41–48

Arukwe A, Goksøyr A (2003) Eggshell and egg yolk proteins in fish: hepatic proteins for the next generation: oogenetic, population, and evolutionary implications of endocrine disruption. Comp Hepatol 2(1):1–21

Bowman CJ, Kroll KJ, Hemmer MJ, Folmer LC, Denslow ND (2000) Estrogen-induced vitellogenin mRNA and protein in sheepshead minnow (Cyprinodon variegatus). Gen Comp Endocrinol 120:300–313

Campbell CM, Idler DR (1980) Characterization of an estradiol induced protein from rainbow trout as vitellogenin by the cross reactivity to ovarian yolk fractions. Biol Reprod 22:605–617

Carnevali O, Carletta R, Cambi A, Vita A, Bromage N (1999) Yolk formation and degradation during oocyte maturation in seabream Sparus aurata: involvement of two lysosomal proteinases. Biol Reprod 60(1):140–146

Ding JL, Hee PL, Lam TJ (1989) Two forms of vitellogenin in the plasma and gonads of male Oreochromis aureus. Comp Biochem Physiol 93B:363–370

Eckelbarger KJ, Davis CV (1996) Ultrastructure of the gonad and gametogenesis in the eastern oyster, Crassostrea virginica. (I) Ovary and oogenesis. Mar Biol 127:79–87

Finn RN, Fyhn HJ (2010) Requirement for amino acids in ontogeny of fish. Aquac Res 41(5):684–716

Finn RN, Kristoffersen BA (2007) Vertebrate vitellogenin gene duplication in relation to the "3R hypothesis": correlation to the pelagic egg and the oceanic radiation of teleosts. PLoS One 2:e169

Fukada H, Haga A, Fujita T, Hiramatsu N, Sullivan CV, Hara A (2001) Development and validation of chemi-luminescent immunoassay for vitellogenin in five salmonid species. Comp Biochem Physiol 130A:163–170

Greeley MS Jr, Calder DR, Wallace RA (1986) Changes in teleost yolk proteins during oocyte maturation: correlation of yolk proteolysis with oocyte hydration. Comp Biochem Physiol 84B:1–9

Hara A, Hiramatsu N, Fujita T (2016) Vitellogenesis and choriogenesis in fishes. Fish Sci 82(2):187–202

Hiramatsu N, Shimizu M, Fukada H, Kitamura M, Ura K, Fuda H, Hara A (1997) Transition of serum vitellogenin cycle in Sakhalin taimen (Hucho perryi). Comp Biochem Physiol 118C:149–157

Hiramatsu N, Matsubara T, Weber GM, Sullivan CV, Hara A (2002) Vitellogenesis in aquatic animals. Fish Sci 68(Suppl I):694–699

Hiramatsu N, Cheek AO, Sullivan CV, Matsubara T, Hara A (2005) Vitellogenesis and endocrine disruption. In: Mommsen TP, Moon TW (eds) Biochemistry and molecular biology of fishes. Elsevier, Amsterdam, p 562

Hiramatsu N, Matsubara T, Fujita T, Sullivan CV, Hara A (2006) Multiple piscine vitellogenins: biomarkers of fish exposure to estrogenic endocrine disruptors in aquatic environments. Mar Biol 149(1):35–47

Kishida M, Specker JL (1993) Vitellogenin in tilapia (*Oreochromis mossambicus*): induction of two forms by estradiol, quantification in plasma and characterization in oocyte extract. Fish Physiol Biochem 12:171–182

LaFleur GJ Jr, Byrne BM, Kanungo J, Nelson LD, Greenberg RM, Wallace RA (1995) Fundulus heteroclitus vitellogenin: the deduced primary structure of a piscine precursor to noncrystalline, liquid-phase yolk protein. J Mol Evol 41:505–521

Larkin P, Knoebl I, Denslow ND (2003) Differential gene expression analysis in fish exposed to endocrine disrupting compounds. Comp Biochem Physiol B: Biochem Mol Biol 136(2):149–161

Lee BH, Lim EH, Lam TJ, Ding JL (1994) Two major groups of vitellogenin cDNA clones from Oreochromis aureus (Steindachner). Biochem Mol Biol Int 34:75–83

Lubzens E, Bobe J, Young G, Sullivan CV (2017) Maternal investment in fish oocytes and eggs: the molecular cargo and its contributions to fertility and early development. Aquaculture 472:107–143

Maitre JL, Le Guellec C, Derrien S, Tenniswood M, Valotaire Y (1985) Measurement of vitellogenin from rainbow trout by rocket immunoelectrophoresis: application to the kinetic analysis of estrogen stimulation in the male. Can J Biochem Cell Biol 63:982–987

Matsubara T, Sawano K (1995) Proteolytic cleavage of vitellogenin and yolk proteins during vitellogenin uptake and oocyte maturation in barfin flounder (Veraspermoseri). J Exp Zool 272:34–45

Matsubara T, Adachi S, Ijiri S, Yamauchi K (1995) Changes of lipovitellin during in vitro oocyte maturation in Japanese flounder, *Paralichthys olivaceus*. Fish Sci 61:478–481

Matsubara T, Ohkubo N, Andoh T, Sullivan CV, Hara A (1999) Two forms of vitellogenin, yielding two distinct lipovitellins, play different roles during oocyte maturation and early development of barfin flounder, *Veraspermoseri*, a marine teleost spawning pelagic eggs. Dev Biol 213:18–32

Nath P, Sundararaj BI (1981) Isolation and identification of female-specific serum lipophosphoprotein (vitellogenin) in the catfish, Heteropneustes fossilis. Gen Comp Endocrinol 43:184–190

Ohkubo N, Mochida K, Adachi S, Hara A, Hotta K, Nakamura Y, Matsubara T (2003) Development of enzyme-linked immunosorbent assays for two forms of vitellogenin in Japanese common goby (Acanthogobius flavimanus). Gen Comp Endocrinol 131(353–364):89

Opresko LK, Karpf RA (1987) Specific proteolysis regulates fusion between endocytic compartments in Xenopus oocytes. Cell 51(4):557–568

Patino-Martinez J, Veiga J, Afonso IO, Yeoman K, Mangas-Viñuela J, Charles G (2022) Light sandy beaches favour hatching success and best hatchling phenotype of loggerhead turtles. Front Ecol Evol 10:823–118

Polzonetti-Magni AM, Mosconi G, Soverchia L, Kikuyama S, Carnevali O (2004) Multihormonal control of vitellogenesis in lower vertebrates. Int Rev Cytol 239(Supplement C):1–46

Prakash O, Goswami SV, Sehgal N (2007) Establishment of ELISA for murrel vitellogenin and choriogenin, as biomarkers of potential endocrine disruption. Comp Biochem Physiol 146C:540–551

Rawat VS, Pipil S, Sharma L, Sehgal N (2013a) Purification, characterization and expression of two vitellogenins in the Indian freshwater murrel Channa punctatus. Gen Comp Endocrinol 189:119–126

Rawat VS, Rani KV, Phartyal R, Sehgal N (2013b) Vitellogenin genes in fish: differential expression on exposure to estradiol. Fish Physiol Biochem 39(1):39–46

Reading BJ, Andersen LK, Ryu YW, Mushirobira Y, Todo T, Hiramatsu N (2018) Oogenesis and egg quality in finfish: yolk formation and other factors influencing female fertility. Aust Fish 3(4):45

Reith M, Munholland J, Kelly J, Finn RN, Fyhn HJ (2001) Lipovitellins derived from two forms of vitellogenin are deferentially processed during oocyte maturation in haddock (Melanogrammus aeglefinus). J Exp Zool 291:58–67

Romano M, Rosanova P, Anteo C, Limatola E (2004) Vertebrate yolk proteins: a review. Mol Reprod Dev 69:109–118

Sawaguchi S, Koya Y, Yoshizaki N, Ohkubo N, Andoh T, Hiramatsu N, Sullivan CV, Hara A, Matsubara T (2005) Multiple vitellogenins (Vgs) in mosquitofish (Gambusia affinis): identification and characterization of three functional Vg genes and their circulating and yolk proteins products. Biol Reprod 72:1045–1060

Sawaguchi S, Kagawa H, Ohkubo N, Hiramatsu N, Sullivan CV, Matsubara T (2006) Molecular characterization of three forms of vitellogenin and their yolk protein products during oocyte growth and maturation in red seabream (Pagrus major), a marine teleost spawning pelagic eggs. Mol Reprod Dev 73:719–736

Shimizu M, Fujiwara Y, Fukada H, Hara A (2002) Purification and identification of a second form of vitellogenin from ascites of medaka (Oryzias latipes) treated with estrogen. J Exp Zool 293:726–735

Sire MF, Babin PJ, Vernier JM (1994) Involvement of the lysosomal system in yolk protein deposit and degradation during vitellogenesis and embryonic development in trout. J Exp Zool 269(1):69–83

Speeker JL, Sullivan CV (1994) Vitellogenesis in fishes: status and perspective [C]. Perspective in comparative endocrinology. National Research Council of Canada, Ottawa, pp 304–315

Sumpter JP, Jobling S (1995) Vitellogenesis as a biomarker for estrogenic contamination of the aquatic environment. Environ Health Perspect 103(Suppl 7):173–178

Sun C, Hu L, Liu S, Gao Z, Zhang SC (2013) Functional analysis of domain of unknown function (DUF) 1943, DUF1944 and von Willebrand factor type D domain (VWD) in vitellogenin2 in zebrafish. Dev Comp Immunol 41:469–476

Trichet V, Buisine N, Mouchel N, Moran P, Pendas AM, Le Pennec JP, Wolff J (2000) Genomic analysis of the vitellogenin locus in rainbow trout (Oncorhynchus mykiss) reveals a complex history of gene amplification and retroposon activity. Mol Gen Genet 263:828–837

Wallace RA, Begovac PC (1985) Phosvitins in Fundulus oocytes and eggs: preliminary chromatographic and electrophoretic analyses together with biological considerations. J Biol Chem 260:11268–11274

Wallace RA, Selman K (1985) Major protein changes during vitellogenesis and maturation of Fundulus oocytes. Dev Biol 110:492–498

Wang H, Yan T, Tan JTT, Gong ZA (2000) Zebrafish vitellogenin gene (vg3) encodes a novel vitellogenin without a phosvitin domain and may represent a primitive vertebrate vitellogenin gene. Gene 256:303–310

Wiegand MD (1982) Vitellogenesis in fishes. In: Reproductive physiology of fish, pp 136–146

Zhang F, Bartels MJ, Brodeur JC, Woodburn KB (2004) Quantitative measurement of fathead minnow vitellogenin by liquid chromatography combined with tandem mass spectrometry using a signature peptide of vitellogenin. Environ Toxicol Chem 23:1408–1415

Zhang S, Dong Y, Cui P (2015) Vitellogenin is an immunocompetent molecule for mother and offspring in fish. Fish Shellfish Immunol 46(2):710–715

Tools for Identification and Characterization of Vitellogenin in Fishes

5

Muthukumar Abinaya and Periyasamy Gnanaprakasam

Abstract

In teleost fishes and other oviparous vertebrates, vitellogenins (Vtg) are the main precursor of the egg yolk proteins that serve as energy reserves for embryonic development. In most cases, endogenous estrogens, such as 17-estradiol (E2), which are released into the bloodstream and then maintained in developing oocytes in adult females, cause the production of Vtg. It has evolved into a significant biomarker in recent decades for determining the estrogenic potency of chemicals and the exposure of animals to estrogenic pollutants found in aquatic environments. Consequently, a variety of different analytical approaches have been used depending on the expertise of diverse research laboratories to assess whether a molecule or effluent is able to generate Vtg. This insights into the molecular mechanism responsible for vitellogenin synthesis and range of techniques are currently in use to measure the Vtg protein as an indication of estrogenic responses in fishes. Thus, more information on the techniques to detect and quantify the vitellogenin is necessary to clarify their endocrine system and to learn more about how vitellogenin protein is utilized as a sensitive biomarker for detecting estrogenic activities in fishes. Hence, this chapter aims to cover the major tools for vitellogenin identification and characterization were explored.

M. Abinaya
Crustacean Molecular Biology and Genomics Division, Biomaterials and Biotechnology in Animal Health Lab, Department of Animal Health and Management, Alagappa University, Karaikudi, Tamil Nadu, India

P. Gnanaprakasam (✉)
Department of Chemistry, Saveetha School of Engineering, Saveetha Institute of Medical and Technical Sciences, Chennai, Tamil Nadu, India

V. Baskaralingam, R. Vanichviriyakit (eds.), *Vitellogenin in Fishes-Diversification, Biological Properties, and Future Perspectives*,
https://doi.org/10.1007/978-981-99-5340-0_5

Keywords

Vitellogenin · Fishes · Estradiol · Identification · Characterization · Aquatic environments

Abbreviations

b-C	Beta-component
CLIA	Chemiluminescent immunoassay
CT	C-terminal
DUF	Domain of unknown function
E2	Estradiol
ELISA	Enzyme-linked immunosorbent assay
ER	Estrogen receptor
LC-MS/MS	Liquid chromatography-tandem mass spectrometry
LPD N	Vitellogenin N/LLT domain
LvH	Lipovitellin heavy chain
LvL	Lipovitellin light chain
Native PAGE	Native polyacrylamide gel electrophoresis
NT	N-terminal
Pv	Phosphorylated serine-rich phosvitin
Q-TOF/MS/MS	Quadrupole time-of-flight tandem mass spectrometer.
RIA	Radioimmunoassay
RT-PCR	Reverse transcriptase-polymerase chain reaction
rtqRT-PCR	Quantitative real-time reverse transcriptase-polymerase chain reaction
SP	Signal peptide
SRID	Single radial immunodiffusion
Vtg	Vitellogenin
VtgR	Vitellogenin receptor
vWD	von Willebrand factor type D domain

5.1 Introduction

Vitellogenin (Vtg) is a lipophosphoglycoprotein complex with a high molecular weight which is generated in oviparous females in response to estrogen signaling with binding to estrogen receptors (Nagler et al. 2010). According to Hiramatsu et al. (2015), Vtg protein is expected to be present at high concentrations only in mature females during the reproductive season, because of its activity as an egg yolk precursor protein. In vertebrates, it is synthetized during vitellogenesis in the liver in response to estradiol-17β from the ovary, which is under control of gonadotropins from the pituitary gland. After secretion into the bloodstream, Vtg is taken up by developing oocytes, where it is cleaved into smaller yolk proteins, including phosvitin, lipovitellin, and β-component

(Matsubara and Sawano 1995; Bergink and Wallace 1974; Mommsen and Walsh 1988; Hara et al. 2016).

In aquatic contexts, the fish vitellogenin (Vtg) protein has been used as a sensitive biomarker for assessing estrogenic activity (Sumpter and Jobling 1995; Matozzo et al. 2008; Robinson and Scott 2012; Tran et al. 2019). The Vtg composed of sugar, lipid, and phosphorus-binding protein that serves as a pre-cursor of vitellin (egg yolk proteins) which interacts with antibodies exerted toward egg extracts. In addition, the ion-binding properties of Vtg serve as a major supply of minerals to the oocytes (Zhong et al. 2014; Hara et al. 2016).

The existence of diverse forms of Vtg in teleosts has been proven by gene cloning and immune-biochemical investigations (Yilmaz et al. 2018; Pan et al. 2019). Relationships between multiple Vtg gene transcripts and their translated Vtg proteins have been confirmed in several species, but procedures for purifying each form of Vtg protein have been developed only for few species due to the close biochemical properties (Hiramatsu et al. 2002a, b, c; Sawaguchi et al. 2005; Amano et al. 2007). Complex combinations of protein precipitation and column chromatography, which include immunosorbent affinity column chromatography using subtype-specific Vtg antibodies each time, are necessary to separate each Vtg from others.

Vtg proteins have been isolated and characterized from diverse fish species (Garnayak et al. 2013). Although the Western blot, enzyme-linked, and RIA remain the utmost popular despite the fact that several assays for quantifying Vtg in fish plasma have indeed been established (Puy-Azurmendi et al. 2013; Flick et al. 2014; Hultman et al. 2015; Dos Santos et al. 2016; Moura Costa et al. 2016; Luna and Coady 2016). Henceforth, the present chapter focused to briefly summarize the major tools for identification and characterization, and molecular mechanisms underlying vitellogenin in fishes were explored.

5.2 Vtg Synthesis

In the yolk of oviparous vertebrates' eggs, vitellogenins are proteins that serve as the main source of nutrients. The amount of Vtg-derived yolk in ovulated eggs with some species can range from 80% to 90% of the total dry mass. Figure 5.1 depicts the formation of egg yolk and synthesis of Vtg.

The production of Vtg is gonadotropin-dependent phenomenon influenced with various parameters like water temperature, nutritional status, and photoperiod. Various environmental factors can stimulate the hypothalamus to secrete the gonadotropin-releasing hormone (GnRH). The pituitary gland secretes gonadotropin hormones (GTHs) in response to GnRH, which increase both the follicle cell E2 synthesis and oocyte vitellogenin absorption. They are produced in the liver under the strict control of E2, secreted into the bloodstream, and are taken up beneath the regulator of GTH by the growing oocytes in the ovary and cleaved into lipovitellins and phosvitins by specific receptors. The binding of estradiol to Vtg receptor present on hepatocytes will lead to the transcription of Vtgs. Vtg's protein backbone is initially generated on membrane-bound ribosomes, followed by posttranslational

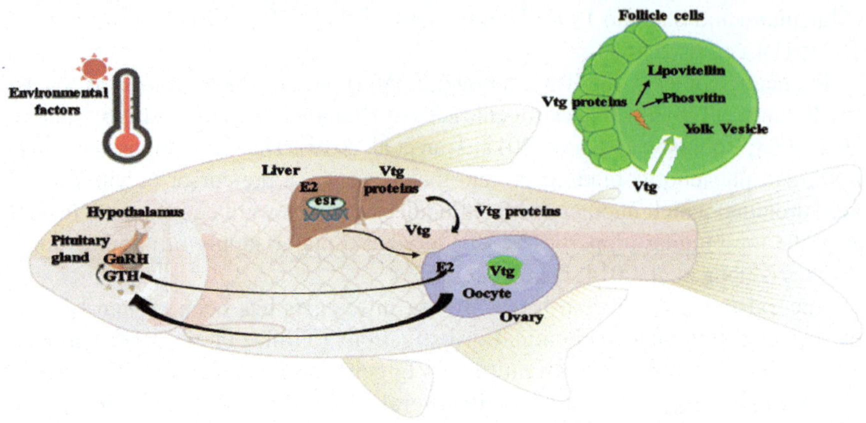

Fig. 5.1 The formation of egg yolk and synthesis of Vtg

modifications such as lipidation, glycosylation, and phosphorylation (Wallace 1985; Bhandari et al. 2003; Arukwe and Goksøyr 2003; Li and Zhang 2017; Tramunt et al. 2021).

5.3 Structure, Expression, and Activation of Vtg

Vtg is a glycolipophosphoprotein homodimer with a molecular size of 250–600 kDa which distributes in the blood/hemolymph. It is encoded by several Vtg genes in various fish species (Tufail and Takeda 2008). Vtg has three conserved domains such as the LPD N-domain which is located at the N-terminus and the DUF 1943 and the vWD at the C-terminus with a variety of proteins (Fig. 5.2). In some Vtg proteins, DUF1944 can be located among DUF1943 and vWD (Dalvin et al. 2011).

A whole fish Vtg starts at the N-terminus and consists of a signal peptide, a LvH, a Pv, a LvL, a b-C, and a CT which makes up the vWD. Furthermore, Pv can be eliminated, as evident in most invertebrate Vtg and zebrafish Vtg3 (Sawaguchi et al. 2005; Wang et al. 2000). Vtg is often made extra-ovarianly and carried to the ovary by the circulatory system. Eventually, it is internalized into receptor-mediated endocytosis for mature oocytes, and aspartic protease cathepsin D cleaves it to produce yolk proteins such as Lv subunits, Pv, and b-C (Opresko and Karpf 1987; Sire et al. 1994; Retzek et al. 1992). Whilst b-C persists in the cytoplasm as a soluble fraction, Lv subunits and Pv are deposited in yolk globules/platelets (Babin 1987; Matsubara et al. 1999; Tyler et al. 1988).

Later, these yolk proteins are turned into nutrients by developing embryos (Arukwe and Goksøyr 2003; Finn and Fyhn 2010). This culminates in the initiation or enhanced transcription of Vtg proteins. As a basis, Vtg has been extensively applied as a biomarker to quantify estrogenic pollution in aquatic habitats (Matozzo et al. 2008).

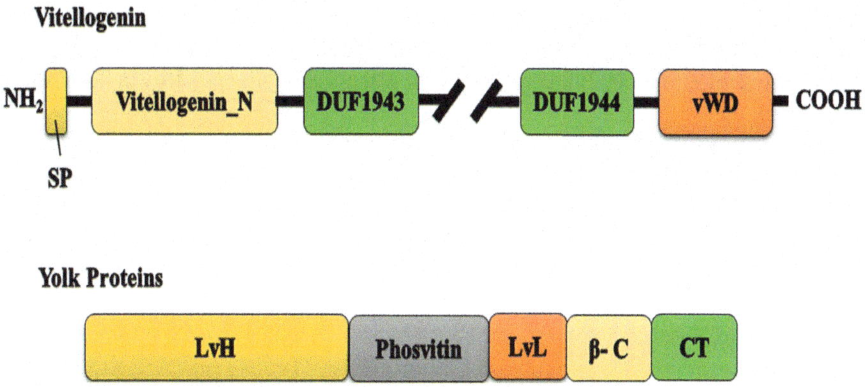

Fig. 5.2 Domain structure of teleost vitellogenin

5.4 Molecular Cloning of Vtg Genes

Estrogen is a crucial transcriptional regulator of Vtg which is triggered via estradiol-17 (E2) with nuclear estrogen receptor (ER) (Hiramatsu et al. 2006; Nelson and Habibi 2013). It is still unclear, though, how Vtg and ER subtypes are related to one another in terms of transcriptional regulation. There have been numerous reports on the molecular characterization and profiles of Vtg and ER subtype expression in fish, with various species using different Vtg production methods. Recently, cDNA cloning investigations and research on molecular and structural analysis have proven the coexistence of certain Vtg variants in the majority of fishes (Hiramatsu et al. 2002a, b, c, 2015; Hara et al. 2016; Li and Zhang 2017; Li et al. 2018; Reading et al. 2011; Sawaguchi et al. 2005; Matsubara et al. 1999, 2003; Patiño and Sullivan 2002; Reith et al. 2001). However, only a few studies have investigated transcriptional regulation of Vtg subtypes in detail. In addition, our understanding of how ER subtypes are involved in the transcriptional regulation of Vtg subtypes remains insignificant.

Multiple ER subtypes have been reported, in addition to Vtg subtypes in teleosts. For example, three ER subtypes, ERα, ERβ1, and ERβ2, have been confirmed in fish such as Atlantic croaker (*Micropogonias undulatus*), Mozambique tilapia (*Oreochromis mossambicus*), largemouth bass (*Micropterus salmoides*), spotted scat (*Scatophagus argus*) and yellowtail (*Seriola quinqueradiata*) (Cui et al. 2017; Davis et al. 2010; Hawkins et al. 2000, 2005; Sabo-Attwood et al. 2004; Mushirobira et al. 2020). ERE-like sequences have been confirmed in promoter regions of Vtg subtypes in blue tilapia, *Oreochromis aureus*, and cutthroat trout, *Oncorhynchus clarkii*, and the activation of the promoter regions has been demonstrated using reporter gene assays in the presence of E2 and ER (Mushirobira et al. 2018; Teo et al. 1998).

In addition, ERE-like sequences have been observed on the erα promoter regions in rainbow trout, *Oncorhynchus mykiss*, and zebrafish, *Danio rerio* (Menuet et al. 2004; Flouriot et al. 1997), suggesting that estrogen regulates ERα, in addition to Vtg subtypes.

The mummichog *Fundulus heteroclitus*, whose cDNAs encode two distinctly different Vgs, VgI and VgII, provided the first molecular evidence of the multiplicity of Vtg genes is found in teleost species (LaFleur et al. 1995, 2005). Following that, full-length cDNAs for two distinctive Vgs were found from haddock, *Melanogrammus aeglefinus* (VgA and VgB; GenBank: AB284035 and AB284034), and medaka, *Oryzias latipes* (Vg1 and VgII; GenBank: AB064320 and AB074891). Dichotomy of Vg proteins, VgA and VgB (Matsubara et al. 1999), and their matching cDNA sequences (VgA and VgB; GenBank: AB181833 and AB181834), was later demonstrated in the barfin flounder, *Verasper moseri*. These various teleost Vgs were classified into two groups, VgA and VgB, based on similarities in their primary structures, constituent yolk protein domains and physiological functions (Hiramatsu et al. 2002b). The VgA group includes mummichog (Fun) VgI, haddock (Had) VgA, medaka (Med) Vg1, and barfin flounder (Bar) VgA. The VgB group includes FunVgII, HadVgB, MedVgII, and BarVgB. The two types of Vtg (VgA and VgB) have similar mass and primary structure. The coding sequences of both the VgA and VgB groups are arranged in linear fashion with respect to yolk protein domains as follows: NH2-LvH (Lv1)-Pv-LvL (Lv2)-bc-C-terminal coding region-COOH (Hiramatsu et al. 2002a, c). Two forms of Vtg protein having considerably different molecular masses have been discovered in two tilapia species, *Oreochromis aureus* (Ding et al. 1989) and *O. mossambicus* (Kishida and Specker 1993; Takemura and Kim 2001). One of the two Vtg proteins has an unusually low native mass and a lower content of phosphorus than the other Vtg (Ding et al. 1989; Kishida and Specker 1993). Wang et al. (2000) identified a Vtg cDNA product of a gene named Vtg3 in the zebrafish, *Danio rerio*, which encodes a novel Vtg without a polyserine domain and with a low degree of similarity to other piscine Vtg cDNAs.

Likewise three Vtg subtypes-VtgAa, VtgAb and VtgC, have been identified through molecular biology and biochemical methods in acanthomorph fishes, notably white perch (*Morone Americana*), mosquitofish (*Gambusia affinis*), red seabream (*Pagrus major*), gray mullet (*Mugil cephalus*), and striped bass (*Morone saxatilis*) (Finn and Kristoffersen 2007; Amano et al. 2007; Hiramatsu et al. 2002a, b, c; Reading et al. 2009; Sawaguchi et al. 2005, 2006; Williams et al. 2014).

From these characteristics, the novel Vtg was designated as a phosvitinless form of Vtg (PvlVg) and is thought to represent an ancient form of the protein. Evidently, Hiramatsu et al. (2002b) identified three forms of Vtg proteins, including VgA, VgB, and PvlVg (VgC) in the white perch, *M. americana*, by immunological and biochemical analyses. Subsequently, a PvlVg protein (Vtg-320; Ohkubo et al. 2003) and its cDNA (GenBank: AB088473) were found in *A. flavimanus* estrogen-treated fish. In mosquitofish, *Gambusia affinis*, we demonstrated the presence of all three forms of Vtg proteins, VgA, VgB, and PvlVg, in plasma from estrogen-treated females and confirmed the presence of their derivative yolk proteins in vitellogenic oocytes (Sawaguchi et al. 2005). Furthermore, full-length cDNAs encoding each type of Vtg (VgA-AB181835, VgB-AB181836, and PvlVg-AB181837) were isolated from a liver cDNA library prepared from estrogen-induced mosquitofish (Sawaguchi et al. 2005).

In barfin flounder, dual Vtgs (VgA and VgB) and their generic yolk proteins are considered to have distinct roles in the regulation of oocyte hydration for the regulation of egg convection (Matsubara et al. 1999). Differential handling of yolk proteins derived from two distinct Vtgs during final maturation has been confirmed in another marine teleost, the haddock (Reith et al. 2001). Such maturation-associated yolk protein degradations have also been discovered in many other marine and brackish species (Greeley Jr et al. 1986; Carnevali and Mosconi 1992; Carnevali et al. 1993; Finn et al. 2002a, b); thus, such a dual-Vtg system is employed among several different taxonomic groups of fishes. The existence of a VgC-formed Pvl yolk protein in oocytes also has been confirmed in various teleosts from wide phylogenetic taxa, including 3-tilapia sp. (Kishida and Specker 1993; Ding et al. 1989; Takemura and Kim 2001), the Japanese goby (Ohkubo et al. 2003), and the mosquito fish (Sawaguchi et al. 2005). In the Japanese common goby, the PvlVtg (Vtg-320) shows no change in its structure after it is taken up by vitellogenic oocytes (Ohkubo et al. 2003). While the PvlVtg of mosquito fish also does not change its native molecular mass after deposition into oocytes, it does receive a "nick" in its primary structure and dissociates into two polypeptides with molecular masses of 112 and 33 kDa (or 26 kDa as a proteolytic variant of the 33 kDa peptide) after separation by SDS-PAGE; these polypeptides represent the LvH and LvL, respectively, of a typical Lv (Sawaguchi et al. 2005).

Little is known about fish vitellogenin genes or their transcripts except two recent studies on vitellogenin mRNA of the brown bullhead, *Ameiurus nebulosus*, and the rainbow trout, *S. gairdneri*. Trout vitellogenin mRNA has 7200 nucleotides, a size comparable to other vertebrate vitellogenin mRNA, whereas the bullhead mRNA is significantly smaller.

5.5 Tools for Identification and Characterization of Vtg

Vtg proteins have been isolated and characterized from diverse fish species (Hiramatsu et al. 2002a, b, c; Mommsen and Walsh 1988; Tyler and Sumpter 1990). Several workers have successfully used chromatographic methods for the purification of Vtg from teleosts (Copeland and Thomas 1988; Jena et al. 2013; Norberg 1995; Roy et al. 2004). The purified sample of Vtg stained positive for carbohydrate (with alcian blue), for lipid (with Sudan black), and for phosphorus (with methyl green) and hence confirmed its glycolipophosphoprotein nature, a characteristic feature of all teleost Vtgs (Nath et al. 2007; Tyler et al. 1996).

The purified Vtg samples exhibited a single band in native PAGE and native gradient PAGE, implying that it could be the main circulating Vtg in *C. batrachus* (Garnayak et al. 2013). Single form of Vtg has been reported in several fish species such as common sole *Solea vulgaris*, smooth flounder *P. putnami*, and African catfish *Clarias gariepinus* (Roy et al. 2004; Manohar et al. 2005). Two molecular mass forms of Vtg were found by protein analysis in tilapia sp. *Oreochromis aureus* (Ding et al. 1989), barfin flounder *Verasper moseri* (Matsubara et al. 1999), medaka *Oryzias latipes* (Shimizu et al. 2002), and Japanese common goby *A. flavimanus* (Ohkubo et al. 2003). Even three molecular weight forms of Vtg have also been reported in white perch *Morone americana* (Hiramatsu et al. 2002a, b, c) and in mosquitofish *Gambusia affinis* (Sawaguchi et al. 2005). A single monomeric polypeptide has been reported in a number of species in SDS-PAGE because the native Vtg molecule is typically a dimer composed of two identical polypeptide subunits in teleosts (Maltais and Roy 2009; Roy et al. 2004; Mosconi et al. 1998; Roubal et al. 1997; Norberg 1995).

Zhang et al. (2004) have proposed a sophisticated quantification technique that combines tandem mass spectrometry and liquid chromatography. Mass spectrometric quantification is based on identification of particular peptides from Vtg protein following enzymatic digestion with synthetic isotope-labeled peptides as standards (Simon et al. 2010).

To date, mass spectrometry-based methods have been successfully established for Vtg quantification in fathead minnow (*Pimephales promaelas*) (Wunschel et al. 2005), zebrafish (*Danio rerio*) (Liang et al. 2015), Greenland halibut (*Reinhardtius hippoglossoides*) (Cohen et al. 2009), Atlantic cod (*Gadus morhua*) (Cohen et al. 2005), Atlantic salmon (*Salmo salar*), and rainbow trout (*Oncorhynchus mykiss*) (Cohen et al. 2006). In addition, LC-MS/MS as an effective method were reported from multiple fish species (fathead minnow, *P. promelas*; largemouth bass, *Micropterus salmoides*; and killifish, *Fundulus heteroclitus*) which use an LC equipped with a high-resolution Q-TOF/MS/MS system (He et al. 2019).

Likewise, the retrieval of Vtg transcripts from hepatic tissues could be made with conventional and evolved molecular techniques which include reverse transcription (RT)-PCR, Northern blotting, quantitative real-time RT-PCR (rtqRT-PCR), differential display RT-PCR, and micro/macro gene arrays (Larkin et al. 2003).

Although different immunological assays for quantifying Vtg in fish plasma have indeed been established (Luna and Coady 2016) such as the Western blot, radioimmunoassay (RIA), single radial immunodiffusion (SRID), chemiluminescent immunoassay (CLIA), and enzyme-linked and radioimmunosorbent assays remain utilized (Flick et al. 2014; Hultman et al. 2015; Moura Costa et al. 2016; Puy-Azurmendi et al. 2013; Dos Santos et al. 2016; Hiramatsu et al. 2006; Fukada et al. 2003). Among these assays, ELISA assays are the most preferred owing to its sensitivity and elimination of the use of radioisotopic agents. Some immunological assays of Vtg fish species were listed in Table 5.1.

Table 5.1 Some immunological assays of Vtg fish species

Fish species	Techniques	Analysis	References
Cyprinus carpio	RIA	Blood plasma	Gimeno et al. (1998a, b)
Misgurnus anguillicaudatus	Western blotting	Blood	Wang et al. (2021)
Pimephales promelas	ELISA	Blood plasma	Tyler et al. (1999), Parks et al. (1999)
Pimephales promelas	RIA	Blood plasma	Tyler et al. (1996)
Pimephales promelas	ELISA	Whole body	Panter et al. (2002)
Oryzias latipes	ELISA	Liver	Seki et al. (2002), Kang et al. (2002)
Oncorhynchus mykiss	RIA	Blood plasma	Thorpe et al. (2000, 2001)
Oncorhynchus mykiss	ELISA	Blood plasma	Schwaiger et al. (2002), Lindholst et al. (2000)
Danio rerio	RIA	Blood plasma	Tyler et al. (1996)
Danio rerio	ELISA	Blood plasma	Van den Belt et al. (2002), Fenske et al. (2001)
Danio rerio	ELISA	Whole body homogenate	Brion et al. (2002)
Paralichthys olivaceus	ELISA	Plasma and liver	Zhang et al. (2019)
Oncorhynchus kisutch & *O. tshawytscha*	ELISA	Plasma and serum	Peck et al. (2011)
Clarias batrachus	ELISA	Plasma	Garnayak et al. (2013)
Gambusia holbrooki	ELISA	Liver and kidneys	Scott et al. (2018)
Pimephales promelas	ELISA	Plasma	Nilsen et al. (2004)
Cyprinus carpio	ELISA	Plasma	Nilsen et al. (2004)
Danio rerio	ELISA	Whole body homogenate	Nilsen et al. (2004)
Oryzias latipes	ELISA	Whole body homogenate	Nilsen et al. (2004)
Liopsetta pinnifasciata	ELISA	Plasma	Shved et al. (2011)
Acanthogobius flavimanus	ELISA	Plasma	Ohkubo et al. (2003)
Oncorhynchus spp.	ELISA		Peck et al. (2011)
Sebastes schlegelii *Pseudopleuronectes yokohamae* *Hexagrammos otakii*	Western blot & ELISA	Plasma	Li et al. (2018)
Pleuronectes vetulus	ELISA	Plasma	Lomax et al. (1998)
Pleuronectes putnami	ELISA	Plasma	Roy et al. (2004)
Misgurnus anguillicaudatus	ELISA	Plasma	Lv et al. (2009)
Oncorhynchus mykiss & *Pimephales promelas*	ELISA	Plasma	Wunschel et al. (2005)
Mugil cephalus	Western blot, CLIA & SRID	Serum	Amano et al. (2019a)
Pleuronectes yokohamae	ELISA	Serum	Amano et al. (2019b)

5.6 Conclusion and Future Perspectives of Vtg

To date, the Vtg protein has been measured using a variety of approaches as an indicator of estrogenic responses in fish. Screening assays could employ ELISA techniques for priority substances suspected of having estrogenic properties. The method has a high level of specificity and sensitivity for relatively low cost. Ultimately, this information is required to assess fully the sensitivity of both the Vtg endpoint and different fish species. Although Vtg is the most commonly employed biomarker in such studies, more research into the biological importance of several Vtgs is required. Possessing compiled data on Vtg as significant proteins involved in oogenesis. Until recently, fish Vtg protein has mainly been assessing estrogenic activity in fish aquatic environments. The knowledge of Vtg receptors is even more limited. Only a few research have looked at the interactions between receptor proteins and their Vtg ligands so far. Future studies should need to focus on the analysis of the entire process of Vtg within a species from a physiological standpoint, the migration of released proteins from the circulation into the egg yolk, which occurs after Vtg production in the liver. To broaden the concept of a "multiple Vtg model," it is also vital to gather Vtg data from a diverse array of fish species.

Acknowledgments The authors appreciatively acknowledge the financial support of DST-INSPIRE fellowship-IF160623, New Delhi, India.

Conflict of Interest The authors declare no conflict of interest.

References

Amano H, Fujita T, Hiramatsu N, Shimizu M, Sawaguchi S, Matsubara T et al (2007) Egg yolk proteins in grey mullet (Mugil cephalus): purification and classification of multiple lipovitellins and other vitellogenin-derived yolk proteins and molecular cloning of the parent vitellogenin genes. J Exp Zool A Ecol Genet Physiol 307(6):324–341

Amano H, Kotake A, Hiramatsu N, Fujita T, Todo T, Aoki JY et al (2019a) Development of specific chemiluminescent immunoassays for three subtypes of vitellogenin in grey mullet (Mugil cephalus). Gen Comp Endocrinol 271:30–38

Amano H, Uno S, Koyama J, Hiramatsu N, Todo T, Hara A (2019b) Development of specific enzyme-linked immunosorbent assays for multiple vitellogenins in marbled sole, Pleuronectes yokohamae. Gen Comp Endocrinol 281:67–72

Arukwe A, Goksøyr A (2003) Eggshell and egg yolk proteins in fish: hepatic proteins for the next generation: oogenetic, population, and evolutionary implications of endocrine disruption. Comp Hepatol 2(1):1–21

Babin PJ (1987) Plasma lipoprotein and apolipoprotein distribution as a function of density in the rainbow trout (Salmo gairdneri). Biochem J 246(2):425–429

Bergink EW, Wallace RA (1974) Precursor-product relationship between amphibian vitellogenin and the yolk proteins, lipovitellin and phosvitin. J Biol Chem 249(9):2897–2903

Bhandari RK, Komuro H, Nakamura S, Higa M, Nakamura M (2003) Gonadal restructuring and correlative steroid hormone profiles during natural sex change in protogynous honeycomb grouper (Epinephelus merra). Zool Sci 20(11):1399–1404

Brion F, Nilsen BM, Eidem JK, Goksøyr A, Porcher JM (2002) Development and validation of an enzyme-linked immunosorbent assay to measure vitellogenin in the zebrafish (Danio rerio). Environ Toxicol Chem 21(8):1699–1708
Carnevali O, Mosconi G (1992) In vitro induction of vitellogenin synthesis in Rana esculenta: role of the pituitary. Gen Comp Endocrinol 86(3):352–358
Carnevali O, Mosconi G, Yamamoto K, Kobayashi T, Kikuyama S, Polzonetti-Magni AM (1993) In-vitro effects of mammalian and amphibian prolactins on hepatic vitellogenin synthesis in Rana esculenta. J Endocrinol 137(3):383–389
Cohen AM, Mansour AA, Banoub JH (2005) 'De novo' sequencing of Atlantic cod vitellogenin tryptic peptides by matrix-assisted laser desorption/ionization quadrupole time-of-flight tandem mass spectrometry: similarities with haddock vitellogenin. Rapid Commun Mass Spectrom 19(17):2454–2460
Cohen AM, Mansour AA, Banoub JH (2006) Absolute quantification of Atlantic salmon and rainbow trout vitellogenin by the 'signature peptide' approach using electrospray ionization QqToF tandem mass spectrometry. J Mass Spectrom 41(5):646–658
Cohen AM, Jahouh F, Sioud S, Rideout RM, Morgan MJ, Banoub JH (2009) Quantification of Greenland halibut serum vitellogenin: a trip from the deep sea to the mass spectrometer. Rapid Commun Mass Spectrom 23(7):1049–1060
Copeland PA, Thomas P (1988) The measurement of plasma vitellogenin levels in a marine teleost, the spotted seatrout (Cynoscion nebulosus) by homologous radioimmunoassay. Comp Biochem Physiol B 91(1):17–23
Cui XF, Zhao Y, Chen HP, Deng SP, Jiang DN, Wu TL et al (2017) Cloning, expression and functional characterization on vitellogenesis of estrogen receptors in Scatophagus argus. Gen Comp Endocrinol 246:37–45
Dalvin S, Frost P, Loeffen P, Skern-Mauritzen R, Baban J, Rønnestad I, Nilsen F (2011) Characterisation of two vitellogenins in the salmon louse Lepeophtheirus salmonis: molecular, functional and evolutional analysis. Dis Aquat Org 94(3):211–224
Davis LK, Katsu Y, Iguchi T, Lerner DT, Hirano T, Grau EG (2010) Transcriptional activity and biological effects of mammalian estrogen receptor ligands on three hepatic estrogen receptors in Mozambique tilapia. J Steroid Biochem Mol Biol 122(4):272–278
Ding JL, Hee PL, Lam TJ (1989) Two forms of vitellogenin in the plasma and gonads of male Oreochromis aureus. Comp Biochem Physiol B 93(2):363–370
Dos Santos DR, Yamamoto FY, Filipak Neto F, Randi MAF, Garcia JE, Costa DDM et al (2016) The applied indicators of water quality may underestimate the risk of chemical exposure to human population in reservoirs utilized for human supply—Southern Brazil. Environ Sci Pollut Res 23(10):9625–9639
Fenske M, van Aerle R, Brack S, Tyler CR, Segner H (2001) Development and validation of a homologous zebrafish (Danio rerio Hamilton–Buchanan) vitellogenin enzyme-linked immunosorbent assay (ELISA) and its application for studies on estrogenic chemicals. Comp Biochem Physiol C Toxicol Pharmacol 129(3):217–232
Finn RN, Fyhn HJ (2010) Requirement for amino acids in ontogeny of fish. Aquac Res 41(5):684–716
Finn RN, Kristoffersen BA (2007) Vertebrate vitellogenin gene duplication in relation to the "3R hypothesis": correlation to the pelagic egg and the oceanic radiation of teleosts. PLoS One 2(1):169
Finn RN, Østby GC, Norberg B, Fyhn HJ (2002a) In vivo oocyte hydration in Atlantic halibut (Hippoglossus hippoglossus); proteolytic liberation of free amino acids, and ion transport, are driving forces for osmotic water influx. J Exp Biol 205(2):211–224
Finn RN, Wamboldt M, Fyhn HJ (2002b) Differential processing of yolk proteins during oocyte hydration in marine fishes (Labridae) that spawn benthic and pelagic eggs. Mar Ecol Prog Ser 237:217–226
Flick RW, Bencic DC, See MJ, Biales AD (2014) Sensitivity of the vitellogenin assay to diagnose exposure of fathead minnows to 17α-ethynylestradiol. Aquat Toxicol 152:353–360

Flouriot G, Pakdel F, Ducouret B, Ledrean Y, Valotaire Y (1997) Differential regulation of two genes implicated in fish reproduction: vitellogenin and estrogen receptor genes. Mol Reprod Dev 48(3):317–323

Fukada H, Fujiwara Y, Takahashi T, Hiramatsu N, Sullivan CV, Hara A (2003) Carp (Cyprinus carpio) vitellogenin: purification and development of a simultaneous chemiluminescent immunoassay. Comp Biochem Physiol A Mol Integr Physiol 134(3):615–623

Garnayak SK, Mohanty J, Rao TV, Sahoo SK, Sahoo PK (2013) Vitellogenin in Asian catfish, Clarias batrachus: purification, partial characterization and quantification during the reproductive cycle by ELISA. Aquaculture 392:148–155

Gimeno S, Komen H, Gerritsen AG, Bowmer T (1998a) Feminisation of young males of the common carp, Cyprinus carpio, exposed to 4-tert-pentylphenol during sexual differentiation. Aquat Toxicol 43(2–3):77–92

Gimeno S, Komen H, Jobling S, Sumpter J, Bowmer T (1998b) Demasculinisation of sexually mature male common carp, Cyprinus carpio, exposed to 4-tert-pentylphenol during spermatogenesis. Aquat Toxicol 43(2–3):93–109

Greeley MS Jr, Calder DR, Wallace RA (1986) Changes in teleost yolk proteins during oocyte maturation: correlation of yolk proteolysis with oocyte hydration. Comp Biochem Physiol B 84(1):1–9

Hara A, Hiramatsu N, Fujita T (2016) Vitellogenesis and choriogenesis in fishes. Fish Sci 82(2):187–202

Hawkins MB, Thornton JW, Crews D, Skipper JK, Dotte A, Thomas P (2000) Identification of a third distinct estrogen receptor and reclassification of estrogen receptors in teleosts. Proc Natl Acad Sci 97(20):10751–10756

Hawkins MB, Godwin J, Crews D, Thomas P (2005) The distributions of the duplicate oestrogen receptors ER-βa and ER-βb in the forebrain of the Atlantic croaker (Micropogonias undulatus): evidence for subfunctionalization after gene duplication. Proc R Soc B Biol Sci 272(1563):633–641

He P, Matich EK, Yonkos LT, Friedman AE, Atilla-Gokcumen GE, Aga DS (2019) Mass spectrometry based detection of common vitellogenin peptides across fish species for assessing exposure to estrogenic compounds in aquatic environments. Sci Total Environ 646:400–408

Hiramatsu N, Hara A, Hiramatsu K, Fukada H, Weber GM, Denslow ND, Sullivan CV (2002a) Vitellogenin-derived yolk proteins of white perch, Morone americana: purification, characterization, and vitellogenin-receptor binding1. Biol Reprod 67(2):655–667

Hiramatsu N, Matsubara T, Hara A, Donato DM, Hiramatsu K, Denslow ND, Sullivan CV (2002b) Identification, purification and classification of multiple forms of vitellogenin from white perch (Morone americana). Fish Physiol Biochem 26(4):355–370

Hiramatsu N, Matsubara T, Weber GM, Sullivan CV, Hara A (2002c) Vitellogenesis in aquatic animals. Fish Sci 68(sup1):694–699

Hiramatsu N, Matsubara T, Fujita T, Sullivan CV, Hara A (2006) Multiple piscine vitellogenins: biomarkers of fish exposure to estrogenic endocrine disruptors in aquatic environments. Mar Biol 149(1):35–47

Hiramatsu N, Todo T, Sullivan CV, Schilling J, Reading BJ, Matsubara T et al (2015) Ovarian yolk formation in fishes: molecular mechanisms underlying formation of lipid droplets and vitellogenin-derived yolk proteins. Gen Comp Endocrinol 221:9–15

Hultman MT, Rundberget JT, Tollefsen KE (2015) Evaluation of the sensitivity, responsiveness and reproducibility of primary rainbow trout hepatocyte vitellogenin expression as a screening assay for estrogen mimics. Aquat Toxicol 159:233–244

Jena B, Mohanty J, Das RC, Garnayak SK, Nandi S (2013) Induction, purification and partial characterization of vitellogenin in an Indian major carp Catla catla (Ham.). Aquac Res 44(12):1901–1911

Kang IJ, Yokota H, Oshima Y, Tsuruda Y, Oe T, Imada N et al (2002) Effects of bisphenol A on the reproduction of Japanese medaka (Oryzias latipes). Environ Toxicol Chem 21(11):2394–2400

Kishida M, Specker JL (1993) Vitellogenin in tilapia (Oreochromis mossambicus): induction of two forms by estradiol, quantification in plasma and characterization in oocyte extract. Fish Physiol Biochem 12(3):171–182

LaFleur GJ Jr, Raldúa D, Fabra M, Carnevali O, Denslow N, Wallace RA, Cerda J (2005) Derivation of major yolk proteins from parental vitellogenins and alternative processing during oocyte maturation in Fundulus heteroclitus. Biol Reprod 73(4):815–824

LaFleur GJ, Byrne BM, Kanungo J, Nelson LD, Greenberg RM, Wallace RA (1995) Fundulus heteroclitus vitellogenin: the deduced primary structure of a piscine precursor to noncrystalline, liquid-phase yolk protein. J Mol Evol 41(4):505–521

Larkin P, Knoebl I, Denslow ND (2003) Differential gene expression analysis in fish exposed to endocrine disrupting compounds. Comp Biochem Physiol B: Biochem Mol Biol 136(2):149–161

Li H, Zhang S (2017) Functions of vitellogenin in eggs. Oocytes:389–401

Li Y, Wang J, Zheng M, Zhang Y, Ru S (2018) Development of ELISAs for the detection of vitellogenin in three marine fish from coastal areas of China. Mar Pollut Bull 133:415–422

Liang YQ, Huang GY, Liu SS, Zhao JL, Yang YY, Chen XW et al (2015) Long-term exposure to environmentally relevant concentrations of progesterone and norgestrel affects sex differentiation in zebrafish (Danio rerio). Aquat Toxicol 160:172–179

Lindholst C, Pedersen KL, Pedersen SN (2000) Estrogenic response of bisphenol A in rainbow trout (Oncorhynchus mykiss). Aquat Toxicol 48(2–3):87–94

Lomax DP, Roubal WT, Moore JD, Johnson LL (1998) An enzyme-linked immunosorbent assay (ELISA) for measuring vitellogenin in English sole (Pleuronectes vetulus): development, validation and cross-reactivity with other pleuronectids. Comp Biochem Physiol B: Biochem Mol Biol 121(4):425–436

Luna LG, Coady K (2016) Quantification of X. laevis vitellogenin by liquid chromatography tandem mass spectrometry. Ecotoxicol Environ Saf 124:296–302

Lv X, Shao J, Zhou Q, Song M, Jiang G (2009) Circannual vitellogenin levels in Chinese loach (Misgurnusanguillicaudatus). Environ Biol Fish 85(1):23–29

Maltais D, Roy RL (2009) Purification and partial characterization of vitellogenin from shorthead redhorse (Moxostoma macrolepidotum) and copper redhorse (Moxostoma hubbsi) and detection in plasma and mucus with a heterologous antibody. Fish Physiol Biochem 35(2):241–254

Manohar D, Rao GD, Sreenivasulu G, Senthilkumaran B, Gupta AD (2005) Purification of vitellogenin from the air breathing catfish, Clarias gariepinus. Fish Physiol Biochem 31(2):235–239

Matozzo V, Gagné F, Marin MG, Ricciardi F, Blaise C (2008) Vitellogenin as a biomarker of exposure to estrogenic compounds in aquatic invertebrates: a review. Environ Int 34(4):531–545

Matsubara T, Sawano K (1995) Proteolytic cleavage of vitellogenin and yolk proteins during vitellogenin uptake and oocyte maturation in barfin flounder (*Veraspermoseri*). J Exp Zool 272(1):34–45

Matsubara T, Ohkubo N, Andoh T, Sullivan CV, Hara A (1999) Two forms of vitellogenin, yielding two distinct lipovitellins, play different roles during oocyte maturation and early development of barfin flounder, Veraspermoseri, a marine teleost that spawns pelagic eggs. Dev Biol 213(1):18–32

Matsubara T, Nagae M, Ohkubo N, Andoh T, Sawaguchi S, Hiramatsu N et al (2003) Multiple vitellogenins and their unique roles in marine teleosts. Fish Physiol Biochem 28(1):295–299

Menuet A, Le Page Y, Torres O, Kern L, Kah O, Pakdel F (2004) Analysis of the estrogen regulation of the zebrafish estrogen receptor (ER) reveals distinct effects of ERalpha, ERbeta1 and ERbeta2. J Mol Endocrinol 32(3):975–986

Mommsen TP, Walsh PJ (1988) 5 Vitellogenesis and oocyte assembly. In: Fish physiology, vol 11. Academic Press, pp 347–406

Mosconi G, Carnevali O, Carletta R, Nabissi M, Polzonetti-Magni AM (1998) Gilthead seabream (Sparus aurata) vitellogenin: purification, partial characterization, and validation of an enzyme-linked immunosorbent assay (ELISA). Gen Comp Endocrinol 110(3):252–261

Moura Costa DD, Bozza DA, Rizzo LE, Garcia J, Costa MDM, de Oliveira Ribeiro CA (2016) Characterization, specificity and sensibility of produced anti-Rhamdia quelen vitellogenin in Brazilian fish species. Fish Physiol Biochem 42(6):1721–1732

Mushirobira Y, Nishimiya O, Nagata J, Todo T, Hara A, Reading BJ, Hiramatsu N (2018) Molecular cloning of vitellogenin gene promoters and in vitro and in vivo transcription profiles following estradiol-17β administration in the cutthroat trout. Gen Comp Endocrinol 267:157–166

Mushirobira Y, Niida M, Hotta T, Fujinami Y, Soyano K (2020) Hepatic expression profiles of three subtypes of vitellogenin and estrogen receptor during vitellogenesis in cultured female yellowtail. Gen Comp Endocrinol 299:113612

Nagler JJ, Davis TL, Modi N, Vijayan MM, Schultz I (2010) Intracellular, not membrane, estrogen receptors control vitellogenin synthesis in the rainbow trout. Gen Comp Endocrinol 167(2):326–330

Nath P, Sahu R, Kabita S, Bhattacharya D (2007) Vitellogenesis with special emphasis on Indian fishes. Fish Physiol Biochem 33(4):359–366

Nelson ER, Habibi HR (2013) Estrogen receptor function and regulation in fish and other vertebrates. Gen Comp Endocrinol 192:15–24

Nilsen BM, Berg K, Eidem JK, Kristiansen SI, Brion F, Porcher JM, Goksøyr A (2004) Development of quantitative vitellogenin-ELISAs for fish test species used in endocrine disruptor screening. Anal Bioanal Chem 378(3):621–633

Norberg B (1995) Atlantic halibut (Hippoglossus hippoglossus) vitellogenin: induction, isolation and partial characterization. Fish Physiol Biochem 14(1):1–13

Ohkubo N, Mochida K, Adachi S, Hara A, Hotta K, Nakamura Y, Matsubara T (2003) Development of enzyme-linked immunosorbent assays for two forms of vitellogenin in Japanese common goby (Acanthogobius flavimanus). Gen Comp Endocrinol 131(3):353–364

Opresko LK, Karpf RA (1987) Specific proteolysis regulates fusion between endocytic compartments in Xenopus oocytes. Cell 51(4):557–568

Pan X, Liu Y, Zhou K, Mu X, Zheng S, Liu C, Hu Y (2019) Tissue expression and bioinformatics analysis of the vitellogenin gene of Asian arowana (Scleropages formosus). J Appl Ichthyol 35(4):970–977

Panter GH, Hutchinson TH, Länge R, Lye CM, Sumpter JP, Zerulla M, Tyler CR (2002) Utility of a juvenile fathead minnow screening assay for detecting (anti-) estrogenic substances. Environ Toxicol Chem 21(2):319–326

Parks LG, Cheek AO, Denslow ND, Heppell SA, McLachlan JA, LeBlanc GA, Sullivan CV (1999) Fathead minnow (Pimephales promelas) vitellogenin: purification, characterization and quantitative immunoassay for the detection of estrogenic compounds. Comp Biochem Physiol C: Pharmacol Toxicol Endocrinol 123(2):113–125

Patiño R, Sullivan CV (2002) Ovarian follicle growth, maturation, and ovulation in teleost fish. Fish Physiol Biochem 26(1):57–70

Peck KA, Lomax DP, Olson OP, Sol SY, Swanson P, Johnson LL (2011) Development of an enzyme-linked immunosorbent assay for quantifying vitellogenin in Pacific salmon and assessment of field exposure to environmental estrogens. Environ Toxicol Chem 30(2):477–486

Puy-Azurmendi E, Ortiz-Zarragoitia M, Villagrasa M, Kuster M, Aragón P, Atienza J et al (2013) Endocrine disruption in thicklip grey mullet (Chelonlabrosus) from the Urdaibai Biosphere Reserve (Bay of Biscay, Southwestern Europe). Sci Total Environ 443:233–244

Reading BJ, Hiramatsu N, Sawaguchi S, Matsubara T, Hara A, Lively MO, Sullivan CV (2009) Conserved and variant molecular and functional features of multiple egg yolk precursor proteins (vitellogenins) in white perch (Morone americana) and other teleosts. Mar Biotechnol 11(2):169–187

Reading BJ, Hiramatsu N, Sullivan CV (2011) Disparate binding of three types of vitellogenin to multiple forms of vitellogenin receptor in white perch. Biol Reprod 84(2):392–399

Reith M, Munholland J, Kelly J, Finn RN, Fyhn HJ (2001) Lipovitellins derived from two forms of vitellogenin are differentially processed during oocyte maturation in haddock (Melanogrammus aeglefinus). J Exp Zool 291(1):58–67

Retzek H, Steyrer E, Sanders EJ, Nimpf J, Schneider WJ (1992) Molecular cloning and functional characterization of chicken cathepsin D, a key enzyme for yolk formation. DNA Cell Biol 11(9):661–672

Robinson CD, Scott AP (2012) 5 Background document: fish vitellogenin (Vtg) as a biomarker of exposure to xenoestrogens. In: Integrated marine environmental monitoring of chemicals and their effects, vol 30

Roubal WT, Lomax DP, Willis ML, Johnson LL (1997) Purification and partial characterization of English sole (Pleuronectes vetulus) vitellogenin. Comp Biochem Physiol B: Biochem Mol Biol 118(3):613–622

Roy RL, Morin Y, Courtenay SC, Robichaud P (2004) Purification of vitellogenin from smooth flounder (Pleuronectes putnami) and measurement in plasma by homologous ELISA. Comp Biochem Physiol B: Biochem Mol Biol 139(2):233–244

Sabo-Attwood T, Kroll KJ, Denslow ND (2004) Differential expression of largemouth bass (Micropterus salmoides) estrogen receptor isotypes alpha, beta, and gamma by estradiol. Mol Cell Endocrinol 218(1–2):107–118

Sawaguchi S, Koya Y, Yoshizaki N, Ohkubo N, Andoh T, Hiramatsu N et al (2005) Multiple vitellogenins (Vgs) in mosquitofish (Gambusia affinis): identification and characterization of three functional Vg genes and their circulating and yolk protein products. Biol Reprod 72(4):1045–1060

Sawaguchi S, Kagawa H, Ohkubo N, Hiramatsu N, Sullivan CV, Matsubara T (2006) Molecular characterization of three forms of vitellogenin and their yolk protein products during oocyte growth and maturation in red seabream (Pagrus major), a marine teleost spawning pelagic eggs. Mol Reprod Dev 73(6):719–736

Schwaiger J, Mallow U, Ferling H, Knoerr S, Braunbeck T, Kalbfus W, Negele RD (2002) How estrogenic is nonylphenol?: a transgenerational study using rainbow trout (Oncorhynchus mykiss) as a test organism. Aquat Toxicol 59(3–4):177–189

Scott PD, Coleman HM, Khan S, Lim R, McDonald JA, Mondon J et al (2018) Histopathology, vitellogenin and chemical body burden in mosquitofish (Gambusia holbrooki) sampled from six river sites receiving a gradient of stressors. Sci Total Environ 616:1638–1648

Seki M, Yokota H, Matsubara H, Tsuruda Y, Maeda M, Tadokoro H, Kobayashi K (2002) Effect of ethinylestradiol on the reproduction and induction of vitellogenin and testis-ova in medaka (Oryzias latipes). Environ Toxicol Chem 21(8):1692–1698

Shimizu M, Fujiwara Y, Fukada H, Hara A (2002) Purification and identification of a second form of vitellogenin from ascites of medaka (Oryzias latipes) treated with estrogen. J Exp Zool 293(7):726–735

Shved N, Kumeiko V, Syasina I (2011) Enzyme-linked immunosorbent assay (ELISA) measurement of vitellogenin in plasma and liver histopathology in barfin plaice Liopsettapinnifasciata from Amursky Bay, Sea of Japan. Fish Physiol Biochem 37(4):781–799

Simon R, Jubeaux G, Chaumot A, Lemoine J, Geffard O, Salvador A (2010) Mass spectrometry assay as an alternative to the enzyme-linked immunosorbent assay test for biomarker quantitation in ecotoxicology: application to vitellogenin in Crustacea (Gammarus fossarum). J Chromatogr A 1217(31):5109–5115

Sire MF, Babin PJ, Vernier JM (1994) Involvement of the lysosomal system in yolk protein deposit and degradation during vitellogenesis and embryonic development in trout. J Exp Zool 269(1):69–83

Sumpter JP, Jobling S (1995) Vitellogenesis as a biomarker for estrogenic contamination of the aquatic environment. Environ Health Perspect 103(Suppl 7):173–178

Takemura A, Kim BH (2001) Effects of estradiol-17β treatment on in vitro and in vivo synthesis of two distinct vitellogenins in tilapia. Comp Biochem Physiol A Mol Integr Physiol 129(2–3):641–651

Teo BY, Tan NS, Lim EH, Lam TJ, Ding JL (1998) A novel piscine vitellogenin gene: structural and functional analyses of estrogen-inducible promoter. Mol Cell Endocrinol 146(1–2):103–120

Thorpe KI, Hutchinson TH, Hetheridge MJ, Sumpter JP, Tyler CR (2000) Development of an in vivo screening assay for estrogenic chemicals using juvenile rainbow trout (Oncorhynchus mykiss). Environ Toxicol Chem 19(11):2812–2820

Thorpe KL, Hutchinson TH, Hetheridge MJ, Scholze M, Sumpter JP, Tyler CR (2001) Assessing the biological potency of binary mixtures of environmental estrogens using vitellogenin induction in juvenile rainbow trout (Oncorhynchus mykiss). Environ Sci Technol 35(12):2476–2481

Tramunt B, Montagner A, Tan NS, Gourdy P, Rémignon H, Wahli W (2021) Roles of estrogens in the healthy and diseased oviparous vertebrate liver. Meta 11(8):502

Tran TKA, Yu RMK, Islam R, Nguyen THT, Bui TLH, Kong RYC et al (2019) The utility of vitellogenin as a biomarker of estrogenic endocrine disrupting chemicals in molluscs. Environ Pollut 248:1067–1078

Tufail M, Takeda M (2008) Molecular characteristics of insect vitellogenins. J Insect Physiol 54(12):1447–1458

Tyler CR, Sumpter JP (1990) The development of a radioimmunoassay for carp, Cyprinus carpio, vitellogenin. Fish Physiol Biochem 8(2):129–140

Tyler CR, Sumpter JP, Bromage NR (1988) In vivo ovarian uptake and processing of vitellogenin in the rainbow trout, Salmo gairdneri. J Exp Zool 246(2):171–179

Tyler CR, Van der Eerden B, Jobling S, Panter G, Sumpter JP (1996) Measurement of vitellogenin, a biomarker for exposure to oestrogenic chemicals, in a wide variety of cyprinid fish. J Comp Physiol B 166(7):418–426

Tyler CR, van Aerle R, Hutchinson TH, Maddix S, Trip H (1999) An in vivo testing system for endocrine disruptors in fish early life stages using induction of vitellogenin. Environ Toxicol Chem 18(2):337–347

Van den Belt K, Wester PW, van der Ven LT, Verheyen R, Witters H (2002) Effects of ethynylestradiol on the reproductive physiology in zebrafish (Danio rerio): time dependency and reversibility. Environ Toxicol Chem 21(4):767–775

Wallace RA (1985) Vitellogenesis and oocyte growth in nonmammalian vertebrates. Oogenesis:127–177

Wang H, Yan T, Tan JT, Gong Z (2000) A zebrafish vitellogenin gene (vg3) encodes a novel vitellogenin without a phosvitin domain and may represent a primitive vertebrate vitellogenin gene. Gene 256(1–2):303–310

Wang W, Lian Q, Chen Y, Hiramatsu N, Wu M (2021) Development and characterization of polyclonal antibodies against subtype specific vitellogenin of the dojo loach, Misgurnus anguillicaudatus. Aquaculture 532:736089

Williams VN, Reading BJ, Amano H, Hiramatsu N, Schilling J, Salger SA et al (2014) Proportional accumulation of yolk proteins derived from multiple vitellogenins is precisely regulated during vitellogenesis in striped bass (Morone saxatilis). J Exp Zool A Ecol Genet Physiol 321(6):301–315

Wunschel D, Schultz I, Skillman A, Wahl K (2005) Method for detection and quantitation of fathead minnow vitellogenin (Vtg) by liquid chromatography and matrix-assisted laser desorption/ionization mass spectrometry. Aquat Toxicol 73(3):256–267

Yilmaz O, Patinote A, Nguyen T, Bobe J (2018) Multiple vitellogenins in zebrafish (Danio rerio): quantitative inventory of genes, transcripts and proteins, and relation to egg quality. Fish Physiol Biochem 44(6):1509–1525

Zhang H, Yan W, Aebersold R (2004) Chemical probes and tandem mass spectrometry: a strategy for the quantitative analysis of proteomes and subproteomes. Curr Opin Chem Biol 8(1):66–75

Zhang Z, Wang J, Gao M, Li X, Cheng Y, Zhang X et al (2019) New methods for purification of Paralichthysolivaceus lipovitellin and immunoassay-based detection of vitellogenin. Ecotoxicol Environ Saf 180:624–631

Zhong L, Yuan L, Rao Y, Li Z, Zhang X, Liao T et al (2014) Distribution of vitellogenin in zebrafish (Danio rerio) tissues for biomarker analysis. Aquat Toxicol 149:1–7

Functional Aspects of Vitellogenin in Oogenesis and Its Regulatory Mechanism

6

Brisca Renuga Ferdinand, X. Venci Candida, Arunthathi Shyla Suganthi, and Jeni Chandar Padua

Abstract

Before the embryo can fend for itself, vitellogenin (Vtg), a key precursor of the yolk proteins, provides the nutrition it needs to survive. The liver produces vitellogenin, the primary precursor of the protein in the yolk, which is then transported to and sequestered inside the eggs. Vitellogenin is crucial for the oocytes' healthy growth and development. Numerous factors such as the kind of egg (pelagic or benthic), the mode of reproduction (placental or nonplacental), and the cleavage pattern (meroblastic or holoblastic) are all impacted by the presence or absence of vitellogenin (Vtg). Eggs and embryos get free amino acids from Vtgs and Yps. The role of Vtgs in the transportation of calcium, phosphorus, lipids, amino acids, and other nutrients to the egg is often associated with its physiological significance. The differential expression of vtg genes typically dictates the kind of spawned eggs. Majority of Vtg genes are expressed in response to the season and reproductive cycle, and their byproducts are released into the bloodstream from the liver. In fish, the nervous and endocrine system coordinate reproduction. The sensory receptors process environmental inputs and the neural signals from these neural signals affect the pituitary gland by chemical messengers called releasing hormones. Gonadotropin-releasing hormones (GnRH) are released by the neurosecretory centers of the hypothalamus stimulate the pituitary to release the gonadotropins, which in turn target the gonads to secret the sex steroids. A hormone of the hypothalamus, anterior pituitary, and gonads is referred to as the hypothalamo-hypophyseal-gonadal axis. The hypothalamus' neurons directly innervate the pars distalis and the pars intermedia of the pitu-

B. R. Ferdinand (✉) · X. Venci Candida · A. S. Suganthi · J. C. Padua
Department of Zoology, Holy Cross College (Autonomous) (Affiliated to Manonmaniam Sundaranar University, Tirunelveli), Nagercoil, Tamil Nadu, India

V. Baskaralingam, R. Vanichviriyakit (eds.), *Vitellogenin in Fishes-Diversification, Biological Properties, and Future Perspectives*, https://doi.org/10.1007/978-981-99-5340-0_6

itary adenohypophysis in teleost, which lack a hypothalamo-hypophyseal portal system. GnRH1 is considered crucial for the release of gonadotropins from the pituitary and gonadal development. Gonadotropins are FSH and LH secreted by the adenohypophysis of the pituitary gland reaches the gonads through blood and stimulate estrogen production in the ovaries. FSH and LH stimulate ovarian follicle cells to synthesis and secrete the steroid hormones, viz., estradiol-17 β (E2), which play an important role in the regulation of vitellogenesis. Estradiol-17 β is carried in plasma and bound to sex-hormone-binding globulin (SHBGs) before entering the liver cells. SHBGs, shield steroids from quick metabolic breakdown, help control the amount of steroid (E2) made accessible to the liver cells and cellular signal transducers. During the reproductive season, estrogen stimulation causes the liver to create Vtg. In the ligand-binding process, the ER dissociates from HSP and goes through dimerization prior to translocation of the complex into the nucleus, involving a complex of coregulator proteins. The dimerized ER/E2 complexes bind estrogen response elements or incomplete ERE sequences in the promoter region of the vitellogenin gene, leading to initiation of gene expression and vitellogenin synthesis. This results in the activation or enhanced transcription of Vtg genes and subsequent increase and stabilization of Vtg messenger RNA. The synthesized Vtgs is then released into the bloodstream. Vitellogenin is produced by the liver and accumulated by the oocyte during the processes of vitellogenesis. The Vtg receptors on the surfaces of oocytes absorb plasma Vtgs, where the enzyme cathepsin cleaves into smaller yolk proteins, where the enzyme cathepsin cleaves into smaller yolk proteins. Vtg are made up of a linear series of five YP domains, namely, amino terminus (N)-lipovitellin heavy chain (LvH), phosvitin (Pv), lipovitellin light chain (LvL), β'-component (β'-c), and C-terminal peptide (Ct)-carboxy terminus-(C). Once cleaved from the parent Vtg, LvH and LvL associate to form lipovitellin, phosvitin, β'-c, and Ct fragments.

Keywords

Embryos · Amino acids · Gonadotropin · Hypothalamus · Adenohypophysis · Estradiol-17 β

Abbreviations

β-c	β-components
C	Carboxy terminus
CT	C-terminal coding regions
E2	Estradiol-17β/estrogen
ER	E2-receptor
ERE	Estrogen response elements
FAAs	Free amino acids
FSH	Follicle-stimulating hormone
GnRH	Gonadotropin-releasing hormones

GTH	Gonadotropic hormone
HSP	Heat shock protein
HUFA	Highly unsaturated fatty acids
LH	Luteinizing hormone
LLTP	Large lipid transfer protein
Lv	Lipovitellins
LvH	Lipovitellin heavy chain
LvL	Lipovitellin light chain
N	Amino terminus
PGCs	Primordial germ cells
POA	Preoptic area of the hypothalamus
PUFA	Polyunsaturated fatty acids
Pv	Phosvitins
Vtg	Vitellogenin
Yps	Yolk platelets

6.1 Introduction

The result of oocyte development and differentiation of an egg, or haploid reproductive cell, which following fertilization grows into a live embryo (Lubzens et al. 2017) typically involves numerous stages, including the production of primordial germ cells (PGCs), their transition into oogonia, and finally their growth into oocytes. Massive amounts of maternal molecules and information needed for early embryonic development, such as RNAs, proteins, lipids, vitamins, and hormones, are deposited during oocyte expansion (Lubzens et al. 2017; Patino and Sullivan 2002). Vitellogenin (Vtg), a significant precursor of the main yolk proteins, is one of the most significant proteins deposited in oocytes called yolk platelets (Yps). In oviparous mammals, the yolk of the eggs serves as a source of nutrients that the embryo needs to survive until it can forage for itself. In those species, vitellogenesis the process through which vitellogenin, the main precursor of the protein in the yolk, is produced by the liver and delivered to and sequestered within the eggs is essential for the successful growth and development of the oocytes. Due to their lower amount of phosphorylation, teleost yolk proteins are more water soluble than those of other vertebrates. In all oviparous animals, this crucial process is regulated by hormones. Two derived lipoproteins, lipovitellin and lipovitellin, are present in the yolk proteins.

In fishes high molecular mass glycolipophosphoprotein, vitellogenins are synthesized in the liver, mainly under estrogen (E2) control (Rosanova et al. 2002), posttranslationally glycosylated, and phosphorylated in the endoplasmic reticulum (ER) and Golgi complexes before being tagged for export, secreted as homodimeric lipoprotein complexes into the blood or circulating body fluid, and transported to the ovary. There, they are internalized during vitellogenesis into developing oocytes through receptor (clathrin)-mediated endocytosis (Amano et al. 2008).

The large lipid transfer protein (LLTP) superfamily includes multidomain proteins called vitellogenins. Vitellogenin has been discovered to be synthesized in male sexually immature animals in smaller proportions (Piulachs et al. 2003), despite being thought to be a protein exclusive to females and an egg yolk protein precursor. At first, vitellogenin was thought to be the embryos' main source of energy. However, a number of investigations have shown that vitellogenin and its derivatives Yps such as Pv and Lv have a number of non-nutritional functions. The vitellogenin has been demonstrated that the cleavage pattern (Finn et al. 2009) and mode of reproduction (Babin et al. 2007; Brawand et al. 2008; Finn et al. 2009) can be determined by the differential expression and deposition of vitellogenins in the developing oocytes. The major purpose of vitellogenin (Vtg) proteins is to provide as a source of yolk nutrients throughout the early stages of development. Multiple vitellogenin genes, though, could serve a variety of purposes. The many roles that vitellogenins play during oogenesis are described here.

6.2 Identifies the Egg Type and Reproduction Mode

Numerous factors, such as the kind of egg (pelagic or benthic), the mode of reproduction (placental or nonplacental), and the cleavage pattern, are all impacted by the presence or absence of vitellogenin (Vtg) (meroblastic or holoblastic). There are several Vtg isoforms produced within a single fish species by a multigene family that codes for vitellogenins. For instance, eight zebrafish Vtg genes have been discovered (Wang et al. 2005). Fish yolking rates are associated with the number of copies of a certain gene, proving that Vtg gene behavior is dose dependent. Fish lay eggs faster than other animals because they contain more copies of the VTG gene than other species (Buisine et al. 2002).

It seems that lineage-specific Vtg gene duplications typically dictate the kind of fish eggs. The differential expression of Vtg genes in *Acanthomorpha teleosts* is correlated with the kind of spawned eggs, either benthic or pelagic (Finn et al. 2009). The quantity of Vtg, Yps, cleavage symmetry, and pattern all correlate with one another. Yps-rich yolks typically stop cleavage. Cleavage happens holoblastically, or throughout the entire egg, in eggs with very few Yps like isolecithal and mesolecithal eggs. Yp-rich eggs rupture meroblastically with only limited cytoplasm discharge.

6.3 Eggs and Embryos Get Free Amino Acids from Vtgs and Yps

The role of Vtgs in the transportation of calcium, phosphorus, lipids, amino acids, and other nutrients to the egg is often associated with its physiological significance. All living things depend on amino acids for survival. They serve as crucial signaling molecules, fuel molecules, and key substrates for the synthesis of a variety of bioactive chemicals and proteins throughout the early fish ontogeny. Because the majority of fish eggs is cleidoic, or closed free-living systems after ovulation and activation, the availability of amino acids by the mother during oogenesis is essential for the early development of fish. The principal means of providing amino acids to the developing oocyte before ovulation are the vitellogenins, even though more than 600 proteins have been found in the maturing oocytes of fish.

Majority of Vtg genes are expressed in response to the season and reproductive cycle, and their byproducts are released into the bloodstream from the liver. In cold-water marine fishes and teleost, the intestine synthesizes Vtg C, which is involved in the transport of highly unsaturated fatty acids (HUFA) and polyunsaturated fatty acids (PUFA) from the gut to the ovary in order to meet the physiological needs of the developing embryos and larvae (Tocher et al. 2008). The majority of Vtg genes are expressed in response to the season and reproductive cycle.

The Vtgs initially deteriorate when they are encapsulated in yolk platelets. Aspartic proteinase cathepsin D breaks down Vtgs into different-sized Vtgs, namely, lipovitellins (Lv), phosvitins (Pv), and C-terminal coding sections (CT) (Komazaki and Hiruma 1999; Retzek et al. 1992). The biggest yolk protein LV is an apoprotein created by the proteolytic processing of Vtgs and delivers phospholipids primarily into oocytes during development. The smallest yolk protein, Pv, is mostly made up of phosphorylated serine residues. Pv is thought to be able to stabilize nascent Vtg structure during lipid loading and improve the solubility of Vtgs in the blood. The tiny -βc and CT are thought to stabilize the Vtg dimer for cellular recognition and receptor binding as well as to guard against premature proteolysis of Vtg or its product Yps. At the end of oocyte growth, just before maturity, an extra proteolysis of Yps takes place in many marine fish species with pelagic eggs (Finn 2007; Reading et al. 2009; Williams et al. 2014).

Fish embryos are entirely dependent on the supply of maternal nutrients and transcripts or signaling factors produced during oogenesis until they reach the mid-blastula transition, at which point they begin to synthesize their own protein. This proteolysis of Vtgs results in the production of smaller peptides and free amino acids (FAAs), which are essential components of the osmotic gradient needed for water absorption during the "hydration phase," which renders the pelagic eggs buoyant (Finn and Fyhn 2010). Lv and Pv are both secondarily degraded during embryonic development by a different enzyme pool, which is expected to contain cysteine proteases. The secondary cleavage of Yps, which is essential for embryonic development, results in the production of small peptides and FAAs. As a result, the Vtg-degraded FAAs are the crucial source of energy for growing embryos. Lysophosphatidic acid, a signaling molecule produced from phosphatidylcholines of Pv, control the development of hemangioblasts and primitive hematopoiesis in zebrafish. These findings imply that Pv and its smaller byproduct molecules play significant roles in embryonic development (Li et al. 2014).

6.4 Vitellogenin Functioning as Lipid and Mineral Transporting Agents

Complete forms of Vtg are made up of a linear series of five YP domains, such as amino terminus (N)-lipovitellin heavy chain (LvH), phosvitin (Pv), lipovitellin light chain (LvL), β'-component (β'-c), and C-terminal peptide (Ct)-carboxy terminus-(C). The LvH polypeptide is composed of secondary and tertiary structures that are amphipathic and create a basket with hydrophobic residues needed to accommodate lipids. Oocyte Vtg receptors are shown to bind to the LvH domain. The lipid basket of Vtg contains the LvL, which is smaller than LvH and shares many chemical and structural traits with LvH. The LvH domain of Vtg also has a site that may bind zinc ion, and both the LvH and LvL domains often include one or more glycosylation sites to which carbohydrate moieties are linked.

The yolk protein products include lipovitellin (Lv) heavy chain (LvH), Lv light chain (LvL), phosvitin (Pv), β′-component (β′-c), and C-terminal component (Ct). Phosphate groups attached to Pv are indicated by the aqua blue circles, and lipids bound to vitellogenin (Vtg) and Lv are indicated by the half-ovals shaded in brown.

The Pv is a metalloprotein that contains more than half of its total residues as serines and can have phosphates covalently bonded before being secreted by the liver. Through ionic interactions that take place in the bloodstream or oocyte, the negatively charged phosphates draw calcium, magnesium, zinc, and other multivalent metal cations (such as ferric iron). Since metal ions are scarcely available for uptake from the environment in freshwater fishes like the masu salmon (*Oncorhynchus masou*) and mosquitofish (Gambusia affinis), Pv's delivery of metal ions is essential for embryo survival. Additionally, the Pv often has a number of possible sites for glycosylation. As a result, Pv helps keep the mostly hydrophobic Vtg particle soluble in water by moving crucial metabolic ions and carbohydrates into the yolk. Additionally, Pv may help to stabilize Vtg structure by interacting with the lipid cargo basket that LvH and LvL create.

The YPs' ct and c′ contain 14 cysteine residues that are known to participate in the disulfide bonds necessary for the complicated folding of the Vtg polypeptide and perhaps for dimerization of native Vtg. When Vtg interacts to its oocyte receptor, the carboxy-terminus of ′-c may play a role in promoting the formation of disulfide bonds during peptide folding and/or adhesion. In addition to having a high-water solubility and the potential for having a glycosylation site, the ′-c also has a lot of hydrophilic residues that are exposed on the surface of the Vtg particle.

6.5 In Eggs and Embryos, Vtgs and Yps Are Maternal Immune-Related Factors

Most fish and aquatic invertebrates release and fertilize their eggs externally, subjecting the developing embryos to a hostile aquatic environment full of potential pathogens that could cause a range of diseases and even death. The ability of their embryos to generate immune-relevant chemicals endogenously is also limited or nonexistent in the early stages of development, and their immune-relevant cells and tissues are not yet fully developed. The aquatic invertebrate embryos withstand pathogenic attacks. Fish and aquatic invertebrates generate eggs that are fully developed fish embryos in an aquatic environment, including all the necessary nutrition and defense mechanisms. It has been established that Vtgs and the proteins they give rise to, Yps, contribute to embryo protection. Vtgs has been shown to work as a multivalent pattern recognition receptor that can bind to lipopolysaccharide, lipoteichoic acid, peptidoglycan, glucan, and virions, as well as a bactericidal molecule that can harm bacterial cell walls and an opsonin that can enhance macrophages' ability to phagocytose bacteria (Garcia et al. 2010; Li et al. 2008).

Lv and Pv produced by the proteolytic cleavage of Vtgs are immune-competent molecule. The Pv functions as a pattern recognition receptor and an antibacterial effector protein, contributing significantly to the immunity of zebrafish embryos. Smaller Pv-derived peptides are also bactericidal and inhibiting the development of the virus-infected cells, thus lowering the virus amounts in the virus-infected cells (Wang et al. 2011; Zhang et al. 2011, 2015). This suggests that Pv is an immune-relevant maternal component that can defend developing embryos against virus infection LvH, and LvL, like Pv, can shield developing embryos and larvae (Liang et al. 2016; Zhang et al. 2011). These demonstrate that Pv and Lv are maternally produced proteins that play a role in immune defense in fish embryos and larvae.

6.6 Vtgs and Yps Are Antioxidant Reagents in Eggs and Embryos

A chemical process called oxidation can generate free radicals, which can set off a cascade of events that seriously harm DNA, proteins, and lipids. Antioxidant defense is therefore regarded to be crucial for all stages of an organism's life. This is also true during the growth and development of the embryo because the strong metabolism of the embryo results in a significant amount of oxidizing chemicals being produced. An intriguing subject in the fields of ecological evolution and animal production is how quickly developing embryos defend themselves from free radical damage (Ebrahimi et al. 2012; Müller et al. 2012). It has been demonstrated that oviparous animals' eggs contain significant amounts of maternally derived antioxidants. Mothers provide a variety of antioxidants, including vitamin A, vitamin E, and beta-carotene, in their eggs. Particular antioxidants found in egg yolk are crucial for embryonic growth (Barim-Oz and Sahin 2016).

Free radicals are produced via a chemical process of oxidation have the potential to damage lipids, proteins, and DNA. Therefore, antioxidant defense is considered to be important during the embryo's growth and development. As a result of the embryo's robust metabolism, sizable amount of oxidizing chemicals are generated. Fish eggs contain a sizable amount of antioxidants such as vitamin A, vitamin E, and beta-carotene. In zebrafish the Pv prevents the oxidation of linoleic acid and scavenges the DPPH radical. It's feasible that Pv's antioxidant action will guard against free radical damage in quickly developing embryos. The antioxidant activity of other Vtg-derived substances such as Lv, -βC, and CT was also reported (Ando and Yanagida 1999).

6.7 Alternative Functions of Vitellogenins

In some fish, Vtgs have also been shown to control their own synthesis. The injection of heterologous Vtg(s) purified from Indian mrigal carp has been shown to promote vitellogenin production in female walking catfish (*Clarias batrachus*) (*Cirrhinus mrigala*). High levels of Vtg(s) inside the oocyte in rainbow trout can change the way Vtg is made in the liver by preventing the ovary from producing E2. The largest YP is the LvH lipoprotein that supplies offspring with amino acids and phospholipids, which can serve as catabolic energy substrates or in anabolic synthesis of membrane or protein structures.

In addition, Vtgs are frequently used to assess exposure of animals in aquatic environments to endocrine-disrupting chemicals (EDCs), specifically to EDCs that mimic the action of estrogens. Since Vtgs are produced in response to endogenous E2, they are potentially ideal markers for assessment of estrogenic EDCs.

6.8 Regulatory Mechanism of Vitellogenin Synthesis

In fishes the principal events responsible for the enormous growth of oocytes are due to the accumulation of yolk proteins within their cytoplasm. Vitellogenin (Vtg), the precursor of the yolk protein rich in phospholipid, is essential for oogenesis, embryonic development, and larval survival. In many teleosts, large amounts of neutral lipids are accumulated in oocyte as lipid droplets during early oocyte growth in a process termed "oocyte lipidation." The occurrence of lipid droplets is generally first observed at the previtellogenic oocyte growth stage, and ongoing accumulation continues as development proceeds through vitellogenesis. Very-low-density lipoprotein (VLDL) is the primary carrier of the neutral lipids into oocytes. After the oocyte completes its growth, it becomes ready for the next phase of oogenesis, that is, the resumption of meiosis, which is accompanied by several maturational processes in the nucleus and cytoplasm of the oocyte. This process, called oocyte maturation, occurs prior to ovulation and is a prerequisite for successful fertilization and is regulated by different hormones.

The main processes in fishes that lead to the massive expansion of oocytes are caused by the buildup of yolk proteins in their cytoplasm. The phospholipid-rich yolk protein precursor known as vitellogenin (Vtg) is crucial for oogenesis, embryonic development, and larval survival. Oocyte lipidation, or the accumulation of substantial amounts of neutral lipids as lipid droplets during early oocyte growth, is a phenomenon seen in many teleosts. Lipid droplets are typically first noticed during the previtellogenic egg growth stage, and they continue to accumulate as vitellogenesis progresses. The next stage of oogenesis, known as oocyte maturation, occurs before ovulation and is a requirement for successful fertilization. It is controlled by many hormones and happens after the oocyte has finished growing.

6.9 Hormonal Regulation of Vitellogenesis in Fishes

In fish, the nervous and endocrine system coordinate reproduction. Vitellogenins are phospholipid-rich yolk protein precursors produced by the liver and accumulated by the oocyte during the processes of vitellogenesis. Vitellogenesis is a seasonal orcyclic process and is closely correlated with several external factors and internal factors which acts on the neuroendocrine system. External factors, such as temperature, photoperiod, stress, and nutrient and internal factors including hormones play key roles in fish reproduction.

6.10 The Hypothalamo-Pituitary-Gonadal (HPG) Axis

The main regulating system is a hormone produced by the hypothalamus, anterior pituitary, and gonads known as the hypothalamo-hypophysial-gonadal axis (Bharadea et al. 2012). The hypothalamus' neurons directly innervate the pars distalis and the pars intermedia of the pituitary adenohypophysis in teleost, which lack a hypothalamo-hypophyseal portal system (Zohar et al. 2010). The sensory receptors process environmental inputs and the neural signals from these neural signals affect the pituitary gland by chemical messengers called releasing hormones. Gonadotropin-releasing hormones (GnRH) are released by the neurosecretory centers of the hypothalamus and stimulate the pituitary to release the gonadotropins, which in turn target the gonads to secret the sex steroids.

6.11 Hypothalamo-Pituitary-Gonadal Axis (HPG)

A hormone of the hypothalamus, anterior pituitary, and gonads is referred to as the hypothalamo-hypophysial-gonadal axis (Bharadwaj et al. 2012). In teleost, the HPG axis lacks a hypothalamo-hypophyseal portal system; however, the neurons of the hypothalamus directly innervate the pars distalis, and the pars intermedia of the pituitary adenohypophysis (Zohar et al. 2010). The external cues are processed by the sensory receptors and these neural signals influence the pituitary gland through the chemical messengers known as releasing hormones.

Gonadotropin-releasing hormones (GnRH) are released by the neurosecretory centres of the hypothalamus stimulates the pituitary to release the gonadotropins, which in turn target the gonads to secret the sex steroids.

6.11.1 Gonadotropin-Releasing Hormones (GnRH)

Upon receiving the seasonal cues, the neurons emanating from the preoptic area of the hypothalamus secretes a short peptide called the gonadotropin-releasing hormone (GnRH) that induces the synthesis of gonadotropins. The GnRH are decapeptides and expressed in multiple forms (Tostivint 2011). In fishes, three variants of GnRH such as GnRH1, GnRH2, and GnRH3 are reported; however, in some teleosts, including Cyprinidae and Salmonidae, the presence of GnRH1is restricted. Among the three variants, GnRH1 is considered crucial for the release of gonadotropins from the pituitary and gonadal development (Nocillado et al. 2007). Neurons that secrete GnRH1 are in the preoptic area of the hypothalamus (POA) that innervates into the pituitary where they regulate reproduction via gonadotropin. GnRH1 regulates feeding and sexual behavior in teleost; GnRH2 and GnRH3 are in the midbrain tegmentum (near III ventricle) and terminal nerve associated with the olfactory region, respectively (Volkoff and Peter 1999). The primary role of GnRH2 is to stimulate the release of LH (Chang et al. 2009), while GnRH3 controls nesting, aggression, and spawning behaviors (Yamamoto et al. 1997; Volkoff and Peter 1999; Ogawa et al. 2006).

While hypothalamic GnRH is a positive regulator of the HPG axis, dopamine, a neurotransmitter released by the neuroendocrine centers of the hypothalamus, exerts negative feedback of HPG activity. Dopamine is synthesized from tyrosine by tyrosine hydroxylase and Dopa decarboxylase (Biran and Levavi-Sivan 2018). Contrasting these effects, dopaminergic inhibition was not observed in several marine species, suggesting that its involvement in reproductive function is mainly important in freshwater fish. Dopamine receptors are GPCRs and are classically divided into two principal subtypes, according to their ability to activate (D1-like subtype) or inhibit (D2-like subtype) adenylyl cyclase, the key enzyme in the conversion of adenosine triphosphate to 30–50 cyclic AMP. Various agonists and antagonists with high specificity to D1 or D2 receptors were employed to demonstrate that the inhibitory effects of dopamine on fish reproduction are due to specific activation of D2-like receptors. The female zebrafish adenohypophysis contains three D2-like receptors in LH cells and that the dopaminergic innervation of the adenohypophysis originates from the hypothalamic POA (Fontaine et al. 2015). These findings provide neuroanatomical support for the existence of dopaminergic inhibition in piscine reproduction.

6.11.2 Gonadotropins (GTH)

The hypothalamus exerts control on pituitary by releasing the neurohormones that diffuses into the pituitary and directs the release of gonadotropins which stimulate growth and development of the gonads.

Gonadotropins are glycoprotein and are heterodimers consisting of two chemically distinct common α subunit (GPα) and a specific β subunit called FSHβ and LHβ secreted by the adenohypophysis of the pituitary gland. Till 1980, it was supposed that GTH was the only hormone that regulates all reproductive activities in fishes. Later, GTH1 and GTHII were discovered and are renamed as LH and FSH due to their sequence identity. Gonadotropin once secreted reaches the gonads through blood and stimulate estrogen production in the ovaries.

6.11.3 Gonadal Hormones

The gonadal hormones, FSH and LH, acts on the ovarian follicle cells to stimulate the synthesis and secretion of the steroid hormones, viz., estradiol-17 β (E2) which play an important role in the regulation of vitellogenesis. The ovarian theca cells synthesize the testosterone which is converted to an estrogen, and granulosa cells secretes E2 (Senthilkumaran et al. 2001). E2 binds to the sex steroid-binding globulin in the blood and is transferred to the hepatocytes to act on the vitellogenin gene to synthesize and secrete vitellogenin (Hara et al. 2016), which subsequently secreted into blood, transported to the ovary, and absorbed into the maturing oocytes. E2 also feeds back to act on the brain and pituitary gland, providing homeostatic control of vitellogenesis and later maturational processes (Ohta et al. 2002).

6.12 Molecular Regulation of Vitellogenesis

As was mentioned earlier in this chapter, the reproductive hypothalamic-pituitary-gonadal neuroendocrine axis controls the seasonal or cyclical process of vitellogenesis. The production of gonadotropin-releasing hormone (GnRH) is induced by a variety of endogenous and environmental factors, including the fish's innate biorhythm, bioenergetic status, seasonal changes in photoperiod, and water temperature. In response, the pituitary releases follicle-stimulating hormone (FSH) (Patino and Sullivan 2002). Both trigger the release of estradiol-17 from the ovarian follicle (E2). Vitellogenin is produced by the liver and accumulated by the oocyte during the processes of vitellogenesis. The ovarian steroid hormone 17-estradiol (E2), which is produced under the control of the hypothalamic-pituitary-gonad axis, is the most significant inducer of vitellogenin (vtg) expression (Fig. 6.1) (Nelson and Habibi 2010).

17β-estradiol; esr: estrogen receptor; Vtg: vitellogenin; VE: vitellin envelope; GTH: gonadotropin; GnRH: gonadotropin releasing hormone.

In most fish species, vitellogenesis begins following exposure to estradiol-17 (E2) and can even be produced in tissues where it is not ordinarily synthesized, such as the hepatocytes in the male liver. The ovarian follicular cells produce estradiol-17 or E2, which is carried in plasma and bound to sex hormone-binding globulin (SHBGs) before entering the liver cells either through diffusion (Hall et al. 2005) or receptor-mediated uptake. These proteins, known as SHBGs, shield steroids from quick metabolic breakdown (Petra 1991) and help control the amount of steroid (E2) that is made accessible to target tissues in the liver. In addition to their role as sex steroid carriers, SHBGs are involved in cellular signal transduction that involves nuclear steroid receptors through specific SHBG membrane receptors in different sex steroid-sensitive tissues. Circulating E2 enters liver cells and binds to an estrogen receptor (ER), which changes conformation and dimerizes.

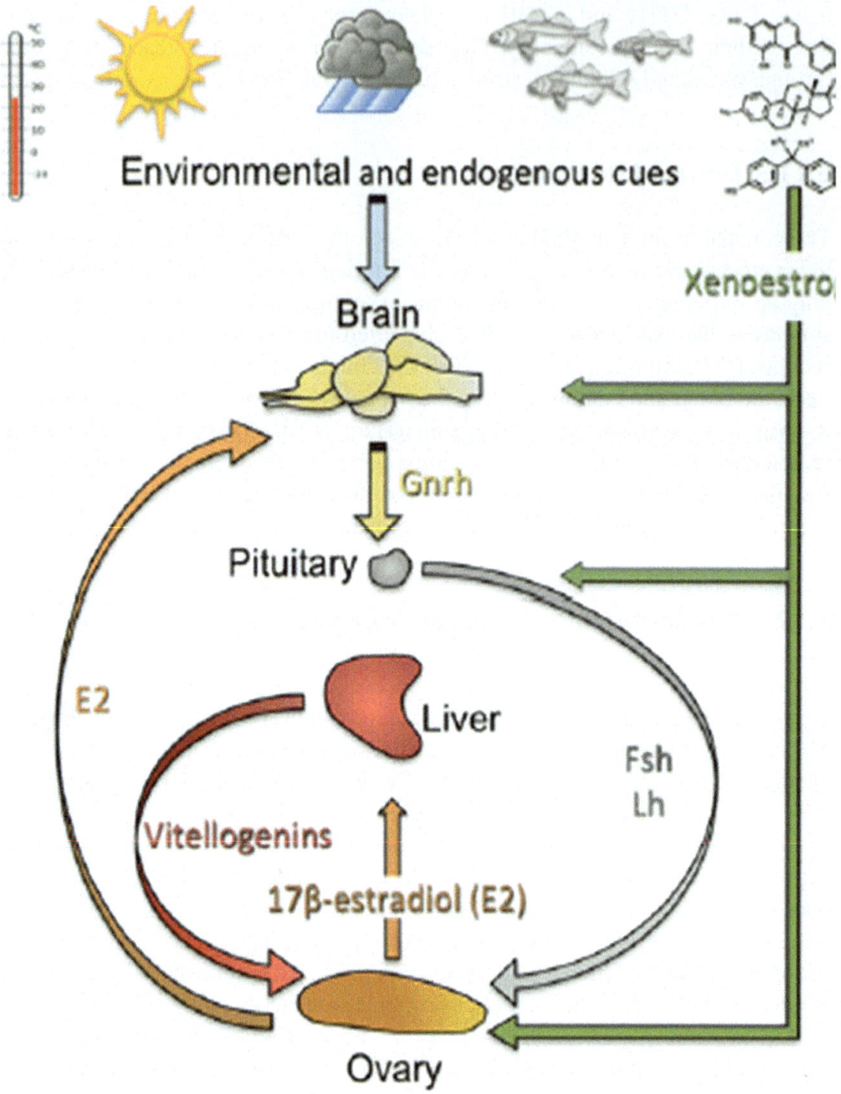

Fig. 6.1 Model regulation of vitellogenesis in teleost fishes via the hypothalamo-pituitary-gonadal (HPG) axis

During the reproductive season, estrogen stimulation causes the liver to create Vtg, which is then released into the bloodstream. The Vtg receptors on the surfaces of oocytes absorb plasma Vtg, which cathepsin then cleaves into smaller yolk proteins. Vitellogenesis, the first step of oocyte formation, consumes a lot of resources and energy. Poor larval development and higher egg mortality are caused by poor vitellogenin oocyte absorption. The deletion of the Vtg genes in female zebrafish causes a sharp decline in the fertility of eggs and the survival of the young. Therefore, fish reproduction success depends on significant Vtg production. Piscine Vtg are classified into three groups: Vtg A, Vtg B, and Vtg C which are encoded by vtgaa, vtgab, and vtgc, respectively.

Estrogen exerts its actions via activating estrogen receptors (ERs). In the liver, E2 is retained in target cells due to its high affinity binding to a specific steroid-receptor protein, the E2-receptor (ER). In the absence of a ligand or E2, the ER is found as a monomer in association with heat shock protein (HSP). In the ligand-binding process, the ER dissociates from HSP and goes through dimerization prior to translocation of the complex into the nucleus, involving a complex of coregulator proteins.

Estrogen receptor 1 (Esr1) and estrogen receptor 2 (Esr 2), also known as ER alpha and ER beta, are the two estrogen receptor (ER) subtypes found in fish. In some species, the Esr 2 subtype has two forms: Esr 2b (formerly ER beta1) and Esr 2 a (formerly ER beta 2 or ER gamma). The estrogen influence on Vtg production in fish liver is thought to be mediated through estrogen receptor 1. During breeding season, Esr1 expression rose. Along with Esr1, the liver also expresses other subtypes of the estrogen receptor type 2 (Esr2). Estrogen receptor 1, estrogen receptor 2b, and estrogen receptor 2a are controlled by gene estrogen receptor 1 (esr1), estrogen receptor 2b (esr2b), and estrogen receptor 2a (esr 2a), respectively.

There is interaction between the development of Vtg and ER subtypes. Leaños and Van Der Kraak (2007) found that esr2b increased Vtg expression in hepatocytes in a dose-dependent manner and mediates estrogen signals to promote esr1 expression, which then sensitizes the hepatocytes to further E2 stimulation and prepares them for vitellogenesis. The receptor protein's (esr2b and esr2a) mRNA expression is also increased by estradiol-17 or E2 (Nelson and Habibi 2010).

Following ligand binding, the ERs form homo or heterodimers that bind to particular palindromic estrogen response elements (ERE) sequences in the promoter region of estrogen-responsive genes, recruiting coactivators or co-repressors to the promoter. This then causes changes in the quantities of mRNA and related protein synthesis, which triggers the physiological response.

The dimerized ER/E2 complexes bind estrogen response elements (EREs) or incomplete ERE (ERE-like) sequences in the promoter region of the vitellogenin gene, leading to initiation of gene expression and vitellogenin synthesis. The hormone-receptor complex binds tightly in the nucleus at estrogen responsive elements (ERE) located upstream of Vtg genes, or within the estrogen-responsive genes in DNA. This results in the activation or enhanced transcription of Vtg genes and subsequent increase and stabilization of Vtg messenger RNA (mRNA). These promoter components and structures differ across fish species. Some of the effects of estrogens are so rapid that they cannot depend on RNA and protein synthesis and are known as non-genomic actions. They involve activating protein-kinase cascades, leading eventually to regulation of gene expression through phosphorylation and activation of transcription factors (TFs) within the nucleus.

Complete forms of Vtg are made up of a linear series of five YP domains, as follows: amino terminus (N)-lipovitellin heavy chain (LvH), phosvitin (Pv), lipovitellin light chain (LvL), β'-component (β'-c), and C-terminal peptide (Ct)-carboxy terminus-(C) . **LvH** is the major component, with an average mass of ~115 kDa, and **LvL** with an average mass of ~25 kDa for (Lubzens et al. 2017). Once cleaved from the parent Vtg, LvH and LvL associate to form lipovitellin

(Lv), which includes 65–100% of the Vtg by mass. Lv is comprised of amphipathic structures that form a basket lined with hydrophobic amino acid residues to accommodate lipids. Lv bears a Vtg receptor-binding peptide (VRBP), which has the special shape and surface charge distribution required to bind to VtgRs on the oocyte surface. Lv is the main source of the polar lipid and amino acid nutrients that serve as catabolic energy substrates, or in anabolic synthesis of membranes or proteins, during early development. Phosvitin is a small (~5–20 kDa) YP made up mainly of long stretches of extensively phosphorylated serine residues, empowering Pv to bind calcium, magnesium, zinc, and iron via ionic interactions. The remaining small YPs (β'-c and Ct) together contain 14 highly conserved cysteine residues known to form disulfide linkages thought to be required for the complicated folding of Vtg polypeptides. The β'-c also contains a CGxC motif implicated in formation of interchain disulfide linkages during Vtg dimerization, which is required for receptor-mediated endocytosis of the Vtgs by the oocyte. When released from Vtg, β'-c forms an ~16 kDa YP, but, with some exceptions, Ct (~13 kDa) may be degraded shortly after cleavage. Various Lv-Pv conjugates (LvH-Pv, Pv-Lvl, LvH-Pv-Lvl) have been detected in the yolk of several species (Reading and Sullivan 2011a). Figure 6.2 shows the pentapartite domain organization of native complete vitellogenin (dimer) in the blood and its corresponding dimeric yolk protein products in growing oocytes. The yolk protein products include lipovitellin (Lv) heavy chain (LvH), Lv light chain (LvL), phosvitin (Pv), β-component, and C-terminal component (Ct) (Reading and Sullivan 2011b).

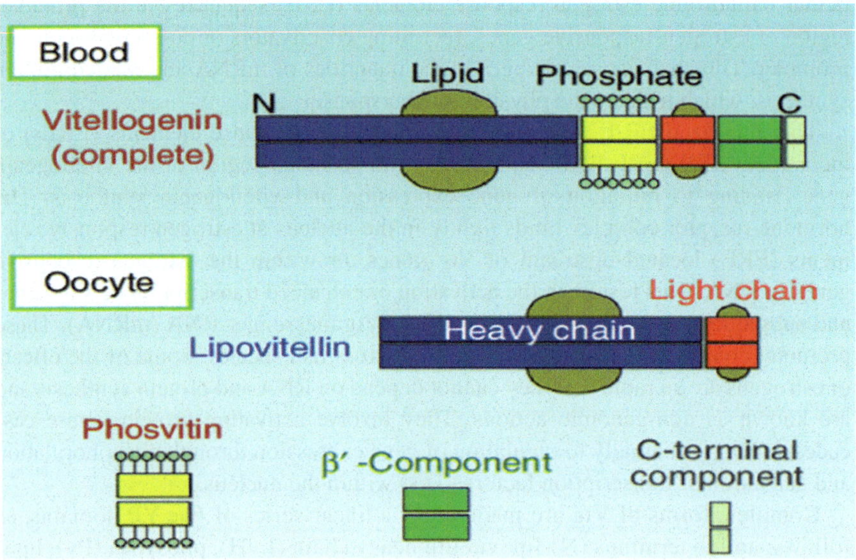

Fig. 6.2 Model showing the pentapartite domain organization of native complete vitellogenin (dimer) in the blood and its corresponding dimeric yolk protein products in growing oocytes

6.13 Conclusion

A major factor influencing egg quality is proper yolk formation, as most fishes are oviparous and the developing offspring are entirely dependent on stored egg yolk for nutritional sustenance. These maternally derived nutrients consist of proteins called vitellogenins in addition to other nutrients. Vitellogenin, the precursor of the yolk protein rich in phospholipid, is essential for oogenesis, embryonic development, and larval survival. Numerous factors, such as the kind of egg, the mode of reproduction, and the cleavage pattern, are all impacted by the presence or absence of vitellogenin. In teleosts, environmental changes, such as photoperiod and water temperature, provide signals that are received by the central nervous system. The main regulating system which is controlling the production of hormone is by hypothalamus, anterior pituitary, and gonads known as the hypothalamo-hypophysial-gonadal axis. Oocyte growth and maturation are regulated by pituitary gonadotropins and ovarian sex steroids. An integral part of hypothalamo-hypophysial-gonadal axis process is the synthesis of the oogenic proteins Vtg. E2 is the major estrogen in female fish. Estrogen exerts its actions via activating estrogen receptors (ERs). E2 stimulates the production of Vtg in the liver, which is then released into the bloodstream. The Vtg receptors on the surfaces of oocytes absorb plasma Vtg, which cathepsin then cleaves into smaller yolk proteins lipovitellin and phosvitin.

References

Amano H, Fujita T, Hiramatsu N, Kagawa H, Matsubara T, Sullivan CV, Hara A (2008) Multiple vitellogenin-derived yolk proteins in gray mullet (Mugil cephalus): disparate proteolytic patterns associated with ovarian follicle maturation. Mol Reprod Dev. 75:1307–1317.

Ando S, Yanagida K (1999) Susceptibility to oxidation of copper-induced plasma lipoproteins from Japanese eel: protective effect of vitellogenin on the oxidation of very low density lipoprotein. Comp Biochem Physiol C Pharmacol Toxicol Endocrinol 123:1–7

Babin PJ, Carnevali O, Lubzens E, Schneider WJ (2007) Molecular aspects of oocyte vitellogenesis in fish. In: Babin PJ, Cerdà J, Lubzens E (eds) The fish oocyte. Springer, Dordrecht. ISBN: 9781402062346

Barim-Oz O, Sahin H (2016) The influence of dietary antioxidant on ovarian eggs and levels of vitamin E, C, A, astaxanthin, β-carotene and oxidative stress in tissues of Astacus leptodactylus (Eschscholtz) during reproduction. Cell Mol Biol 62:1–10

Bhardwaj A, Nayan V, Yadav P, De S, Datta TK, Goswami SL. Heterologous expression and characterization of indian sahiwal cattle (Bos indicus) alpha inhibin. Anim Biotechnol 2012;23(2):71–88

Biran J and Levavi-Sivan, B. (2018). Endocrine Control of Reproduction, Fish. In M. K. Skinner (Ed.), Encyclopedia of Reproduction. vol. 6, pp. 362–368. Academic Press: Elsevier. https://dx.doi.org/10.1016/B978-0-12-809633-8.20579-7. ISBN: 9780128118993

Brawand D, Wahli W, Kaessmann H (2008) Loss of egg yolk genes in mammals and the origin of lactation and placentation. PLoS Biol l6:e63

Buisine N, Trichet V, Wolff J (2002) Complex evolution of vitellogenin genes in salmonid fishes. Mol Gen Genomics 268:535–542

Chang JP, Johnson JD, Sawisky GR, Grey CL, Mitchell G, Booth M (2009) Signal transduction in multifactorial neuroendocrine control of gonadotropin secretion and synthesis in teleosts-studies on the goldfish model. Gen Comp Endocrinol 161:42–52. https://doi.org/10.1016/j.ygcen.2008.09.005

Ebrahimi MR, Ahangari YJ, Zamiri MJ, Akhlaghi A, Atashi H (2012) Does preincubational in ovo injection of buffers or antioxidants improve the quality and hatchability in long-term stored eggs? Poult Sci 91:2970–2976

Finn RN (2007) Vertebrate yolk complexes and the functional implications of phosvitins and other subdomains in vitellogenins. Biol Reprod 76:926–935

Finn RN, Fyhn HJ (2010) Requirement for amino acids in ontogeny of fish. Aquac Res 41:684–716

Finn RN, Kolarevic J, Kongshaug H, Nilsen F (2009) Evolution and differential expression of a vertebrate vitellogenin gene cluster. BMC Evol Biol 9:2

Fontaine P, Wang N, Hermelink B (2015) Broodstock management and control of the reproductive cycle. Biology and culture of percid fishes: principles and practices: 103–122

Garcia J, Munro ES, Monte MM, Fourrier MC, Whitelaw J, Smail DA, Ellis AE (2010) Atlantic salmon (Salmo salar L.) serum vitellogenin neutralises infectivity of infectious pancreatic necrosis virus (IPNV). Fish Shellfish Immunol 29:293

Hall G, Phillips TJ. Estrogen and skin: the effects of estrogen, menopause, and hormone replacement therapy on the skin. Journal of the American Academy of Dermatology. 2005;1;53(4):555–68.

Hara A, Hiramatsu N, Fujita T (2016) Vitellogenesis and chroriogenesis in fish. Fish Sci 82(2):187–202

Komazaki S, Hiruma T (1999) Degradation of yolk platelets in the early amphibian embryo is regulated by fusion with late endosomes. Develop Growth Differ 41:173–181

Leaños-Castañeda O, Van Der Kraak G. Functional characterization of estrogen receptor subtypes, ERα and ERβ, mediating vitellogenin production in the liver of rainbow trout. Toxicol Appl Pharmacol 2007;224(2):116–125

Li Z, Zhang S, Liu Q (2008) Vitellogenin functions as a multivalent pattern recognition receptor with an opsonic activity. PLoS One 3:e1940

Li H, Yue R, Wei B, Gao G, Du J, Pei G (2014) Lysophosphatidic acid acts as a nutrient-derived developmental cue to regulate early hematopoiesis. EMBO J 33(12):1383–1396

Liang X, Hu Y, Feng S, Zhang S, Zhang Y, Sun C (2016) Heavy chain (LvH) and light chain (LvL) of lipovitellin (Lv) of zebrafish can both bind to bacteria and enhance phagocytosis. Dev Comp Immunol 63:47–55

Lubzens E, Young G, Bobe J, Cerda J (2017) Oogenesis in teleosts: how eggs are formed. Gen Comp Endocrinol 165:367

Müller W, Vergauwen J, Eens M, Blount JD (2012) Environmental effects shape the maternal transfer of carotenoids and vitamin E to the yolk. Front Zool 9:1–11

Nelson ER, Habibi HR (2010) Functional significance of nuclear estrogen receptor subtypes in the liver of goldfish. Endocrinology 151(4):1668–1676

Nocillado JN, Levavi-Sivan B, Carrick F, Elizur A (2007) Temporal expression of G-protein-coupled receptor 54 (GPR54), gonadotropin releasing hormones (GnRH), and dopamine receptor D2 (drd2) in pubertal female grey mullet, Mugil cephalus. Gen Comp Endocrinol 150:278–287

Ogawa S, Akiyama G, Kato S, Soga T, Sakuma Y, Parhar IS (2006) Immunoneutralization of gonadotropin-releasing hormone type-III suppresses male reproductive behavior of cichlids. Neurosci Lett 403:201–205

Ohta K, Yamaguchi S, Yamaguchi A, Gen K, Okuzawa K, Kagawa H, Matsuyama M (2002) Biosynthesis of steroid in ovarian follicles of red sea bream, Pagrus major (Sparidae, Teleostei) during final oocyte maturation and the relative effectiveness of steroid metabolites for germinal vesicle breakdown in vitro. Comp Biochem Physiol B 133:45–54

Patino R, Sullivan CV (2002) Ovarian follicle growth, maturation, and ovulation in teleost fish. Fish Physiol Biochem 26:57–70

Petra PH (1991) The plasma sex steroid binding protein (SBP or SHBG). A critical review of recent developments on the structure, molecular biology and function. J Steroid Biochem Mol Biol 40(4–6):735–753

Petersen DN, Tkalcevic GT, Koza-Taylor PH, Turi TG, Brown TA (1998). Identification of estrogen receptor β2, a functional variant of estrogen receptor β expressed in normal rat tissues. Endocrinology.1;139(3):1082–92.

Piulachs MD, Guidugli KR, Barchuk AR, Cruz J, Simoes ZLP, Belles X (2003) The vitellogenin of the honey bee, *Apis mellifera*: structural analysis of the cDNA and expression studies. Insect Biochem Mol Biol 33:459–465

Reading BJ, Sullivan CV (2011a) The reproductive organs and processes: vitellogenesis in fishes. https://doi.org/10.1016/B978-0-12-374553-8.00257-4

Reading BJ, Sullivan CV (2011b) Chapter 257: vitellogenesis in fishes. In: Ferrell AP (ed) Encyclopedia of fish physiology: from genome to environment. The reproductive organs and processes. Elsevier, Maryland Heights, MI, pp 635–646. 2272 pp. http://dx.doi.org.prox.lib.ncsu.edu/10.1016/B978-0-12-374553-8.00257-4

Reading BJ, Hiramatsu N, Sawaguchi S, Matsubara T, Hara A, Lively MO, Sullivan CV (2009) Conserved and variant molecular and functional features of multiple egg yolk precursor proteins (vitellogenins) in white perch (Morone americana) and other teleosts. Mar Biotechnol 11:169–187

Retzek H, Steyrer E, Sanders EJ, Nimpf J, Schneider WJ (1992) Molecular cloning and functional characterization of chicken cathepsin D, a key enzyme for yolk formation. DNA Cell Biol 11:661–672

Rosanova P, Romano M, Marciano R, Anteo C, Limatola E (2002) Vitellogenin precursors in the liver of the oviparous lizard, Podarcis sicula. Mol Reprod Dev 63:349

Senthilkumaran B, Okuzawa K, Gen K, Kagawa H. (2001). Effects of serotonin, GABA and neuropeptide Y on seabream gonadotropin releasing hormone release in vitro from preoptic-anterior hypothalamus and pituitary of red seabream, Pagrus major. J Neuroendocrinol 13:395–400

Tocher DR, Bendiksen EÅ, Campbell PJ, Bell JG (2008) The role of phospholipids in nutrition and metabolism of teleost fish. Aquaculture 280:21–34

Tostivint H (2011) Evolution of the gonadotropin-releasing hormone (GnRH) gene family in relation to vertebrate tetraploidizations. Gen Comp Endocrinol 170:575–581. https://doi.org/10.1016/j.ygcen.2010.11.017

Volkoff H, Peter RE (1999) Actions of two forms of gonadotropin releasing hormone and a GnRH antagonist on spawning behavior of the goldfish Carassius auratus. Gen Comp Endocrinol 116(3):347–355. https://doi.org/10.1006/gcen.1999.7377

Wang H, Tan JT, Emelyanov A, Korzh V, Gong Z (2005) Hepatic and extrahepatic expression of vitellogenin genes in the zebrafish, Danio rerio. Gene 356:91–100

Wang S, Wang Y, Ma J, Ding Y, Zhang S (2011) Phosvitin plays a critical role in the immunity of zebrafish embryos via acting as a pattern recognition receptor and an antimicrobial effector. J Biol Chem 286:22653–22664

Williams VN, Reading BJ, Hiramatsu N, Amano H, Glassbrook N, Hara A, Sullivan CV (2014) Multiple vitellogenins and product yolk proteins in striped bass, Morone saxatilis: molecular characterization and processing during oocyte growth and maturation. Fish Physiol Biochem 40:395–415

Yamamoto T, Ikeda K, Unuma T (1997) Apparent availabilities of amino acids and minerals from several protein sources for fingerling rainbow trout. Fish Sci 63(6):995–1001

Zhang S, Wang S, Li H, Li L (2011) Vitellogenin, a multivalent sensor and an antimicrobial effector. Int J Biochem Cell Biol 43:303–305

Zhang S, Dong Y, Cui P (2015) Vitellogenin is an immunocompetent molecule for mother and offspring in fish. Fish Shellfish Immunol 46:710

Zohar Y, Munoz-Cueto JA, Elizur A, Kah O (2010) Neuroendocrinology of reproduction in teleost fish. Gen Comp Endocrinol 165:438–455. https://doi.org/10.1016/j.ygcen.2009.04.017

Role of Vitellogenin as Immunocompetent Molecule

7

SavariyarAdimy Prakash Shoba, Johnson Vinoliya Josephine Mary, Chellathangam Anitha, and Amirtha Mani Punitha

Abstract

Vitellogenin a glycolipoprotein present in the yolk of eggs is found circulating in the blood of vertebrates and plays a critical role in defense against microbes. Diverse vitellogenin molecules have been identified of which the cleaved product of Vtg—phosvitin and lipovitellin—displays antimicrobial activity functioning as an immune-relevant molecule. Though vitellogenin is primarily involved in yolk protein formation, it has been proved to be an efficient immunocompetent molecule and also possess antioxidant properties. Fishes which live in an aquatic environment are susceptible to different pathogens in its vicinity and can cause mass mortality of young ones. A highly immunocompetent female fish transfers the immune factors to its offspring, and high level of IgM has been observed in the larval forms. This circulating antibody incorporates into the vitellogenic oocytes, thus getting transferred to the larva via the yolk sac. The maternal antibodies are metabolized during the different larval stages and totally disappears in the later developmental stage. Vtg functions as a pattern recognition molecule by its capability to recognize pattern associated molecular patterns like lipopolysaccharide, lipoteichoic acid, peptidoglycans, and glucans found on the cell surface of microbes. Research on fish embryos has documented the role of Vtg in the immunity of various species like carp, zebra fish, and rosy barb and its function as a bactericidal molecule. The role of Vtg as an opsonin and its ability to phagocytose bacteria by macrophages, its antiviral and antioxidant property, has been elaborated through various studies on fishes.

S. P. Shoba (✉) · J. V. J. Mary · C. Anitha · A. M. Punitha
Department of Zoology, Holy Cross College (Autonomous) (Affiliated to Manonmaniam Sundaranar University, Tirunelveli), Nagercoil, Tamil Nadu, India

V. Baskaralingam, R. Vanichviriyakit (eds.), *Vitellogenin in Fishes-Diversification, Biological Properties, and Future Perspectives*,
https://doi.org/10.1007/978-981-99-5340-0_7

Keywords

Glycolipoprotein · Phosvitin · Lipovitellin · Antioxidant · Opsonin · Pattern recognition receptors

Abbreviations

DEAE	Diethylaminoethyl
DNA	Deoxy ribonucleic acid
IgM	Immunoglobulin M
LCDV	Lymphocystis disease virus
LLTP	Large lipid transfer protein
LPS	Lipopolysaccharide
LTA	Lipoteichoic acid
Lv	Lipovitellin
nsSNPs	Non-synonymous single-nucleotide polymorphisms
PAMP	Pathogen-associated molecular pattern
PGN	Peptidoglycan
Pv	Phosvitin
RNAs	Ribonucleic acid
rPv	Recombinant phosvitin
Vg	Vitellogenin
Vtg	Vitellogenin
Yps	Yolk proteins

7.1 Introduction

Oocyte growth and embryonic development of oviparous animals depends on maternal information and molecules like proteins, ribonucleic acids, lipids, vitamins, and hormones (Lubzens et al. 2010; Patino and Sullivan 2002). Vitellogenin (Vtg) is an important glycolipoprotein present in the yolk of eggs, and it aids in the yolk formation. It is a homodimer formed of sugar, protein, and fat, and many sugar-specific transporter proteins are found in the blood of vertebrates. Vitellogenin belongs to the lipid transfer protein superfamily, which includes microsomal triglyceride transfer protein and apolipoprotein. Vtg separates into yolk protein, namely, phosvitin (Pv) and lipovitellin (Lv) that provides the nutrition for the developing embryos (Li and Zhang 2017). Vitellogenin is an immune-competent molecule that recognizes the microbes by the sugar moieties expressed on the cell walls of bacteria and involves in the host defense against pathogens, thus demonstrating as bactericidal molecule (Zhang et al. 2015; Liu et al. 2009; Shi et al. 2006; Sattar Khan et al. 2000). The role of vitellogenin as an immune molecule and its capacity to recognize PAMP has been documented in several fish species (Sun et al. 2013; Hu et al 2015; Sun and Zhang 2015). The cleaved products of Vtg that is Pv and Lv display an antibacterial role and possess antioxidant activity (Li and Zhang 2017). The pattern recognition receptors via the PAMP binds to the bacteria, where the recognition domain has a leucine-rich repeat which carries nsSNPs, moderating the ability to recognize pathogens. Pv has a lead role in the immunity of embryos of zebrafish through a pattern recognition receptor and an antimicrobial effector molecule (Zhang et al. 2015).

7.2 Classification

Vtg is a reproductive protein, made up of yolk proteins—phosvitin (Pv) and lipovitellin (Lv), which are stored in egg yolks and nourish early embryo during development. It is an immunocompetent factor that has the capacity to protect the host against the invading microbes that include bacteria and virus. The proteolytically cleaved products of maternal Vtg—Pv, Lv, and Pv-derived small peptides—demonstrate an antibacterial role and possess antioxidant activity that is capable of protecting embryonic cells from impairment by free radicals. Thus, Vtg helps not only in yolk protein formation but also in non-nutritional roles such as immune-molecule and antioxidant reagents (Li and Zhang 2017).

7.3 Occurrence of Vitellogenin

Vitellogenin (Vtg), a large precursor of the major yolk protein, deposited in oocytes is found in the hemolymph and liver of invertebrates, fishes, amphibians, reptiles, birds, and egg-laying mammals (Arukwe and Goksoyr 2003). Vtg, a homodimer, is a high-molecular-mass glycolipophosphoprotein found in the blood of vertebrates and hemolymph of invertebrates. In oviparous animals, under estrogen stimulation, vitellogenins are synthesized and secreted extra-ovarianly by the precursor of egg yolk, circulated in the blood stream and delivered to the ovary. It is absorbed by the oocytes and embryos through receptor-mediated endocytosis and cleaved into yolk proteins and are stored in the ooplasm (Rosanova et al. 2002; Wallace 1985).

7.4 Maternal Immunity in Fishes

Immunity is of prime importance right from the initial development of an organism, and this is achieved through maternal immunity. An immunocompetent female fish transfers the immune factors, both innate and adaptive, to a naive neonate that is immunocompromised. The immune factors are immunoglobulin, complement, protease inhibitor, lectin, lysozyme, and serine protease. Of the different immunoglobulin types, IgM is observed in most of the teleosts and is transferred from mother to the offsprings. IgM is a circulating antibody found in the larval forms of fishes. Analysis of larval tilapia and *Oreochromis mossambicus* by ELISA test showed low level of IgM in the blood of the prelarval stage, while comparatively higher IgM level was noted in the larval homogenate, suggesting that the IgM is of maternal origin and is found in the yolk sac of the larva. However, in the post-larval stage, an increased IgM was found in the blood of the larval stage similar to the IgM level in the larval homogenate suggesting that the maturation of immune system starts during the post-larval stages. Similarly, when bovine serum albumin was injected in the female brood fish, there was an increase in antibody in serum, egg homogenate, and sera of pre-larvae. Takemura and Takano (1997) suggested that the antibody that is produced in the maternal sera enters the larval circulation by incorporating into the vitellogenic oocytes and are transferred via larval yolk sac. Once the larva transforms into post-larvae, the level of antibody decreases to normal, suggesting the metabolization of maternal antibody throughout the different larval stages. Thus, the maternal IgM remains for a short time and reduces as the yolk is completely absorbed and totally disappears during later developmental stages. Brood fishes are subjected to stress factors in the environment like pollution, overcrowding, and adverse climatic conditions which affect the breeding performance, seed production, and the quality of young ones. These factors may cause high mortality among the fry and fingerlings causing a reduction in the healthy fishes. Thus, the immunity of the brood fish during vitellogenesis and oogenesis is vital to reduce the mortality, and the transfer of immunity to the embryo depends on the health of the brood fish (Swain and Nayak 2009).

7.5 Vitellogenin: An Immunocompetent Protein

Vitellogenin has been identified as an immunocompetent protein owing to its capacity to protect the host against the microbial attack. Pv-, Lv-, and Pv-derived peptides of maternal Vtg exhibits an antibacterial effect in embryos. Vtg and Pv also possess antioxidant property that protects cell damage by free radicals (Li and Zhang 2017). Vitellogenin performs multiple functions that include immune defense reaction, and fish Vtg has an important part in the regulation of innate immunity by recognition of specific moieties expressed on microbial cell wall and PAMPs, thus inducing macrophage phagocytosis (Hu et al 2015). Immunocompetent properties of vitellogenin were given in Fig. 7.1.

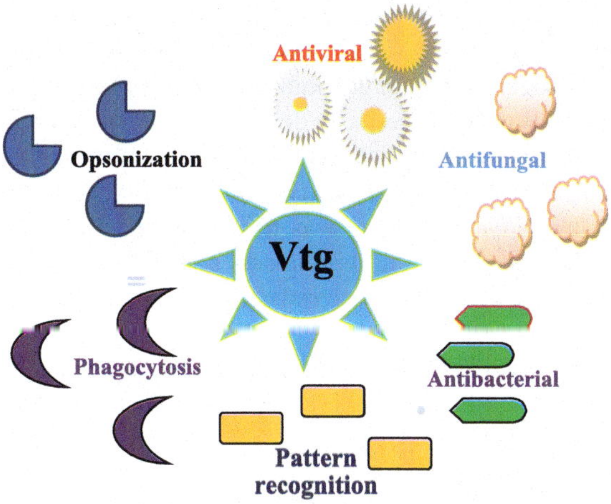

Fig. 7.1 Immunocompetent properties of vitellogenin

7.6 Vitellogenin as PAMP Recognition Molecule

Vitellogenin is a pattern recognition protein which can bind to diverse pathogens like bacteria (gram positive and negative), viruses, and fungi by recognizing the pathogen-associated molecular patterns (PAMPs) found on the cell walls of the pathogens like lipopolysaccharide (LPS), peptidoglycans, lipoteichoic acids, and glucans and initiating host response. Brood fishes rely on Vg during oocyte development and help to transfer maternal immunity (Hu et al 2015).

7.7 Vitellogenin as Defense Molecule

Immune response is pivotal for every living organism where recognition and elimination of nonself are vital processes of innate immunity mediated by germline-encoded proteins called pattern recognition receptors (Swain et al. 2006; Van Loon et al. 1981) that identify the PAMPs of bacterial and fungal cell wall (Mor and Avtalion 1990; Takemura and Takano 1997). Phagocytosis is performed by phagocytes such as macrophages and dendritic cells in vertebrates (Kanlis et al. 1995). Humoral molecules like antibody fragments and complements can adhere to the cell wall of pathogen and act as an opsonin thus promoting phagocytosis (Picchietti et al. 2001; Breuil et al. 1997). The phagocytosis is also enhanced by scavenger receptors and C-type lectins (Picchietti et al. 2004). Fish embryos are exposed to microbial attack, when they develop in an external environment and few pathogens are capable of causing mortality. In order to overcome the pathogenic stress, the embryos have to develop immunocompetent molecules. Though there is an early appearance of lymphoid organs and lymphocytes during embryogenesis (Zapata et al. 2006), the fish embryos have inadequate capacity to synthesize immune molecules endogenously (Magnadottir et al. 2004; Ellis 1988). Hu et al. (2008) explored the influence of vitellogenin (Vg) in the immunity of carp. Vitellogenin was purified from carp by gel filtration combined with diethylaminoethyl (DEAE) chromatography and was tested for antimicrobial activity. A strong antibacterial response was observed against *E. coli* and *S. aureus* and weak inhibitory activity against *S. cerevisiae*. When Vg-treated macrophage was treated with *E. coli* and *S. aureus*, it resulted in phagocytosis suggesting its role as a pattern recognition molecule and opsonin. Hemagglutinating and antibacterial activities of Vg has been demonstrated in the rosy barb, *Puntius conchonius* (Shi et al. 2004) which was able to produce Vtg on induction with *E. coli* as an infection-induced response (Table 7.1).

Table 7.1 Immunocompetent molecules in fishes

S. no	Name of the fish	Immune molecule identified	Interaction with bacteria/virus	Function	References
1.	*Oreochromis mossambicus*	IgM	–	Antibody transfer to larval forms	Takemura and Takano (1997)
2.	*Puntius conchonius*	Lv	*E. coli*	Hemagglutinating and antibacterial	Shi et al. (2004)
3.	Zebrafish	Vg and Pv	*E. coli, S. aureus*	Antibacterial, phagocytosis, and opsonization	Garcia et al. (2010), Hu et al. (2008)
4.	Zebrafish	Pv	Lymphocystis disease virus	Cytopathic effect	Garcia et al. (2010)
5.	Zebrafish	Vtg gene	Immuno-challenge with *Citrobacter freundii*	Expression of vitellogenin genes in the skin of fish	Lu et al. (2012)
6.	Carp and zebrafish	Vtg	*E. coli* and *S. aureus*	Antibacterial	Liu et al. (2009), Tong et al. (2010)
7.	Carp	Vtg	*E. coli, S. aureus,* and *S. cerevisiae*	Antibacterial, phagocytosis, and opsonization	Hu et al. (2008)
8.	*Hexagrammos otakii*	Vtg (E2-inducible protein)	*E. coli, S. aureus, Pichia pastoris*	Multivalent pattern recognition receptor—aids in phagocytosis and opsonization of bacteria	Jordan and Starks (1895)

(continued)

Table 7.1 (continued)

S. no	Name of the fish	Immune molecule identified	Interaction with bacteria/virus	Function	References
9.	Rosy barb *Puntius conchonius*	Lipovitellin	*Escherichia coli* and *Staphylococcus aureus*	Function as opsonin	Zhang and Zhang (2011)
10	Atlantic salmon	Vtg	Pancreatic necrosis virus	Antiviral	Sun et al. (2013)
11.	*Labeo rohita*	Serum antibody	Immune challenge with *Aeromonas hydrophila*	Augmentation of antibody	Swain et al. (2006)
12.	Rosy barb *P. conchonius*	LPS, LTA, and PGN	*E. coli* and *S. aureus*	Function as opsonin	Sun and Zhang (2015)
13.	Zebrafish	Pv (Pt5)	*A. hydrophila*	Antimicrobial agent, immune modulator, suppressing the pro-inflammatory cytokine gene expression	Ma et al. (2013)
14.	Tiger grouper brood fish *Epinephelus fuscoguttatus*	IgM	Vaccination of inactivated *V. harveyi*	Production and transfer of maternal antibodies increase IgM	Azman et al. (2019)

7.8 Antibacterial Potential of Vitellogenin

Fishes breed in water and the larval forms are subjected to microbes and can cause mass mortality. Vitellogenins are bactericidal molecules that contribute to the antimicrobial defense by its capacity to recognize the lipopolysaccharide, lipoteichoic acid, and peptidoglycan and bind to the bacterial cell wall. It acts as an opsonin by augmenting the phagocytosis of bacteria by macrophages (Garcia et al. 2010) as observed with *E. coli* and *S. aureus* (Hu et al 2015). Extract of zebrafish embryos presented antimicrobial potential against different pathogens including *Aeromonas hydrophila,* and phosvitin (Pv), a protein rich in eggs, was associated to the antimicrobic effect. Wang et al. (2011) developed a recombinant phosvitin (rPv) that can recognize the LPS, lipoteichoic acid, and peptidoglycan of gram-negative and positive microbes *E. coli*, *A. hydrophila*, and *S. aureus* and was proficient in killing the microbes. The C-terminal 55 residues (Pt5) with the sites Arg^{242} and Ala^{201}/Ile^{203} were vital for the antimicrobial effect of Pv and microinjection of rPv or Pt5 into early embryos boosted their resistance on challenge with *A. hydrophila*, and the resistance was lessened by co-injection of anti-Pv antibody and rPv or Pt5 but not by injection of anti-actin antibody and rPv. Pv partakes in the defense of early stages of embryos against microbial bouts by binding to the pathogens and destroying them. Vg from fish *Hexagrammos otakii* (Jordan and Starks 1895) functions as a multivalent pattern recognition receptor by binding to LPS, LTA, PGN, glucan, and laminarin, thus promoting macrophage phagocytosis. Mortality in larval stages of Indian major carp is due to the lack of immunocompetent molecules, and studies on immune challenge of brood fish *Labeo rohita* with *Aeromonas hydrophila* showed a dramatic increase in antibody levels, and this was transferred to the larval forms via the egg. During the first and second week of post hatch, there was no difference in the antibody levels, but an increase was observed from third week fry as they were naturally exposed to *A. hydrophila* in the aquatic environment. Swain et al. (2006) observed a reduction in mortality in larvae and fry that were immunized when compared to the control suggesting the protective role of maternally derived antibodies in the protection of young ones against pathogens. Immunochallenge of zebrafish with gram-negative bacteria *Citrobacter freundii* resulted in the expression of vitellogenin genes in the skin of fish (Lu et al. 2012). Similarly, Tong et al. (2010) observed the production of Vtg on induction with LPS and LTA in male zebrafish, and antibacterial activity against *E. coli* and *S. aureus* was noted in carp, zebrafish (Hu et al 2015; Tong et al. 2010), and fish *Hexagrammos otakii* (Li et al. 2008). The opsonin function of lipovitellin was documented by Zhang and Zhang (2011). Lv purified from the eggs of rosy barb *P. conchonius* could interact with the lipopolysaccharide, lipoteichoic acid, and peptidoglycan of bacteria *E. coli* and *S. aureus* (Sun and Zhang 2015). Pt5 a Pv-derived peptide increased the survival rate of zebrafish challenged with *A. hydrophila* suggesting its role as an antimicrobial agent. Function of Pt5 as an immune modulator and effector was evident when it was able to regulate the immune response of host by suppressing the pro-inflammatory cytokine gene expression and augmenting the anti-inflammatory cytokine genes (Ma et al. 2013). Vaccination of inactivated *V. harveyi* to tiger grouper brood fish *Epinephelus fuscoguttatus* resulted in the production and transfer of maternal antibodies to both eggs and larvae with a marked increase in IgM levels (Azman et al. 2019).

7.9 Antiviral

Phosvitin (Pv), a yolk protein of vitellogenin, has demonstrated antiviral activity in fishes. Purified Pv was able to inhibit the cytopathic effect in Lymphocystis disease virus (LCDV)-infected cells of zebrafish and neutralize the virus. Garcia et al. (2010) demonstrated the neutralizing effect of Vtg in the Atlantic salmon exposed to pancreatic necrosis virus, signifying that Vtg is active in targeting anti-viral resistance (Sun et al. 2013).

7.10 Antioxidant

Antioxidants, have gained attention in recent past owing to their roles in deterrence of chronic diseases. The recombinant phosvitin (rPv) is effective in inhibiting the oxidation of the linoleic acid and scavenging the 2,2-diphenyl-1-picrylhydrazyl radical. Zebrafish rPv is a cellular antioxidant adept of protecting radical-mediated oxidation of biomolecules (Sun and Zhang 2015). Toxic analysis of zebrafish rPv to murine macrophage RAW 264.7 cells indicated the noncytotoxic nature of Pv, suggesting it as a vital antioxidant which can be used as a supplementary mediator for diseased condition (Hu et al. 2015). In addition, due to the presence of serine and phosphorous content in Pv, the chelation of ion takes place, thus preventing DNA damage (Ishikawa et al. 2004).

7.11 Conclusion

Proteins are crucial for the growth and survival of living organisms; protein variants have been identified that contribute to the immune response of organisms. Vitellogenin, a female reproductive protein which is cleaved into phosvitin and lipovitellin, is stored in eggs and utilized for the development of early embryos. Vtg and its yolk proteins (Pv and Lv) has been identified as an immunocompetent molecule that displays an antimicrobial role in the developing embryos through opsonization, and the phagocytic potential of Vtg has been documented. In addition, it has been studied for its antioxidant property which can be used to protect cells from damage due to free radicals and oxidative stress. Thus, the research on immune-relevant molecule Vtg of fishes can deepen the understanding of Vtg and be applied for human health.

Acknowledgments The authors are grateful to the management, Holy Cross College (Autonomous), for the constant support and encouragement.

Conflict of Interest The authors have no conflicts of interest to declare.

References

Arukwe A, Goksoyr A (2003) Eggshell and egg yolk proteins in fish: hepatic proteins for the next generation: oogenetic, population, and evolutionary implications of endocrine disruption. Comp Hepatol 2:4

Azman M, Nain H, Rafidah O, Ching F, Senoo S, Zamri-Saad M (2019) Passive maternal antibody transfer to eggs and larvae of tiger grouper (*Epinephelus fuscoguttatus*). J Phys Conf Ser 1358:012017. IOP Publishing. https://doi.org/10.1088/1742-6596/1358/1/012017

Breuil G, Vassiloglou B, Pepin JF, Romestand B (1997) Ontogeny of IgM bearing cells and changes in the immunoglobulin M-like protein level (IgM) during larval stages in sea bass (*Dicentrarchus labrax*). Fish Shellfish Immunol 7:29–43

Ellis AE (1988) Ontogeny of the immune system in teleost fish. In: Ellis AE (ed) Fish vaccination. Academic Press, London, pp 20–31

Garcia J, Munro ES, Monte MM, Fourrier MC, Whitelaw J, Smail DA, Ellis AE (2010) Atlantic salmon (*Salmo salar* L.) serum vitellogenin neutralises infectivity of infectious pancreatic necrosis virus (IPNV). Fish Shellfish Immunol 29:293–297

Ishikawa S, Yano Y, Arihara K, Itoh M (2004) Egg yolk phosvitin inhibits hydroxyl radical formation from the fenton reaction. Biosci Biotechnol Biochem 68:1324–1331

Jordan DS, Starks EC (1895) The fishes of Puget Sound. Proc Calif Acad Sci (Ser 2) 5:785–855

Kanlis G, Suzuki Y, Tauchi M, Numata T, Shirojo Y et al (1995) Immunoglobulin in oocytes, fertilized eggs, and yolk sac larvae of red sea bream. Fish Sci 61:787–790
Li H, Zhang S (2017) Functions of vitellogenin in eggs. Results Probl Cell Differ. https://doi.org/10.1007/978-3-319-60855-6_17
Li Z, Zhang S, Liu Q (2008) Vitellogenin functions as a multivalent pattern recognition receptor with an opsonic activity. PLoS One 2008(3):e1940
Hu l, Sun C, Luan J, Lu L, Zhang S. (2015) Zebrafish phosvitin is an antioxidant with non-cytotoxic activity, Acta Biochimica et Biophysica Sinica, 47(5):349–354, https://doi.org/10.1093/abbs/gmv023
Liu QH, Zhang SC, Li ZJ, Gao CR (2009) Characterization of a pattern recognition molecule vitellogenin from carp (*Cyprinus carpio*). Immunobiology 2009(214):257–267
Lu A, Hu X, Xue J, Zhu J, Wang Y, Zhou G (2012) Gene expression profiling in the skin of zebrafish infected with *Citrobacter freundii*. Fish Shellfish Immunol 32:273–283
Lubzens E, Young G, Bobe J, Cerda J (2010) Oogenesis in teleosts: how eggs are formed. Gen Comp Endocrinol 165:367
Ma J, Wang H, Wang Y, Zhang S (2013) Endotoxin-neutralizing activity of hen egg phosvitin. Mol Immunol 2013(53):355–362
Magnadottir B, Lange S, Steinarsson A, Gudmundsdottir S (2004) The ontogenic development of innate immune parameters of cod (*Gadus morhua* L.). Comp Biochem Physiol B Biochem Mol Biol 139:217–224
Mor A, Avtalion RR (1990) Transfer of antibody activity from immunised mother to embryos in tilapias. J Fish Biol 37:249–255
Patino R, Sullivan CV (2002) Ovarian follicle growth, maturation, and ovulation in teleost fish. Fish Physiol Biochem 26:57–70
Picchietti S, Scapigliati G, Fanelli M, Barbato F, Canese S et al (2001) Sex related variations of serum immunoglobulin during reproduction in gilthead sea bream and evidence for a transfer from the female to the eggs. J Fish Biol 59:1503–1511
Picchietti S, Taddei AR, Scapigliati G, Buonocore F, Fausto AM et al (2004) Immunoglobulin protein and gene transcripts in ovarian follicles throughout oogenesis in the teleost *Dicentrarchus labrax*. Cell Tissue Res 315:259–270
Rosanova P, Romano M, Marciano R, Anteo C, Limatola E (2002) Vitellogenin precursors in the liver of the oviparous lizard, *Podarcis sicula*. Mol Reprod Dev 63:349
Sattar Khan MA, Nakamura S, Ogawa M, Akita E, Azakami H, Kato A (2000) Bactericidal action of egg yolk phosvitin against Escherichia coli under thermal stress. J Agric Food Chem 2000(48):1503–1506
Shi X, Zhang S, Sun Y, Pang P, Sawant MS (2004) Purification, characterization and antigenic species-specific reactivity of vitellogenin of rosy barb (Puntius conchonius Hamilton). Indian J Biochem Biophys 41:216–220
Shi X, Zhang S, Pang Q (2006) Vitellogenin is a novel player in defense reactions. Fish Shellfish Immunol 20:769
Sun C, Zhang S (2015) Immune-relevant and antioxidant activities of vitellogenin and yolk proteins in fish. Nutrients 7(10):8818–8829. https://doi.org/10.3390/nu7105432
Sun C, Li H, Liu S, Hu G, Zhang S (2013) Antiviral activity of phosvitin from zebrafish *Danio rerio*. Dev Comp Immunol 40(1):28–34
Swain P, Nayak S (2009) Role of maternally derived immunity in fish. Fish Shellfish Immunol 27(2):89–99
Swain P, Dash S, Bal J, Routray P, Sahoo PK et al (2006) Passive transfer of maternal antibodies and their existence in eggs, larvae and fry of Indian major carp, *Labeo rohita*. Fish Shellfish Immunol 20:519–527
Takemura A, Takano K (1997) Transfer of maternally-derived immunoglobulin (IgM) to larvae in tilapia, *Oreochromis mossambicus*. Fish Shellfish Immunol 7:355–363
Tong Z, Li L, Pawar R, Zhang S (2010) Vitellogenin is an acute phase protein with bacterial-binding and inhibiting activities. Immunobiology 215:898–902

Van Loon JJA, van Oosterom R, van Muiswinkel WB (1981) Development of the immune system in carp (*Cyprinus carpio*). Asp Comp Dev Immunol 1:469–470

Wallace RA (1985) Vitellogenesis and oocyte growth in nonmammalian vertebrates. Dev Biol 1:127–177

Wang S, Wang Y, Ma J, Ding Y, Zhang S (2011) Phosvitin plays a critical role in the immunity of zebrafish embryos via acting as a pattern recognition receptor and an antimicrobial effector. J Biol Chem 286(25):22653–22664. https://doi.org/10.1074/jbc.M111.247635

Zapata A, Diez B, Cejalvo T, Frias CG, Cortes A (2006) Ontogeny of the immune system of fish. Fish Shellfish Immunol 20:126–136

Zhang J, Zhang S (2011) Lipovitellin is a non-self-recognition receptor with opsonic activity. Mar Biotechnol 2011(13):441–450

Zhang S, Dong Y, Cui P (2015) Vitellogenin is an immunocompetent molecule for mother and offspring in fish. Fish Shellfish Immunol 46(2):710–715

Vitellogenesis and Reproductive Strategies in Fishes

8

Anjugam Mahalingam and Perumal Santhanam

Abstract

Vitellogenesis is a hormonally mediated process in fishes in which the protein vitellogenin is being secreted by reproductive hormones gonadotropins and facilitated by estrogen. Sexual activity of some fish is generally seasonal and fertilization is external. After fertilization, fishes produce haploid eggs and undergo the process of differentiation. Oocyte development occurs at the stage of differentiation. For proper embryonic development, gathering of yolk in oocytes is a vital process which leads the reproduction in a successful manner. Sequentially, there are three main steps involved in oocyte production which are (1) generation of primordial germ cells (PGCs), (2) transformation of PGCs to oogonia, and finally (3) the production of oocytes. Gonadotropins are produced into the bloodstream during the oogenesis process and then delivered to the ovaries to enhance the growth of the oocyte, otherwise known as ovulation. In addition to this, follicle cells are triggered to synthesize estrogen hormone also named as primary estradiol. Sequentially, the estradiol is introduced into the serum of the animal where it is bound by steroid-binding protein or albumin. In fact, the exogenous synthesis of vitellogenin is initiated by gonadotropins and regulated by the hormone estrogens. Estradiol or estrogen plays the key role in triggering vitellogenin production.

Keywords

Vitellogenesis · Vitellogenin · Reproductive strategy · Spermatogenesis · Oogenesis

A. Mahalingam · P. Santhanam (✉)
Department of Marine Science, School of Marine Sciences, Bharathidasan University, Thiruchirapalli, Tamil Nadu, India

V. Baskaralingam, R. Vanichviriyakit (eds.), *Vitellogenin in Fishes-Diversification, Biological Properties, and Future Perspectives*,
https://doi.org/10.1007/978-981-99-5340-0_8

8.1 Introduction

Studies on vitellogenesis and fish reproduction assist the aquaculture industry in meeting the ever-increasing demand for fish, by refining protocols for good quality of egg production and superior viability of progeny. Since the vast majority of fishes only reproduce periodically and a few numbers do so continuously, they have developed a variety of reproductive techniques. Whatever their method of reproduction, there is a universal rule for successful reproduction. According to Sundararaj and Vasal (1976) and Davis et al. (1999), the hypothalamo-hypophyseal-gonadal axis appears to be a neuroendocrine regulatory system that controls reproduction in nature. The major neuroendocrine connection in the conversion of environmental stimuli into the series of hormonal signals that control changes in reproductive activities is the hypothalamic neuropeptide gonadotropin-releasing hormone (GnRH). Majorly, GnRH work to cause the pituitary to release gonadotropin hormone (GTH). According to biochemical tests and molecular cloning studies, fish have two GTHs, GTH-I (similar to follicle-stimulating hormone, or FSH) and GTH-II (similar to luteinizing hormone, or LH). Through the generation of steroid hormones such E2 and 17- and 20-dihydroxy-pregnene-3-one in females and 11-ketotestosterone in males, both GTHs have a role in regulating fish reproductive strategies (Weltzien et al. 2004). The family of yolk proteins known as vitellogenins is by far the most prevalent in oviparous animals. Vitellogenins are produced by the process of vitellogenesis. Vitellogenins are yolk proteins that are essential for the development of offspring or progeny. Vitellogenins are glycoproteins expressed in vertebrate liver. After being absorbed by yolk platelets in vertebrates, vitellogenins are also broken down by cathepsins to create lipovitellin and phosvitins. Phosvitins are smaller hydrophobic subunits with a high level of phosphorylation, whereas lipovitellin are larger hydrophobic subunits that contain lipids. Vitellogenin has a molecular weight of about 325 kDa, but when it was subjected to an SDS-PAGE examination, it broke up into two main peptides with molecular weights of 190 and 190 kDa. Oocytes in the ovary that are mature collect the yolk through a process called vitellogenesis. The majority of fish species have cyclical vitellogenesis processes, and spawning only occurs once each cycle. Therefore, any intrinsic or extrinsic condition that disrupts the vitellogenic cycle can significantly reduce the reproductive success (number of offspring produced).

8.2 Vitellogenesis

In fish, a process known as vitellogenesis produces the precursors of the egg yolk protein (vitellogenins), which are then secreted to the plasma and carried to the oocytes for uptake. Many metabolic changes, including a large rise in liver weight, lipid deposition, RNA content, glycogen depletion, plasma protein, magnesium, calcium, and phosphoprotein quantity, take place in mature female fish throughout the process of vitellogenesis (Wiegand 1996; Arukwe and Goksøyr 2003). The gonadotropins hormone is necessary for the seasonal or cyclical process of vitellogenesis. The hypothalamus, a region of the brain, secretes the gonadotropin-releasing hormone (GnRH), which is controlled by both endogenous and exogenous variables such as innate biorhythms, nutritional status, seasonal variations, and water temperature. Followed by the release of GnRH, sequentially, follicle-stimulating hormone (FSH) and luteinizing hormones are secreted. Both hormone secretions are regulated by hypothalamic nerve fibers. In addition, GnRH stimulates the expression and secretion of glycoprotein hormone alpha (GPalpha), FSHbeta, and LHbeta. Secretion of FSH stimulates the theca and granulosa cells of the ovarian follicle to secrete estradiol-17β (E2) which inspires the liver to synthesize vitellogenins and secrete them into the bloodstream. Certain species may also induce hepatic vitellogenesis by growth hormone, thyroid hormone, cortisol, androgens, and other estrogenic steroids like estrone. In hagfish (*Eptatretus stoutii*), vitellogenesis is generally less responsive to E2 and might be regulated by other factors, for example, feeding.

8.3 Vitellogenin

In the 1900s, in the study of sex determination in fishes, a specific antigen was identified and named as vitellogenin. Vitellogenin is the precursor of major yolk protein in all oviparous animals including fishes, amphibians, reptiles, invertebrates, and birds. Vitellogenin is a female-specific phospholipoglycoprotein; nourish protein and lipid-rich nutrients to growing embryos and larvae. Apart from vitellogenin, choriogenin is another protein which was identified as a unique ancestor of egg envelope proteins that is secreted into the bloodstream like vitellogenin. Vitellogenin is confined with covalently linked carbohydrates and phosphates and non-covalently bound lipids. Vitellogenin is a high-molecular-weight (350 kDa) and female-specific protein, but some quantum of reports stated that male and juvenile fishes also produce vitellogenin (Wallace 1985; Hara et al. 1980; Wallace and Selman 1982; Chang et al. 1994; Purdom et al. 1994; Tyler and Sumpter 1990; Harries et al. 1997). Vitellogenin get fragmented into two major peptides with molecular weight of 190 and 160 kDa when subjected to SDS-PAGE analysis. Stimulation of estrogen hormone or substances similar to estrogens are the major factor for producing vitellogenin in both male and juvenile fishes. Extraovarian tissues of female animals produce vitellogenin under the hormonal control of estrogen and travelled to the ovary with the help of the bloodstream. At the site of ovary, budding oocytes internalized the vitellogenin and proteolytically cleaved to form yolk proteins and that are nourished to the developing embryos as nutrients (Tyler and Lancaster 1993; Wallace 1978; Ng and Idler 1983; Mommsen and Walsh 1988). In a fact, fish vitellogenins do not only give nourishment to fish offsprings but are also involved in other multifaceted functions such as providing an osmotic gradient for initiating good egg development and metabolic water for embryonic growth. Moreover, young ones received a sufficient quantity of protein, carbohydrate, and lipid to the grown-up embryos and yolk sac larvae at their diverse developmental stages. They also provide a programed provisioning of protein, carbohydrate, and lipid nutrition to developing embryos and yolk sac larvae at different developmental time points. Different vitellogenin type has been observed in a variety of fish species to date including grey mullet (*Mugil cephalus*) (Amano et al. 2007), barfin flounder (*Veraspermoseri*) (Matsubara et al. 1999; Sawaguchi et al. 2008), striped bass (*Morone saxatilis*) (Williams et al. 2014a, b), white perch (Reading et al. 2009;Schilling et al. 2015), mosquitofish (*Gambusia affinis*) (Sawaguchi et al. 2005), red seabream (*Pagrus major*), goldsinny wrasse (*Ctenolabrus rupestris*) (Kolarevic et al. 2008), haddock (*Melanogrammus aeglefinus*) (Reith et al. 2001), Atlantic halibut (*Hippoglossus hippoglossus*) (Finn 2007a, b; Finn et al. 2002), and zebrafish (*Danio rerio*) (Yilmaz et al. 2018). It has been mentioned that, at the final growth stage, the egg yolk composition may be altered from species to species which displays distinct requirements for egg buoyancy, larval nutrition, or time frame until first feeding. Vitellogenin is normally synthesized in liver parenchymal cells of female oviparous vertebrates including fish under the estrogen control post-translationally modified in the liver (phosphorylation, glycosylation, and lipidation), secreted into the bloodstream and transported to the ovary, where it is internalized by the growing oocytes and proteolytically cleaved to form the yolk proteins which are later used as the nutrient material by the developing embryos and larvae. Furthermore, while vitellogenin does not participate in the oogenesis process, it is used as an effective biomarker for assessing the impact of estrogen-like endocrine-disrupting chemicals (environmental hormones) in aquatic ecosystems (Hara et al. 2016). Vitellogenins such as protein, carbohydrates, lipids, minerals, and vitamins are vital materials for the process of embryogenesis. Generally, yolk protein or vitellogenin consists of five components including heavy-chain, lipovitellin, light-chain lipovitellin, phosvitin, β-component, and carboxy-terminal component. Based on the presence or absence of these five components, vitellogenins are divided into complete and incomplete vitellogenin. In incomplete vitellogenin, there are two types by which having phosvitin and phosvitinless (Pv-less) domain. Vitellogenin C is another type of incomplete vitellogenin having heavy-chain lipovitellin and light-chain lipovitellin but not having phosvitin, β-component, and carboxy-terminal component. In female fishes, vitellogenin is one of the important factors to identify the process of puberty and gonad maturation. Interestingly, in the fish which doesn't have sexual dimorphism, vitellogenin is the only factor in the blood to identify the gender of the fishes in aquaculture.

8.4 Vitellogenin as a Defense Molecule

Recognition of self from non-self is one of the inevitable processes in immune reactions. Majorly, identification of foreign invaders and activation of defense reactions is carried out by a set of proteins named as pattern recognition proteins (PRPs). PRPs have the ability to recognize the external intruders by recognizing the pathogen-associated molecular patterns (PAMPs) such as lipopolysaccharides, β-glucan, peptidoglycans, and lipoteichoic acid which are found on the surface of invaders (Anjugam et al. 2016). Humoral proteins and complement components of vertebrates can bind to the surface moieties of pathogens and trigger the phagocytic mechanism in resemblance to opsonin. In invertebrates, plasma proteins and membrane proteins also stimulate the phagocytic mechanism against external invaders. Just like in vertebrates, especially fish vitellogenin is involved in defense reactions. Li et al. (2008) reported that fish vitellogenin regulates innate immunity by recognizing microbial cell wall constituents known as PAMPs and enhancing macrophage phagocytosis. Furthermore, it has been documented that the clotting protein which is the member of Vg family participated in the defense mechanism of nematode, crayfish, and sea urchin (Soderhal 1997). This is one of the key points which denote that Vg is not only responsible for yolk protein synthesis but also has the role in defense reactions of both male and female fish. The statement was supported by Zhang et al. (2005), who stated that Vg isolated from amphioxus possess antibacterial and hemagglutinating activity. In addition, Sun and Zhang (2015) stated that Vg actively participated in host immune response and play multifunctional roles in defense mechanism. For example, Vg has recently been shown to have both hemagglutinating and antibacterial activities in the protochordate amphioxus (*Branchiostoma belcheri*) as well as the bony fish rosy barb (*Puntius conchonius*) (Shi et al. 2006; Zhang et al. 2005; Amdam et al. 2004). Moreover, the male rosy barb produces Vg as a defense molecule when infected with *Escherichia coli*, indicating that Vg may be connected to an infection-resistant response. Yet, it is still unclear how Vg contributes to an anti-infectious response. The Vg purified from *H. otakii* can bind with PAMPS such as lipopolysaccharide (LPS) of Gram-negative bacteria, lipoteichoic acid (LTA) from Gram-positive bacteria, peptidoglycan (PGN) from both Gram-positive and Gram-negative bacteria, and β-glucan from fungi and laminarin from brown algae. Hence, it has been universally proposed that both vertebrate and invertebrate Vg have antibacterial property (Sun and Zhang 2015). Generally, invertebrates only have the tendency to recognize the β-glucan molecule, since β-glucan-binding protein has been reported only in invertebrates. In the case of vertebrates, some of the immune proteins like Dectin-1 and scavenger receptors recognize the surface moieties of fungi and act as β-glucan-binding protein. Fish Vg is the first plasma β-glucan-binding protein discovered in vertebrates to date. In addition, fish Vg is able to bind with *E. coli*, *Staphylococcus aureus*, and fungus *Pichia pastoris* by recognizing their surface markers known as PAMPs. The preceding statement clearly demonstrated that Vg is a novel pattern recognition receptor with a broad specificity capable of identifying non-self components such as LPS, LTA, PGN, glucan, and laminarin (Li et al. 2008). Along with the recognition of non-self molecules, fish Vg also triggers the macrophage phagocytosis, thus acts as opsonin (Sun and Zhang 2015). Besides, credit to the immune-related functions of Vg, it possesses an antioxidant property, in which they regulate reactive oxygen species (ROS) or free radical generation to reduce the oxidant stress of the host. These functions disclose the physiological role of the Vg and also give the knowledge platform on the potential application of the fish Vg in human health.

8.5 Fish Reproduction

Reproduction is a basic feature of all known life in which any organism produce an individual or new offspring to fix their race on the earth. Reproduction may be classified into sexual reproduction and asexual reproduction. In the case of fishes, they produce their young ones by sexual reproduction. Briefly, female fishes release their eggs into the water for fertilization, hence named as external fertilization. When a female fish reproduces, there are two main physiological processes that take place: (1) the maturation, ovulation, and spawning of yolky oocytes and (2) the gradual expansion of the ovaries caused by the development of yolky oocytes, a process commonly referred to as vitellogenesis. GTHs control both of these processes; GTH-I is engaged in vitellogenesis, while GTH-II initiates maturation and ovulation (Nagahama 2000).

8.6 Reproductive Strategies in Fishes

A reproductive strategy is a collection of traits that are predominantly heritable, such as size and age at first reproduction, spawning seasonality and frequency, and size- or age-specific fecundity. These phenotypic characteristics usually occur together and most likely originated through natural selection (Fig. 8.1).

The teleosts are one of the most diverse vertebrate groups. Teleosts exhibit a wide range of morphological, physiological, and behavioral traits. Because many of the world's fisheries are routinely overfished, there is an increasing need to understand the dynamics of fish populations and the underlying mechanisms that control their capacity to rebound (Boreman et al. 1997; Food and Agricultural Organization 2002; Worm et al. 2009). For fisheries biologists, the process of reproduction and the subsequent recruitment of juveniles into the fishery are extremely important. Their success has largely been attributed to their ability to use various reproduction methods to fill a range of ecological niches. These reproductive systems have both behavioral and biological components. Several methods of fertilization, the number of spawning cycles, and gender differentiation are all characteristics of biological systems. Examples of behavioral systems include parental care and mating systems.

Fish have ovarian development that is organized in one of three ways: synchronous, group synchronous, or asynchronous. The majority of semelparous species have synchronized spawners, in which all of the oocytes develop simultaneously. At least two distinct oocyte populations, such as one with larger oocytes and another with a mixed population of smaller oocytes, are present in group synchronous spawners. This is typical of iteroparous species, which have short spawning seasons and a uniform ground fish structure. The ultimate organization, known as asynchronous, occurs when oocytes from all embryonic stages are present. This usually occurs in species with lengthy spawning seasons because yolk accumulation depends mainly on the availability of food sources during these times (Murua and Saborido-Rey 2003). Organization of the spawning pattern is similar to ovarian development. Total and batch spawning are the two spawning patterns. All of the

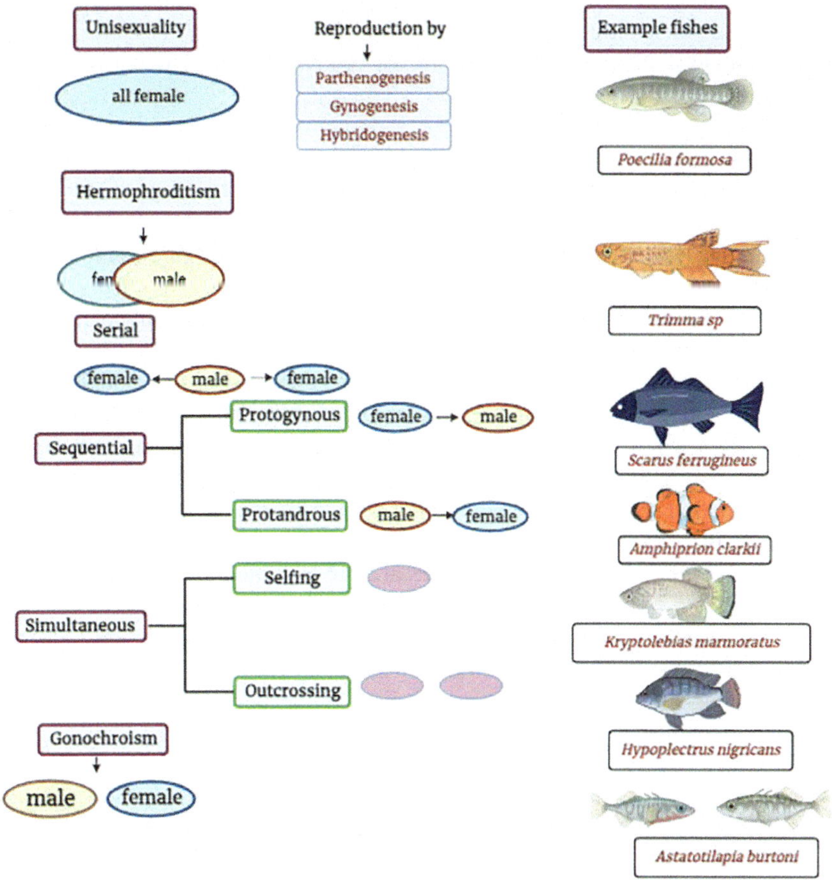

Fig. 8.1 Reproductive strategy in fishes

eggs were rapidly shed by total spawners. Throughout the spawning season, different batches of eggs are released by batch spawners. With this method, a species can release eggs over a long period of time and increase the chance that the offspring will survive. Iteroparous fish frequently cycle through different states, including resting, developing, ripe, and running. This cycle of maturational stages is regulated by endogenous factors acting in concert with external environmental stimuli. According to Mommsen and Korsgaard (2008), this regulation ensures that gamete production and spawning take place simultaneously for both sexes. In fish, whether it occurs during a single spawning event or over the course of numerous spawning seasons, spawning is the most taxing activity (Nielsen et al. 2011). The ocean sunfish (*Mola mola*, Molidae), which has an estimated maximum oocyte production of 300 million from a single female, may have the highest fecundity. The pace of egg production turnover is extremely impressive. For instance, the enormous dolphin fish *Coryphaena hippurus* of the Coryphaenidae family has a yearly egg production capacity of around 100 million. Even very tiny species lay between 1000 and 2000 eggs during their lifetimes, such as the 3-cm medaka (*Oryzias latipes*, Adrianichthyidae). According to Grier et al. (2009), the ovary is the most active organ in an adult fish.

8.7 Hormones Involved in Fish Reproduction

Hormones are chemical messengers that are necessary for the communication between various kinds of cells, distinguish their responsibility and function through receptors. Receptors are nothing but protein structures specialized in molecular recognition. In all living being, biochemical reactions are happened only after the interaction between hormone and receptor, frequently termed as hormonal receptor interaction (Yada and Nakanishi 2002). Generally, in all vertebrates (from mammals to fishes), reproduction is regulated by variety of hormones. The major hormones involved in fish reproduction are follicle-stimulating hormone (FSH) and luteinizing hormone (LH) collectively called as gonotropin hormones released by the pituitary gland. Both FSH and LH are glycoproteins and necessary for the development of gonad and reproductive function. In females, FSH stimulates the growth of ovarian follicles, and LH stimulates the process of ovulation. In males, FSH regulates spermatogenesis, and LH induces the process of spermatogenesis (Marshall et al. 1986; Levavi-Sivan et al. 2010; Hollander-Cohen et al. 2021). Sequential secretion pattern of GnRH indicates the reproductive state of the animal and their importance in maintaining a healthy reproductive state (Stamatiades and Kaiser 2018).

8.8 Factors Affecting Reproduction in Fishes

Undoubtedly, the environmental parameters like temperature, current velocity, photoperiod, water quality, pH, food convenience, and meteorological conditions are responsible factors for fish reproduction. Other than these factors, overcrowding of fishes, handling, and poor management can also decline the rate of quality young one production. Availability of food is one of the major factors in fish reproduction. Deficiency of food directly affects the fecundity rate of the fish which leads to produce very less quantity of egg. Overcrowding of fish also leads to inhibition in vitellogenesis. Overpopulation of fish diminishes the growth rate and cardiac activity and stimulates embryonic mortality. Furthermore, man-made hazardous factors including air, water, and land pollution negatively affected the spawning of fishes. Industrialization causes the pollution on land, water, and air which leads high mortality in fishes. Moreover, freshwater habitat fishes receives considerable amount of waste products from both industrial and domestic areas. The sewage discharged from industries contains many harmful chemicals which resemble endogenous hormones, hence directly affect the fish reproduction. Those hazardous factors collapse the reproductive steps such as gametogenesis, oocyte maturation, ovulation, and spermiation. At the process of gametogenesis, these unfavorable environmental factors may delay the progression of initiation, completion, fecundity, and egg superiority. Fishes can adapt themselves according to the changes happened in environment throughout the year. In a fact, some fishes can lead their life even at sub-zero temperatures and high pressure in the oceanic environment.

8.8.1 Spermatogenesis

Spermatogenesis is a highly organized and developmental process, in which diploid spermatogonial stem cells undergo the process of proliferation, meiosis, and differentiation to form mature spermatozoa (Fig. 8.2). It is a well-known fact that the two hormones, namely, pituitary gonadotropins follicle-stimulating hormone (FSH) and luteinizing hormone (LH), play pivotal role in the regulation of spermatogenesis in all vertebrates including fishes (Thackray et al. 2010). Spermatogenesis is a developmental process during which a small number of diploid spermatogonial stem cells produce a large number of highly differentiated spermatozoa carrying a haploid, recombined genome. Fish sperms are broadly divergent and had a wide range of shapes, sizes, and structures, hence could not exactly predict the spermatic model of fish as like mammals and snakes (Mattei 1991). They differ from aflagellate to biflagellate and had wide variations in number and location of organelles (Baccetti et al. 1984; Baccetti 1986; Jones and Butler 1988).

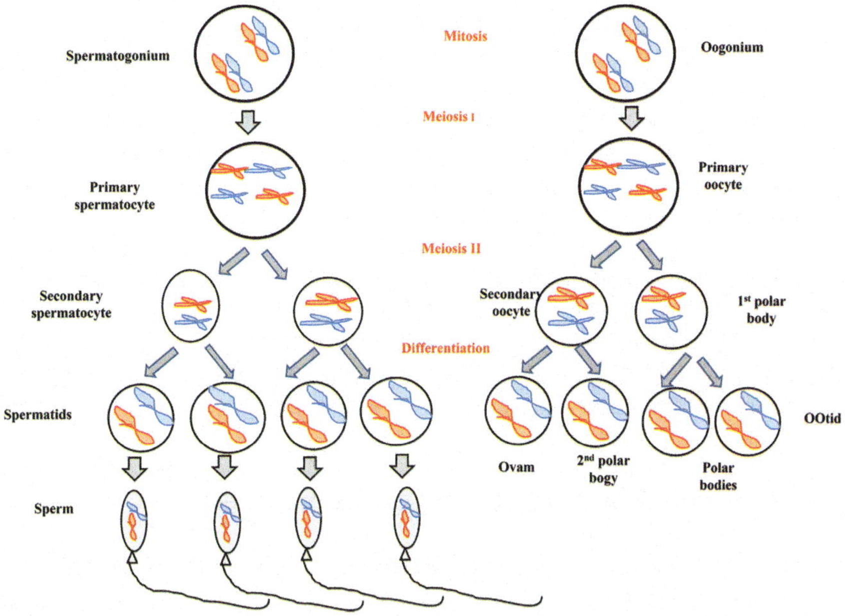

Fig. 8.2 Spermatogenesis and oogenesis

8.8.2 Sperm Quality

Sperm quality and quantity are an important parameter for successful offspring production in both natural and artificial spawning. Sperm quality plays the key role in aquaculture practice for ensuring their successful production of offsprings (Rurangwa et al. 2004). Furthermore, sperm quality indicates the environmental status of that particular water. Quality and healthiness of the sperm may be screened by various characters including sperm motility, concentration of sperm, volume, seminal plasma pH, and osmolality. Off these, sperm motility is one of the significant features of the male gamete, since the speed of the sperm only fix their ability to reach and penetrate the female gamete for fertilization (Islam and Akhter 2011; Kowalski and Cejko 2019). Additionally, the length of the flagellum and the number of mitochondria found in the sperm also determine the motility of the sperm. Furthermore, after releasing into the water, sperms are metabolically active and move less than 2 min. In marine fishes, high CO_2 concentration in semen sustains the intracellular pH at 7.2. The parameters like concentration, volume, seminal plasma pH, osmolality, and ionic composition are the major factors which trigger the sperm maturation and have been linked with the capability of sperm to fertilize eggs. Apart from that, the membrane stability of the sperm, DNA integration, enzymatic activity, and mitochondrial concentration also responsible for monitoring the sperm superiority. Moreover, in artificial hatcheries, the quantity and quality of sperm is inadequate which leads to the lack of successful fertilization. Hence, screening of those significant characters of sperm delivers the familiarity about the sperm quality and male fertility and also provides the information about how to improve and maintaining the brood stock for successful fertilization and healthiest young one production. Furthermore, those data help in handling the gametes in a proper manner and fertilization process. Some species release a very least amount of sperm and is not sufficient for artificial fertilization because many batches of eggs can be released during artificial spawning which displayed that males are frequently subjected for collecting all their milt. Furthermore, many species belong to the endangered category in aquaculture. Therefore, for safeguarding the endangered species, their sperm or egg should be preserved in a proper manner, a technique known as cryopreservation. Though cryopreservation technique fills the space in sperm availability, their potency may be diminished during the process of cryopreservation. Thus, the sperm lost their efficiency to fertilize the egg. In addition to this, evaluation of sperm quality of endangered species may help to store the high-quality sperms through cryopreservation for gene banking.

8.9 Factors Affecting Sperm Quality in Fishes

Sexual activity of some fish is generally seasonal and fertilization is external. Sperm, once differentiated in the gonad, remain there completely quiescent until they are released into the external medium, which is either freshwater or sea water. Various parameters such as ion concentrations (K^+, Na^+, Ca^{2+}), osmotic pressure, pH, and temperature affect motility. Studying the effects of these factors on teleost sperm can help establish good activation and/or immobilizing media for improving either artificial fertilization or cryopreservation.

8.9.1 Oogenesis

The oogenesis is a very dynamic process in the ovaries, in which the oocyte passes through various phases of the development that are very similar in different fish species. Vitellogenesis and zonagenesis are two essential oogenesis stages which are involved in oocyte growth. Zonagenesis is the process by which a thick, highly differentiated acellular zone grows around the developing oocyte and is referred to as the zona radiata, zona pellucida, vitelline envelope, or chorion. Conversely, during vitellogenesis in response to GTH-induced ovarian E2, hepatocytes produce and release the yolk precursor protein vitellogenin (Vg), which is then absorbed and incorporated into the growing oocytes through a process known as receptor-mediated endocytosis (Arukwe and Goksøyr 2003). Before being deposited in the ooplasm, Vg(s) is proteolytically cleaved into yolk proteins (lipovitellin (Lv), phosvitin (Pv), and ′-component (′-C)). Although vitellogenin is thought to be a protein that only exists in females, comparable proteins have been found in the males of various fish and can be stimulated by E2. The precise function of Vg in males has not yet been determined. During the oogenesis process, the oocyte develops in the follicle. The oocyte is the only component of the follicle, which is supported by a thin, vascularized theca and is encircled by a single layer of follicular cells (Grier et al. 2009). Chromatin-nucleolus, previtellogenesis, and vitellogenesis are the three stages of the oogenesis. In the chromatin-nucleolus stage of meiosis, when oogonia become oocytes, the germinal vesicle (nucleus) exhibits the lampbrush chromosomes.

The oocyte accumulates the organelles and components of the ooplasm during previtellogenesis. The oocyte acquires the complex and varied nutrients contained in the yolk during vitellogenesis (Aranzábal et al. 2009; Grier et al. 2009). There are several characteristics identify previtellogenesis. The number of lipid droplets, cortical alveoli, and the diameter of the oocyte all increase, while the solitary nucleolus of the germinal vesicle (nucleus) multiplies into several nucleoli. When yolk globules are deposited in the ooplasm, vitellogenesis is observed; the yolk becomes fluid, and some lipid droplets may be visible around the oocyte periphery. When the follicular epithelium becomes columnar, it shows that the oocyte is being fed heavily (Arcanjo et al. 2014). At the end of vitellogenesis, the oocyte reaches its maximum size, and the germinal vesicle migrates to the periphery of the ooplasm at the animal pole (Grier et al. 2009; Arcanjo et al. 2014).

8.9.2 Ovarian Maturation and Yolk Formation

Oocytes enter a phase known as post-vitellogenesis after vitellogenesis is finished, which is typically characterized by translational quiescence. In this stage, multiple number of oocyte ribosomes disorganizes from the ooplasm. At this time, ovarian follicles are ready for ovulation by gaining maturation competence. To fully mature an egg and ovulate in response to a progesterone signal, maturity competence is required (Sullivan et al. 2003; Patiño and Sullivan 2002). The luteinizing hormone (LH), which is secreted by the pituitary in response to gonadotropin-releasing hormone (GRH) released by the hypothalamus, then induces the ovarian follicle to secrete progesterone, typically either 17-, 20-, 21-trihydroxy-4-pregnen-3-one (20-S) or 17-, 20-dihydroxy-4-pregnen-3-one (17-, 20-DHP). Fishes of different species respond differently to these progesterone's effects on ovarian maturation and ovulation.

8.9.3 Egg Yolk Composition

The egg yolk nourishes the nutrients to developed embryo and is being considered as an essential constituent for the proper embryonic development (Reading et al. 2017). Vitellogenin hold the components like phosphate, lipid, carbohydrate, and protein components like low-density lipoprotein and VLDL. Furthermore, 11–12% of vitellogenin polypeptide residues was occupied by an amino acid such alanine and is being considered as the most abundant amino acid. The large lipid transfer protein superfamily, which also contains other serum lipoproteins like low-density lipoprotein (LDL) and VLDL, includes vitellogenins, which are composed of phosphate, lipid (approximately 20% by weight), carbohydrate, and protein components (Smolenaars et al. 2007). Alanine is the most abundant amino acid comprising about 11–12% of the total residues of vitellogenin polypeptides since it acts as the intermediator for carbohydrate metabolism in particular for embryonic gluconeogenesis. Alanine may serve as an important intermediary of carbohydrate metabolism, especially for embryonic gluconeogenesis. Generally, sufficient quantities of metal ions are not existed in freshwater environment. Fortunately, vitellogenins serve as the carrier and transporter of different metal ions, including calcium, magnesium, iron, zinc, and copper, as well as different minerals and vitamins, including retinoids, and carotenoids (Specker and Sullivan 1994; Finn 2007b). In the case of marine fishes, the least amount of metal ions is present in their egg yolk, since marine water provides satisfactory quantity of metal ions for good embryonic development. It reveals that vitellogenins may play a fewer role in the transportation and production of those metal ions to embryos of marine fishes (Reading et al. 2017, 2018).

8.9.4 Quality of Egg

The fish farming industry has been more focused toward the quality of eggs and larvae rather than that of sperm, even though the sperm quality of male brood stock also affects the production of healthy larvae. The quality of an egg may governed by both intrinsic and extrinsic factors. Furthermore, for an efficient captive propagation and fruitful employment, the quality of egg is the major parameter. Since the majority of fish are oviparous, the only source of nutrition for their developing offspring is stored egg yolk. The ingredients found in the egg yolk are proteins, carbohydrates, lipids, vitamins, ions, and minerals. These nutrients are essential for proper development of offspring, and these nutrients are shifted from the liver to the ovary by the protein named as vitellogenin. Other than these, some of the factors like gene transcripts are also responsible factor in determining the quality of egg (Reading et al. 2018). Any changes or abnormal functions in gene transcription or expression may cause the production of poor-quality eggs and failed in fertilization. Moreover, stress can affect fish behavior, fertility, and ovulation rate as well as cause ovarian atresia or reproductive failure. Last but not least, postovulatory aging can happen when fish fail to spawn in a timely manner and the eggs get overripe, resulting in decreased fertility, which is frequently seen during manual strip spawning of fish.

8.9.5 Factors Affecting the Egg Quality

Egg quality can be defined as the capability of an egg to be fertilized and constantly develop into a normal embryo (Bobe and Labbé 2010). The quality of egg determines the development of aquaculture. Egg quality is extremely inclined by environmental factors and husbandry practices. Egg quality is a difficult to measure phenotype and is therefore not currently included in most selection programs. In aquaculture, reduced egg quality can lead to several types of difficulties such as lack of fertilization, problems of egg activation, development arrests, embryonic deaths, and embryonic abnormalities (Bobe and Labbé 2010; Migaud et al. 2013; Bobe 2015). The major factors affecting egg quality is water quality, temperature, and light intensity. Since fishes are poikilothermic animals, water temperature directly affects the reproductive cycle and their growth. Among the various phases in oocyte development, final oocyte maturation stage is strictly affected by water temperature. In addition, photoperiodic cues also responsible for changes occurs in reproductive cycle. In many species, sexual maturation (i.e., development of the gonad, ultimately leading to gamete production) is influenced by photoperiod. Furthermore, fish nutrition also play a key role in determining the egg quality, because the egg yolk will be frequently utilized for the good embryonic development. It is therefore important that broodstock diets are adjusted to ensure good larval survival and early development (Izquierdo et al. 2001). Thus, broodstock diets should be framed to certify all crucial nutrient requirements are met for the species being cultured (Migaud et al. 2013). The deficiency of vitamins and some key components in the diet leads major reproductive problems in newly cultured species (Izquierdo et al. 2001). If the brooders receive nutritional diets as they required, then problems in egg quality had minimum impact in young one production. Apart from these environmental factors, man-made stress also affects the quality of egg. Generally poor broodstock management techniques may be the reason for poor egg quality problems. The overall impact on egg quality is highly dependent on the species, the techniques used, and the physiological status of the fish, including stress level.

8.10 Conclusion and Future Aspects

In conclusion, to attain a great success in fish aquaculture, the quality of egg and sperm is most significant. The quality of both male and female gametes is determined by the factors including age, management, feeding, chemical and physical factors, water quality, etc. These parameters possess the strong influence on the lifetime of embryos, larvae, and/or fry in the short or long term.

Acknowledgments MA gratefully acknowledges the Science and Engineering Research Board (SERB), New Delhi, India, for the financial assistance rendered (Ref: PDF/2021/001905).

References

Amano H, Fujita T, Hiramatsu N, Shimizu M, Sawaguchi S, Matsubara T, Kagawa H, Nagae M, Sullivan CV, Hara A (2007) Egg yolk proteins in grey mullet (*Mugil cephalus*): purification and classification of multiple lipovitellins and other vitellogenin-derived yolk proteins and molecular cloning of the parent vitellogenin genes. J Exp Zool A Ecol Genet Physiol 2007(307):324–341

Amdam GV, Hartfelder K, Norberg K, Hagen A, Omholt SW (2004) Altered physiology in worker honey bees (Hymenoptera: Apidae) infested by the mite Varroa destructor (Acari: Varroidae): a factor in colony loss during over-wintering? J Econ Entomol 97:741–747

Anjugam M, Iswarya A, Vaseeharan B (2016) Multifunctional role of β-1, 3 glucan binding protein purified from the haemocytes of blue swimmer crab *Portunus pelagicus* and in vitro antibacterial activity of its reaction product. Fish Shellfish Immunol 48:196–205

Aranzábal MC, Grier HJ, De la Rosa CG, Alarcón AG (2009) Modifications in ovarian and testicular morphology associated with viviparity in teleosts. Reproductive biology and phylogeny of fishes (Agnathans and bony fishes). phylogeny, reproductive system, Viviparity, Spermatozoa 3:85

Arcanjo RB, de Souza LP, Rezende CF, Silva JR (2014) Embryonic development and nourishment in the viviparous fish Poecilia vivipara (Cyprinodontiformes: Poeciliidae). Acta Zool 95:493–500

Arukwe A, Goksøyr A (2003) Eggshell and egg yolk proteins in fish: hepatic proteins for the next generation: oogenetic, population, and evolutionary implications of endocrine disruption. Comp Hepatol 2:1–21

Baccetti B (1986) Evolutionary trends in sperm structure. Comp Biochem Physiol A 85:29–36

Baccetti B, Burrini AG, Callaini G et al (1984) Fish germinal cells. I. Comparative spermatology of seven cyprinid species. Gamete Res 10:373–396

Bobe J (2015) Egg quality in fish: present and future challenges. Anim Front 5:66–72

Bobe J, Labbé C (2010) Egg and sperm quality in fish. Gen Comp Endocrinol 165:535–548

Boreman J, Nakashima BS, Wilson JA, Kendall RL (eds) (1997) Northwest Atlantic groundfish: perspectives on a fishery collapse. American Fisheries Society, Bethesda, MD

Chang CF, Lee FY, Huang YS, Hong TH (1994) Purification and characterization of the female specific protein (vitellogenin) in mature female hemolymph of the prawn, *Penaeus monodon*. Invertebr Reprod Dev 25:185–192

Davis B, Bromage N, Swanson P (1999) The brain-pituitary-gonadal axis of female rainbow trout, Oncorhynchus mykiss: effects of photoperiod manipulation. Gen Comp Endocrinol 115:155–166

Finn RN (2007a) The maturational disassembly and differential proteolysis of paralogous vitellogenins in a marine pelagophil teleost: a conserved mechanism of oocyte hydration. Biol Reprod 76:936–948

Finn RN (2007b) Vertebrate yolk complexes and the functional implications of phosvitins and other subdomains in vitellogenins. Biol Reprod 76:926–935

Finn RN, Østby GC, Norberg B, Fyhn HJ (2002) In vivo oocyte hydration in Atlantic halibut (*Hippoglossus hippoglossus*); proteolytic liberation of free amino acids, and ion transport, are driving forces for osmotic water influx. J Exp Biol 205:211–224

Food and Agricultural Organization (2002) The state of world fisheries and agriculture. FAO, Rome

Grier HJ, Uribe-Aranzábal MC, Patiño R (2009) The ovary, folliculogenesis, and oogenesis in teleosts. Reproductive biology and phylogeny of fishes (agnathans and bony fishes) 8(Part A):25–84

Hara A, Yamauchi K, Hirai H (1980) Immunochemical identification of female-specific serum protein vitellogenin and egg yolk proteins in Japanese eel *Anguilla japonica*. Comp Biochem Physiol 65B:315–320

Hara A, Hiramatsu N, Fujita T (2016) Vitellogenesis and choriogenesis in fishes. Fish Sci 82:187–202

Harries JE, Shehan DD, Jobling S, Matthiessen P, Neall M, Sumpter JP et al (1997) Estrogenic activity in five United Kingdom rivers detected by measurement of vitellogenesis in caged male trout. Environ Toxicol Chem 16:534–542

Hollander-Cohen L, Golan M, Levavi-Sivan B (2021) Differential regulation of gonadotropins as revealed by transcriptomes of distinct LH and FSH cells of fish pituitary. Int J Mol Sci 22:6478

Islam MS, Akhter T (2011) Tale of fish sperm and factors affecting sperm motility: a review. Adv Life Sci 1:11–19

Izquierdo MS, Fernandez-Palacios H, Tacon AG (2001) Effect of broodstock nutrition on reproductive performance of fish. Aquaculture 197:25–42

Jones PR, Butler RD (1988) Spermatozoon ultrastructure of Platichthys flesus. J Ultrastruct Mol Struct Res 98:71–82

Kolarevic J, Nerland A, Nilsen F, Finn RN (2008) Goldsinny wrasse (Ctenolabrus rupestris) is an extreme vtgAa-type pelagophil teleost. Mol Reprod Dev 75:1011–1020

Kowalski RK, Cejko BI (2019) Sperm quality in fish: determinants and affecting factors. Theriogenology 135:94–108

Levavi-Sivan B, Bogerd J, Mañanós EL, Gómez A, Lareyre JJ (2010) Perspectives on fish gonadotropins and their receptors. Gen Comp Endocrinol 165:412–437

Li Z, Zhang S, Liu Q (2008) Vitellogenin functions as a multivalent pattern recognition receptor with an opsonic activity. PLoS One 3:1940

Marshall GR, Bint Akhtar F, Weinbauer GF, Nieschlag E (1986) Gonadotrophin-releasing hormone (GnRH) overcomes GnRH antagonist-induced suppression of LH secretion in primates. J Endocrinol 110:145–150

Matsubara T, Ohkubo N, Andoh T, Sullivan CV, Hara A (1999) Two forms of vitellogenin, yielding two distinct lipovitellins, play different roles during oocyte maturation and early development of barfin flounder, *Veraspermoseri*, a marine teleost that spawns pelagic eggs. Dev Biol 213:18–32

Mattei X (1991) Spermatozoa ultrastructure and its systematic implications in fishes. J Zool 69:3038–3055

Migaud H, Bell G, Cabrita E, McAndrew B, Davie A, Bobe J, Herraez MP, Carrillo M (2013) Gamete quality and broodstock management in temperate fish. Rev Aquac 5:S194–S223

Mommsen T, Korsgaard B (2008) Vitellogenesis. In: Rocha MJ, Arukwe A, Kapoor BG (eds) Fish reproduction. Science Publishers, Enfield, NH, pp 113–169

Mommsen TP, Walsh PJ (1988) Vitellogenesis and oocyte assembly. In: Hoar WS, Randall VJZ (eds) Fish physiology, XIA. Academic Press, San Diego, pp 347–406

Murua H, Saborido-Rey F (2003) Female reproductive strategies of marine fish species of the North Atlantic. J Northwest Atl Fish Sci 33:23–31

Nagahama Y (2000) Gonadal steroid hormones: major regulators of gonadal sex differentiation and gametogenesis in fish. In: International symposium on the reproductive physiology of fish

Ng TB, Idler DR (1983) Yolk formation and differentiation in teleost fishes. In: Hoar WS, Randall DJ, Donaldson EMZ (eds) Fish physiology, IX. Academic Press, New York, pp 373–404

Nielsen JL, Turner SM, Zimmerman CE (2011) Electronic tags and genetics explore variation in migrating steelhead kelts (Oncorhynchus mykiss), Ninilchik River, Alaska. Can J Fish Aquat Sci 68(1):1–6

Patiño R, Sullivan CV (2002) Ovarian follicle growth, maturation, and ovulation in teleost fish. Fish Physiol Biochem 26:57–70

Purdom CE, Hardiman PA, Bye VJ, Eno NC, Tyler CR, Sumpter JP (1994) Estrogenic effects of effluents from sewage treatment works. Chem Ecol 8:275–285

Reading BJ, Hiramatsu N, Sawaguchi S, Matsubara T, Hara A, Lively MO, Sullivan CV (2009) Conserved and variant molecular and functional features of multiple egg yolk precursor proteins (vitellogenins) in white perch (*Morone americana*) and other teleosts. Mar Biotechnol 11:169–187

Reading BJ, Sullivan CV, Schilling J (2017) Vitellogenesis in fishes. In: Reference module in life sciences. Elsevier, Amsterdam

Reading BJ, Andersen LK, Ryu YW, Mushirobira Y, Todo T, Hiramatsu N (2018) Oogenesis and egg quality in finfish: yolk formation and other factors influencing female fertility. Aust Fish 3:45

Reith M, Munholland J, Kelly J, Finn RN, Fyhn HJ (2001) Lipovitellins derived from two forms of vitellogenin are differentially processed during oocyte maturation in haddock (Melanogrammus aeglefinus). J Exp Zool 291:58–67

Rurangwa E, Kime DE, Ollevier F, Nash JP (2004) The measurement of sperm motility and factors affecting sperm quality in cultured fish. Aquaculture 234:1–28

Sawaguchi S, Koya Y, Yoshizaki N, Ohkubo N, Andoh T, Hiramatsu N, Sullivan CV, Hara A, Matsubara T (2005) Multiple vitellogenins (Vgs) in mosquitofish (*Gambusia affinis*): identification and characterization of three functional Vg genes and their circulating and yolk protein products. Biol Reprod 72:1045–1060

Sawaguchi S, OhkubO N, Amano H, Hiramatsu N, Hara A, Sullivan CV, Matsubara T (2008) Controlled accumulation of multiple vitellogenins into oocytes during vitellogenesis in the barfin flounder, *Verasper moseri*. Cybium Int J Ichthyol 32:262

Schilling J, Loziuk PL, Muddiman DC, Daniels HV, Reading BJ (2015) Mechanisms of egg yolk formation and implications on early life history of white perch (*Morone americana*). PLoS One 10:0143225

Shi XD, Zhang SC, Pang QX (2006) Vitellogenin is a novel player in defense reactions. Fish Shellfish Immunol 20:769–772

Smolenaars MM, Madsen O, Rodenburg KW, Van der Horst DJ (2007) Molecular diversity and evolution of the large lipid transfer protein superfamily. J Lipid Res 48:489–502

Soderhal K (1997) Defence reactions in a crustacean. Dev Comp Immunol 21:137

Specker JL, Sullivan CV (1994) Vitellogenesis in fishes: status and perspectives. In: Davey KG, Peter RE, Tobe SS (eds) Perspectives in comparative endocrinology. National Research Council of Canada, Ottawa, ON, pp 304–315

Stamatiades GA, Kaiser UB (2018) Gonadotropin regulation by pulsatile GnRH: signaling and gene expression. Mol Cell Endocrinol 463:131–141

Sullivan CV, Hiramatsu N, Kennedy AM, Clark RW, Weber GM, Matsubara T, Hara A (2003) Induced maturation and spawning: opportunities and applications for research on oogenesis. Fish Physiol Biochem 28:481–486

Sun C, Zhang S (2015) Immune-relevant and antioxidant activities of vitellogenin and yolk proteins in fish. Nutrients 7:8818–8829

Sundararaj BI, Vasal S (1976) Photoperiod and temperature control in the regulation of reproduction in the female catfish *Heteropneustes fossilis*. J Fish Res Board Can 33:959–973

Thackray VG, Mellon PL, Coss D (2010) Hormones in synergy: regulation of the pituitary gonadotropin genes. Mol Cell Endocrinol 314:192–203

Tyler CR, Lancaster P (1993) Isolation and characterization of the receptor for vitellogenin from follicles of the rainbow trout. J Comp Physiol 163B:225–233

Tyler CR, Sumpter JP (1990) The purification and partial characterization of carp, *Cyprinus carpio*, vitellogenin. Fish Physiol Biochem 8:111–120

Wallace RA (1978) Oocyte growth in non-mammalian vertebrates. In: Jones REZ (ed) The vertebrate ovary. Plenum Press, New York, pp 469–502

Wallace RA (1985) Vitellogenin and oocyte growth in nonmammalian vertebrates. In: Browder LW (ed) Developmental biology. Plenum Press, New York, pp 127–177

Wallace RA, Selman K (1982) A new procedure for the isolation of intact vitellogenin from teleosts. In: Richter CJJ, Goos HJT (eds) Reproductive physiology of fish. Pudoc, Wageningen

Weltzien FA, Andersson E, Andersen Ø, Shalchian-Tabrizi K, Norberg B (2004) The brain–pituitary–gonad axis in male teleosts, with special emphasis on flatfish (Pleuronectiformes). Comp Biochem Physiol A Mol Integr Physiol 137(3):447–477

Wiegand M (1996) Composition, accumulation and utilization of yolk lipids in teleost fish. Rev Fish Biol Fish 6:259–286

Williams VN, Reading BJ, Amano H, Hiramatsu N, Schilling J, Salger SA, Islam Williams T, Gross K, Sullivan CV (2014a) Proportional accumulation of yolk proteins derived from multiple vitellogenins is precisely regulated during vitellogenesis in striped bass (*Morone saxatilis*). J Exp Zool A Ecol Genet Physiol 321:301–315

Williams VN, Reading BJ, Hiramatsu N, Amano H, Glassbrook N, Hara A, Sullivan CV (2014b) Multiple vitellogenins and product yolk proteins in striped bass, *Morone saxatilis*: molecular characterization and processing during oocyte growth and maturation. Fish Physiol Biochem 40:395–415

Worm B, Hilborn R, Baum JK, Branch TA, Collie JS, Costello C, Fogarty MJ, Fulton EA, Hutchings JA, Jennings S, Jensen OP (2009) Rebuilding global fisheries. Science 325(5940):578–585

Yada T, Nakanishi T (2002) Interaction between endocrine and immune systems in fish. Int Rev Cytol 220:35–92

Yilmaz O, Patinote A, Nguyen T, Bobe J (2018) Multiple vitellogenins in zebrafish (Danio rerio): quantitative inventory of genes, transcripts and proteins, and relation to egg quality. Fish Physiol Biochem 44:1509–1525

Zhang SC, Sun YN, Pang QX, Shi XD (2005) Hemagglutinating and antibacterial activities of vitellogenin. Fish Shellfish Immunol 19:93–95

Vitellogenin: As a Hormone

9

V. Ramasubramanian and V. Brindha Priyadarisini

Abstract

Animal protein is a major component of human diets derived from fishing. In India, about 35% of the population consumes fish. The typical Indian's protein consumption is estimated to be approximately 30%, with both marine and freshwater fish accounting for 20%. As a result, aquaculture has greatly focused on producing and growing freshwater fish. As in other animals, fish growth is regulated by both intrinsic (sex hormones, growth hormone (GH), insulin-like growth factor-I (IGF-I), and leptin) and extrinsic (photoperiod, temperature, and food availability) variables. Though fish may grow throughout their lifetimes, due to energy partitioning, body weight increase slows dramatically during gonadal development and gamete generation.

Keywords

Growth hormone · Insulin · Photoperiod · Vitellogenin · Endocrine · Yolk sac

9.1 Introduction

In the early 1900s, research on sex discrimination in fish, a particular antigen in the blood of gravid females, was discovered using immunological techniques. The principal precursor of egg yolk protein, vitellogenin, is currently identified as this

V. Ramasubramanian (✉)
Unit of Aquatic Biotechnology, Department of Zoology, Bharathiar University, Coimbatore, Tamil Nadu, India

V. B. Priyadarisini
Clinical Biotechnology Lab, Department of Microbial Biotechnology, Bharathiar University, Coimbatore, Tamil Nadu, India

V. Baskaralingam, R. Vanichviriyakit (eds.), *Vitellogenin in Fishes-Diversification, Biological Properties, and Future Perspectives*,
https://doi.org/10.1007/978-981-99-5340-0_9

particular antigen and is synthesized in the female liver before being secreted into the blood and absorbed into the egg. Recent protein and gene analyses have led to the identification of several vitellogenin variations. A unique precursor of egg envelope proteins secreted into the blood in response to estrogen stimulation is choriogenin, which was also found in the 1980s with vitellogenin. These two proteins are utilized as efficient biomarkers for determining the effects of estrogen-like endocrine-disrupting substances (environmental hormones) in aquatic environments and playing important roles in the process of oogenesis.

The growth hormone (GH)/insulin-like growth factor-I (IGF-I) axis is a fundamental regulator of vertebrate growth that also regulates reproduction and development. Although the processes driving somatic development and reproductive cross-talk in fish systems are not fully known, gonadal steroids are critical for puberty and sexually dimorphic growth (Meinhardt and Ho 2006). In fish, gonadal steroid hormones have been demonstrated to affect growth hormone receptors in several ways (Jiao et al. 2006). IGF-I, the major mitogenic factor, has been demonstrated to promote somatic growth in fish (Duan 1997). It has also been found that E2 stimulates vitellogenesis by upregulating numerous vg genes and decreasing IGF-I levels when plasma GH levels increase or remain unchanged (Davis et al. 2007; Riley et al. 2002). During sexual maturity, plasma levels of E2 and T increase and correspond with changes in plasma GH levels, showing that sex hormones may regulate pituitary GH production (Canosa et al. 2007; Makino et al. 2007; Melamed et al. 1998). E2 increases the hepatic production and release of vitellogenin (Vg), a yolk-precursor protein integrated into developing oocytes to generate yolky oocytes during vitellogenesis in female teleosts (Nath and Maitra 2001; Mommsen and Walsh 1988). The number of forms present, native and sub-unit molecular weight, and degree of post-translational modification of many fish vitellogenins in circulation vary more. Vg has recently been isolated and characterized in various teleosts, and one, two, or three types of Vg in circulation have been demonstrated to contribute to yolk protein synthesis (Mahapatra et al. 2017; Maitra et al. 2007).

Fish Vg has been widely used as a biomarker for EDC exposure in addition to its traditional function as a nutrient reserve during embryonic development. It may also play a role in immune functions and act as a carrier molecule for ions like calcium, magnesium, and iron (Hara et al. 2016; Nath et al. 2007; Mommsen and Walsh 1988). Reis-Henriques et al. (2000) showed that Vg in the oocyte could modify the production of ovarian E2, which in turn affects Vg synthesis in the liver.

Additionally, different Vg types produced at different phases of oogenesis may have different functions throughout oocyte maturation and embryonic development (Pousis et al. 2011; Sawaguchi et al. 2008; Hiramatsu et al. 2002). In the past, we showed that semi-purified conspecific Vg might cause full vitellogenesis in *C. batrachus*, an Indian walking catfish (Vg synthesis and integration into oocytes) (Nath et al. 1997). Exogenous mrigal, *Cirrhinus mrigala*, Vg administration affects Vg production and the integration of Vg into developing oocytes for development and conversion into yolky oocytes in female catfish, *C. batrachus*. Furthermore, fish's early embryonic development is caused by yolk protein until the yolk sac is digested.

The vertebrate neuroendocrine system controls important processes, including development, growth, metabolism, and reproduction. It is now recognized that many chemical substances created during the past century can alter it. Environmental toxicology now includes a significant amount of study on the identification and consequences of such compounds, and this work has drawn considerable public interest. Endocrine disrupters, also known as endocrine-disrupting chemicals, are generally defined as substances that either imitate or oppose the functions of naturally occurring hormones (EDCs). These include estrogenic EDCs (environmental estrogens), which function similarly to endogenous estradiol-17 to trigger an estrogenic response (E2), as well as synthetic estrogens like ethinylestradiol (EE2) and diethylstilbestrol; examples of estrogenic EDCs include the biodegradation products of pesticides like dichlorodiphenyltrichloroethane (DDT), polychlorinated biphenyls (PCBs), and alkylphenol poly ethoxylates (DES).

In bioassays of animal exposure to estrogenic EDCs, measurements of proteins controlled by estrogen are frequently utilized. For these evaluations, vitellogenin (Vg), a yolk precursor protein, has been employed most frequently. Vg is a useful biomarker for fish exposure to estrogenic EDCs in aquatic habitats because of the following characteristics: (1) Fish have been used extensively in field studies of EDCs and aquatic health. (2) Induction of vitellogenesis is a specific physiological response of fishes to estrogen or estrogenic chemicals. (3) Induction of Vg synthesis by estrogen is dose-dependent within broad limits. (4) Vg appears naturally in maturing females but not in males or immature fish. (5) Vitellogenesis is induced in males and juveniles exposed to estrogen or estrogenic EDCs.

Fish Vg screenings that are sensitive and precise are useful techniques for detecting the presence of environmental estrogens. The "single Vg" model of piscine oogenesis is still widely accepted, whether on purpose or due to a failure to keep up with recent research on vitellogenin multiplicity. However, this assumption is based on most current approaches to assessing and interpreting the effects of estrogenic EDCs using Vg as a biomarker. Recent cDNA cloning and immunohistochemical studies (Hiramatsu et al. 2002; Patino and Sullivan 2002; Matsubara et al. 2003; Hiramatsu et al. 2006) have demonstrated that the existence of several forms of Vg is the norm in the majority of fish species investigated so far. Vg-based assays will be misapplied to evaluate the impact of EDCs because the target type of Vg gene or protein, and its sensitivity to estrogen induction is usually unknown for the species under investigation and may differ between assay systems and even laboratories performing assays on the same species. This concise review will concentrate on our current understanding of the size, distribution, and function of various teleost Vgs in connection to potential breakthroughs in developing and interpreting Vg-based bioassays of EDC exposure in fish. Vitellogenin and the vitellogenesis process, "vitellogenin" was coined by Pan et al. (1969) to describe a female-specific protein discovered in the hemolymph of the Cecropia moth.

The trout FSSP was later purified and recognized as Vg in a teleost for the first time (Hara and Hirai 1978). Numerous biochemical and immunological techniques have been used to purify, identify, and characterize Vg in various fish species (Wallace and Begovac 1985; Mommsen and Walsh 1988; Patino and Sullivan 2002).

All Vg proteins share the following traits in general: (1) they are female-specific serum or plasma proteins, (2) they are precursors to yolk proteins, (3) they are induced by estrogen, (4) phosphoprotein with molecular masses ranging from 300 to 600 kDa, and (5) they are carrier proteins with both a lipid and ionic component (e.g., calcium, zinc, cadmium, iron). A relatively novel paradigm that has significantly impacted recent studies on the reproductive physiology of fishes is the structural and functional multiplicity of Vgs. Even for discovering several Vgs within a single species, the nomenclature and categorization of dual or multiple Vg proteins and their related genes have become fairly muddled. Two varieties of Vgs have a whole structure for the yolk protein domain ("complete" Vg: NH_2-LvH-Pv-LvL-′-c-C-terminal peptide-COOH).

9.2 Biological Actions of Growth Hormone in Fish

Model depicting oogenesis in a temperate perciform fish, the white perch (*Morone americana*) (Fig. 9.1). The oocyte along with its surrounding somatic tissues (the granulosa cells and theca layer) is called a follicle. The previtellogenic, primary growth oocyte accumulates neutral lipids that are stored in the ooplasm as lipid droplets. During vitellogenic growth, lipid deposition continues and the oocyte accumulates yolk proteins, which are stored in yolk granules. After vitellogenic growth is completed, the follicle undergoes maturation, which includes resumption of meiosis and cytoplasmic maturation by the oocyte, culminating in ovulation of an egg that is competent to undergo fertilization (Fig. 9.2).

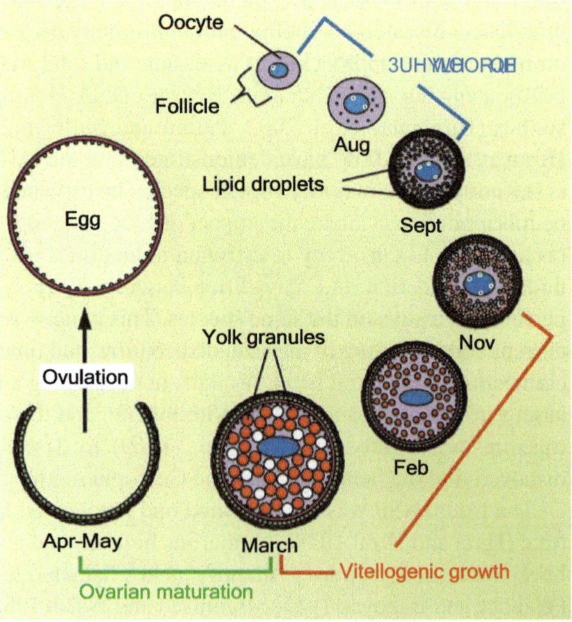

Fig. 9.1 Model depicting oogenesis in a temperate perciform fish, the white perch (*Morone americana*)

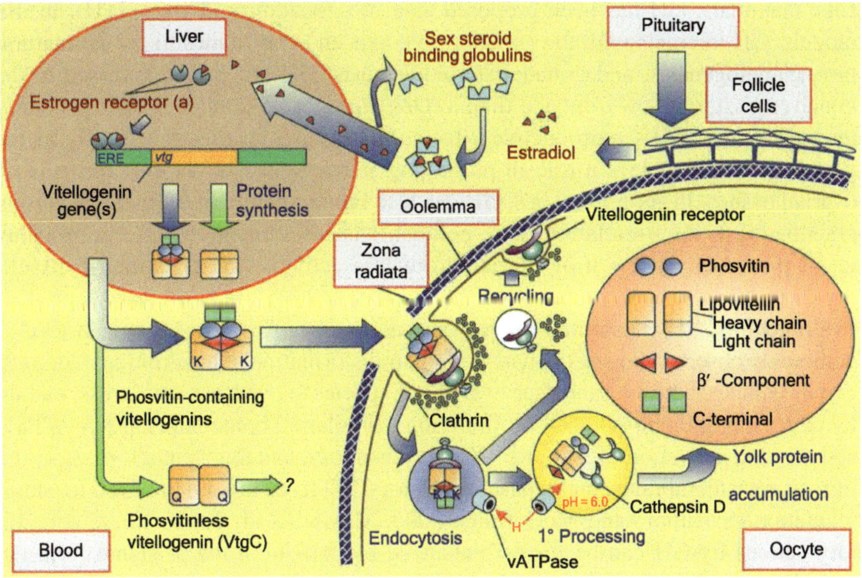

Fig. 9.2 Model of vitellogenesis and oocyte growth. (Redrawn from Hiramatsu N, Cheek AO, Sullivan CV, Matsubara T, and Hara A (2005) Vitellogenesis and endocrine disruption. In: Mommsen TP and Moon TW (eds). Biochemistry and Molecular Biology of Fishes, vol. 6, pp. 431–471. Amsterdam: Elsevier)

GH release from the anterior pituitary is a major regulator of body growth and metabolism in fish models. Fish and human GH treatment increase somatic and linear development in fish species such as rainbow trout, salmon, and carp. These growth-promoting effects have been further confirmed by transgenesis studies using mammalian GH (e.g., common carp (*Cyprinus carpio*)) and fish GH transgenes (e.g., salmon and tilapia species) (Zbikowska 2003). During the growth boost mediated by GH, a decline in condition factor (measured as body weight 100/length^3) is usually seen, suggesting that the fish become comparatively "leaner" with a contemporaneous gain in body weight. GH therapy can boost de novo protein synthesis in tissues in representative species, such as rainbow trout. Lipolysis is also triggered throughout the process, as evidenced by significant increases in free fatty acid and glycerol levels in circulation. This catabolic effect probably is caused by the differential actions of GH on hepatic triacylglycerol lipase and acetyl-coenzyme A carboxylase activities (Bjornsson 1997). In fish models, GH is also known to modulate the behavior pattern during foraging. In salmonids (e.g., trout), GH treatment can increase the appetite and dominant feeding behavior with a drop in avoidance responses to predators. Similar behaviors are also observed in transgenic fish with GH overexpression, suggesting that the metabolic demand for growth enhancement may increase the risk-taking behavior during foraging (Sundstrom et al. 2004). These behavioral changes are suspected of GH modulation of dopaminergic activities/neuronal circuitry within the central nervous system (Bjornsson et al. 2002).

Like mammals, GH has been proposed as a "co-gonadotropin" (co-GTH). In fish models, GH interacts with the gonadotropic axis and contributes to sexual maturation, gametogenesis, and gonadal steroidogenesis. GH receptors expressed in the gonad (e.g., in rainbow trout and tilapia (*Oreochromis mossambicus*)) are primarily responsible for GH's reproductive effects. GH is also synthesized locally at the gonadal level and plays a role in promoting steroidogenesis via direct effects on ovarian tissues. In certain species, GH can also indirectly affect steroid synthesis by potentiating GTH stimulation (e.g., goldfish). These stimulatory effects probably result from GH induction of ovarian aromatase activity via activation of cAMP-dependent cascades (Kajimura et al. 2004; Li et al. 2005). In addition to reproductive functions, GH is essential for seawater adaptation. A rise in serum GH level is commonly observed during the parr–a smolt transformation of anadromous salmons.

GH enhances the tolerance/survival of fish species to hyperosmotic stress, mainly by increasing gill chloride cell proliferation, stimulating branchial Naþ/Kþ-ATPase activity, and activation of Naþ, Kþ, 2Cl-cotransporter, and ion channels (e.g., cystic fibrosis transmembrane conductance regulator (CFTR) channels) involved in osmoregulation (Sakamoto and McCormick 2006; Makino et al. 2007). These stimulatory actions by GH can be further enhanced by cortisol, a major signal from the hypothalamic-pituitary-adrenal (HPA) axis in fish during osmotic stress. This synergism is partly caused by GH-induced cortisol receptor expression in the gills (Pelis and McCormick 2001). Given that (1) IGF-I can mimic GH induction of chloride cell proliferation and Naþ/Kþ-ATPase activity, (2) a rise in IGF-I level in circulation is observed in fish during hyperosmotic stress, and (3) both IGF-I mRNA and IGF-I binding sites can be detected in the gill epithelium; it is commonly accepted that both endocrine and autocrine/paracrine components of IGF-I are involved in the osmoregulatory functions of GH in fish models (Sakamoto and McCormick 2006).

The increase in GH release during hyperosmotic stress in euryhaline fish occurs concurrently with immune function activation, similar to the immunomodulatory effects of GH observed in mammals. Hypophysectomy can reduce immunological responses in fish models (e.g., channel catfish (*Ictalurus punctatus*) and rainbow trout), which can be partially reversed by GH replenishment (Yada 2007). Furthermore, GH injection can improve fish survival against bacterial infection and artificial vibriosis (Sakai et al. 1997). (1) GH's immunoprotective effects are due to its stimulation of antibody production and immune cell proliferation, (2) activation of phagocytic and nonspecific cytotoxic activities in leukocytes, (3) induction of superoxide production and lysozyme activity, and (4) anti-inflammatory actions via ceruloplasmin production (Yada 2007). The increase in GH release during hyperosmotic stress in euryhaline fish occurs concurrently with immune function activation, similar to the immunomodulatory effects of GH observed in mammals. Hypophysectomy in fish models (e.g., channel catfish (*Ictalurus punctatus*) and rainbow trout) can suppress immune responses (e.g., by reducing Ig-secreting leukocytes), which can be partially reversed by GH replacement (Yada 2007). Likewise, GH injection can improve fish biota's resistance to bacterial infection and artificial vibriosis (Sakai et al. 1997). GH's immunoprotective effects are due to its stimulation of antibody production and immune cell proliferation, (2) activation of phagocytic and nonspecific cytotoxic activities in leukocytes, (3) induction of superoxide production and lysozyme activity, and (4) anti-inflammatory actions via ceruloplasmin production (Yada 2007).

9.3 Nomenclature and Classification of Vitellogenin

Fish Vtgs have a complex evolutionary history, and they exhibit considerable variation in structure and function. The currently accepted Vtg nomenclature developed by R.N. Finn and associates is based on multiple types of Vtg having arisen via whole-genome duplications (WGDs) during the evolutionary history of vertebrates. According to this scenario, the ancestral chordate Vtg (VtgABCD, present in silver lamprey, *Ichthyomyzon unicupsis*) gave rise to two forms of Vtg, VtgAB (chondrostean vitellogenin; present in chondrostean fishes and amphibian and avian species), and VtgCD. The VtgAB subsequently gave rise to VtgA (universally present in teleosts) and to VtgB, which is extinct. VtgCD gave rise to teleost vitellogenin type C (VtgC; also called Vtg3 or phosvitinless Vtg) (present in teleosts from diverse phylogenetic lineages) and to VtgD, which is extinct. Additional Vtg gene duplications, not always involving WGD, have occurred within different lineages of fishes. Dual or multiple A-type Vtgs are present in *Protacanthopterygii* (salmonids, VtgAsa, and VtgAsb), Ostariophysi (Ostariophysian vitellogenin type 1 (VtgAo1) and type 2 (VtgAo2)), and Elopomorpha (Elopomorph vitellogenin type 1 (VtgAe1), type 2 (VtgAe2), and type 3 (VtgAe3)). In a late round of WGD, the VtgA gave rise to two paralogous forms of Vtg (acanthomorph vitellogenin type Aa (VtgAa; also called Vtg1 or VtgI) and type Ab (VtgAb; also called Vtg2 or VtgII)), which are present in *Acanthomorphteleosts.*

Incomplete Vtgs are subdivided into those that contain Pv and those that lack or have severely reduced Pv domains (Pv-less). The VtgC is incomplete, being Pv-less and also lacking β9-c and Ct domains, and consists only of lipovitellin (LvH and LvL). Incomplete Pv-containing Vtgs occur only in Ostariophysian fishes (VtgAo1) and contain three YP domains: LvH, Pv, and LvL. A protein dendrogram shows the classification and domain structure of various types of Vtg present in diverse fish species.

9.4 Chemistry of Vitellogenins and Egg Yolk

Vitellogenins are dimeric proteins, consisting of two identical subunits with phosphate, lipid, carbohydrate, and protein components. They are members of the large lipid transfer protein superfamily, which includes other serum lipoproteins such as low-density lipoprotein (LDL). The Vtgs are large lipoproteins (350–600 kDa) and 20% lipid by weight. They also are specialized carriers of important ions, such as calcium, magnesium, iron, zinc, copper, and various minerals and vitamins, such as retinoids and carotenoids. In addition, regulatory compounds found in fish egg yolk, including lipid-soluble steroid and thyroid hormones, may be transported in part by Vtgs.

For fishes that spawn demersal (sinking) eggs that lack prominent oil droplets, Vtgs are major carriers of lipids into growing oocytes, most of which (80%) are phospholipids, often phosphatidylcholines. In marine fishes that spawn pelagic (floating) eggs with no large oil droplet(s), phospholipids, triacylglycerides, and wax or steryl esters can account for >70%, 8–12%, and 4% of total egg lipids, respectively. Other teleosts produce pelagic eggs with large oil droplets comprised of neutral lipids (e.g., triacylglycerides and wax or steryl esters) that can occupy >50% of the ooplasm. As the deposition of neutral lipids is initiated before and can occur independently from, accumulation of Vtg-derived YPs, Vtg is not a major source of neutral lipids in oocytes of these species. The neutral lipids must be delivered to the oocyte via a different mechanism, perhaps by LDL or related lipoproteins.

Alanine is the most abundant amino acid comprising Vtg polypeptides (11–12% of total residues). In addition, Vtgs in some species (e.g., anguillid eels) contain polyalanine regions. It is suggested that lanine serves as an important intermediary of carbohydrate metabolism, especially embryonic gluconeogenesis.

9.5 Vitellogenin Structure and Product YPs

Native Vtg monomers consist of covalently linked YP domains that are proteolytically processed into corresponding YPs after being taken up by the oocyte. Complete Vtg molecules consist of five linear YP domains from the amino-terminus: lipovitellin heavy chain (LvH), phosvitin (Pv), lipovitellin light chain (LvL), β9-component (β9-c), and C-terminal peptide (Ct).

Structure wise, the A type of complete Vg (VgA) is similar to mummichog VgI (Genbank: T43141), barfin flounder (*Verasper moseri*) VgA (Genbank: AB181833), and haddock (*Melanogrammus aeglefinus*) VgA (Genbank: AAK15158), and its constituent LvH is severely degraded during oocyte hydration associated with final maturation (see Dual Vitellogenin Model, below). Complete Vg (VGB) of the B type shares structural similarities with the barfin flounder (VgB; Genbank: AB181834), the haddock (VgB; Genbank: AAK15157), and mummichog 6 (VgII; T43144). During oocyte maturation, LvH is either not degraded or is only partially hydrolyzed.

A Vg that is most comparable to zebrafish (*Danio rerio*) vg3 (Genbank: AAG30407), Japanese goby (*Acanthogobius flavimanus*) Vg-320 (Genbank: BAC06191), insect Vgs, or chicken VgIII should be regarded as a C type Vg since it lacks a Pv domain or has a significantly shorter Pv domain (VgC or phosvitinless Vg). A lower concentration of phosphorus or serine residues, as well as a much smaller molecular mass than either VgA or VgB, might be used to classify a VgC.

Further studies of the scope and distribution of numerous Vgs across teleost species taken from various phylogenetic groups are now being conducted on the basis of this categorization approach. Matsubara et al. published these trials' preliminary findings (Matsubara et al. 2003). In mosquitofish (*Gambusia affinis*), white perch (*Morone americana*), red seabream (*Pagrus* major), white-edged rockfish (*Sebastes taczanowskii*), mummichog, and striped mullet, all three types of Vg (A–C) transcripts have been found (*Mugil cephalus*).

The largest YP is the LvH lipoprotein that supplies offspring with amino acids and phospholipids, which can serve as catabolic energy substrates or in anabolic synthesis of membrane or protein structures, respectively. Teleost LvH has an average mass of 114 kDa predicted from deduced amino acid sequences and resolves close to this position by acrylamide gel electrophoresis in most species. The LvH polypeptide consists of largely amphipathic secondary and tertiary structures that form a basket with a lumen of hydrophobic residues required to accommodate lipids. This characteristic structure is similar to vertebrate apolipoprotein B, the primary protein scaffold of other lipid-transporting particles such as LDL. Almost universally included in the LvH domain is a short sequence known to bind oocyte Vtg receptors. The LvL domain, which is smaller than LvH (25 kDa), also forms part of the lipid-basket of Vtg and has chemical and structural characteristics similar to LvH. The LvH and LvL domains of Vtg usually contain one or more glycosylation sites to which carbohydrate moieties are attached, and the LvH also contains a site that may bind zinc ion. The Pv is a metalloprotein, consisting largely of serines (>50% of total residues), to which phosphates may be covalently attached prior to secretion by the liver. The negatively charged phosphates attract calcium, magnesium, zinc, and other multivalent metal cations (e.g., ferric iron) via ionic interactions occurring in the bloodstream or oocyte. In freshwater fishes, such as masu salmon (*Oncorhynchus masou*) and mosquitofish (*Gambusia affinis*), metal ions delivered by Pv are crucial for embryo survival, since they are not abundantly available for uptake from the environment. The egg yolk of marine fishes, such as barfin flounder (*Veraspermoseri*), red seabream (*Pagrus major*), and Pacific herring (*Clupea pallasii*), contains less of the metal ions abundantly present in seawater (e.g., calcium and magnesium), suggesting that Pv plays a less significant role in maternal provision of these ions in marine fishes. The Pv also usually contains several potential glycosylation sites. Thus, Pv acts to transport important metabolic ions and carbohydrates into the yolk and, in doing so, helps maintain aqueous solubility of the largely hydrophobic Vtg particle. In addition, Pv may aid in stabilizing Vtg structure through interaction with the basket enclosing lipid cargo formed by LvH and LvL.

It is technically difficult to precisely resolve Pv by acrylamide gel electrophoresis, since it does not stain by traditional protein dyes and all species of Pv do not migrate to the same position in the gel (size range 6–20 kDa) due to high and varying degrees of phosphorylation and to the presence of various Pv-Lv conjugates in the yolk. The size of Pv also is highly variable among fishes, and the Pv domain may even be lacking in some incomplete forms of Vtg. The Pv-containing Vtgs from *Ostariophysian* and *Protacanthopterygian* fishes generally contain shorter Pvs with fewer serines (24–34) than those from *Acanthomorphteleosts*, which contain 50 serines. In contrast, Japanese eel (*Anguilla japonica*) Vtgs have massive Pvs with 87–91 serines.

The remaining two small YPs (β9-c and Ct) are devoid of lipid and phosphate but together contain 14 highly conserved cysteine residues, which are known to be involved in disulfide linkages required for complex folding of the Vtg polypeptide and possibly for dimerization of native Vtg. This carboxy-terminal end of Vtg is highly similar in structure to cysteine-rich von Willebrand factor (vWF)-type D2 domain located in the short pro-peptide of vertebrate vWF. In addition, a conserved CGxC motif found in both vWF-type D2 domain and the carboxy terminus of β9-c may be involved in facilitation of disulfide bond formation during peptide folding and/or adhesion when Vtg binds its oocyte receptor. The β9-c may have a glycosylation site and, like Pv, also has high aqueous solubility with many of its hydrophilic residues exposed on the surface of the Vtg particle. Although β9-c is known to be released from the carboxy-terminal end of Vtg as an 16 kDa YP in a number of fish species, the presence of Ct as a bona fide YP has only been verified in barfin flounder and deduced in Atlantic herring. The molecular weight of β9-c in the oocytes of several fishes indicates that Ct is cleaved from its carboxy-terminus; however, it may be degraded as opposed to store as yolk, since a Ct YP typically cannot be detected (predicted size 13 kDa).

9.6 Regulation of Vitellogenesis

Major environmental cues regulating vitellogenesis include seasonal changes in water temperature and daylength, but almost anything predictive of successful reproduction can serve as a cue. Examples include rainy weather (e.g., "monsoon season"), which may flood preferred spawning areas, and the presence of conspecific males. The converse is also applicable, as the perceived absence of positive cues is inhibitory. An example in some farmed fishes that spawn every few days is the response to removal of males from the tank, which causes females to cease reproductive processes, terminate vitellogenesis, and undergo massive pre-ovulatory atresia. Internal cues signaling adequate nutritional status or fat reserves also influence vitellogenesis and may be especially important where seasonal cues are lacking. Information imparted by received cues is integrated in the brain, ultimately influencing the activity of neurons emanating from the preoptic area that course to the anterior lobe of the pituitary gland and terminate on the gonadotropic cells. These neurons secrete a short peptide called gonadotropin releasing-hormone (Gnrh) that induces the gonadotrophs to synthesize and secrete gonadotropins. Fish possess two gonadotropins similar to those in higher vertebrates, follicle-stimulating hormone (Fsh) and luteinizing hormone (Lh). Depending on species and reproductive status, one or both of these gonadotropins act on responsive ovarian follicle cells to stimulate their synthesis and secretion of the estrogen, estradiol-17b (E2). Circulating E2 then induces hepatocytes in the liver to synthesize and secrete Vtgs. E2 also feeds back to act on the brain and pituitary gland, providing homeostatic control of vitellogenesis and later maturational processes. Xenoestrogens present in the environment can "short-circuit" the brain-pituitary-gonad axis and act directly on the liver to induce abnormal vitellogenesis in male and juvenile fish, sometimes resulting in the appearance of "intersex" males bearing previtellogenic oocytes in their testes, and in reproductive failure. Vitellogenesis is highly responsive to synthetic and natural estrogens, and fish Vtgs have been globally adopted as biomarkers of the presence of estrogenic chemical contaminants in aquatic environments, which is indicated by elevated Vtg levels in the blood or tissues of male and/or and juvenile fish (Dang 2016; Adeel et al. 2017).

9.7 Endocrine Control of Vitellogenesis

Vitellogenesis is a seasonal or cyclic process depending on gonadotropins. The production of gonadotropin-releasing hormone (GnRH) by the brain (hypothalamus) is mediated by a variety of endogenous and environmental factors, including innate biorhythms, nutritional status, and seasonal changes in day length and water temperature. In response to GnRH, follicle-stimulating hormone (FSH) is secreted by pituitary gonadotrophs and induces the theca and granulosa cells of the ovarian follicle to secrete estradiol-17β (E2), which instructs the liver to synthesize Vtgs and secrete them into the bloodstream. In some species, luteinizing hormone produced by the pituitary in response to GnRH also elicits follicular production of E2. Growth hormone, thyroid hormone, cortisol, androgens, and other estrogenic steroids, such as estrone, may also contribute to the induction of hepatic vitellogenesis in some species. Endocrine regulation of the induction of Vg synthesis by the liver has been extensively studied both in vivo and in vitro (Wallace and Begovac 1985; Mommsen and Walsh 1988). Briefly, hepatic Vg synthesis naturally ensues from activation of the hypothalamus-pituitary-gonad neuroendocrine axis by environmental and endogenous signals. Data obtained for some teleosts (e.g., salmonids) indicate that increased levels of follicle-stimulating hormone (FSH) in the blood induce follicular production of E2, which triggers hepatic synthesis of Vg. In other species, ovarian E2 production may be regulated by luteinizing hormone. Other estrogens, such as estrone, may contribute to the induction of vitellogenesis by priming the liver to respond more strongly to endogenous E2 rhythms. Although the liver is recognized as the organ responsible for Vg synthesis in fishes, ovarian connective tissues or the oocyte itself are possible sites for Vg (vitellin) synthesis in some bivalve species. Following synthesis, Vg is released into the bloodstream, taken up by the growing oocytes, and enzymatically processed into yolk proteins that are stored throughout the ooplasm in yolk granules or globules. Growing fish oocytes, as well as those in the chicken (Shen et al. 1993) and in *Xenopus laevis* (Opresko and Wiley 1987), selectively accumulate Vg via receptor-mediated endocytosis (Stifani et al. 1990). A membrane receptor on the oocyte surface with a high affinity for Vg, called the Vg receptor (VgR), mediates the endocytosis process. Vitellogenin bound to the VgR is clustered in clathrin-coated pits that invaginate to form coated vesicles. These endocytosed vesicles fuse with lysosomes in the peripheral ooplasm, forming multivesicular bodies. The lysosomes contain the proteolytic enzyme cathepsin D, and possibly other enzymes (such as cathepsin B), that process Vg into its product yolk proteins (Carnevali et al. 2006; Hiramatsu et al. 2002; Mosconi et al. 2002; Romano et al. 2004). The yolk of an ovulated egg consists largely of materials that have been deposited in the growing oocyte to be later utilized by the nascent embryo as a source of nutrition and energy to support development. In salmonid species, Vg is a major carrier of egg phospholipids, and the yolk proteins derived from Vg can account for 80–90% of the dry mass of ovulated eggs (Selman and Wallace 1989; Tyler et al. 1999). However, in some marine or anadromous fishes that spawn heavily lipidated eggs, Vg is not a major source of the neutral lipids commonly found in the oil droplet of ovulated eggs (Patino and Sullivan 2002; Sullivan et al. 2003).

9.8 Novel Functions of Vitellogenins

In addition to their fundamental functions as primary precursors of YPs and as transporters of lipids and other nutritional resources into the egg yolk, teleost Vtgs have become objects of curiosity for their novel, non-nutritive functions. It is increasingly evident that Vtg is an immunocompetent factor capable of protecting the host against attack by microbes including bacteria and viruses (Sun and Zhang 2015; Li and Zhang 2017). Both serum Vtg and its product YPs, Lv and Pv, are immunologically active via their ability to recognize and bind to pathogen-associated molecular patterns (PAMPs). These properties of multivalent pattern recognition receptors were associated with the ability of Lv and Pv to bind Gram-positive and Gram-negative bacteria through recognition of their conserved lipopolysaccharide, peptidoglycan, lipoteichoic acid, and glucan components. Pv was shown to be an effector molecule capable of killing bacteria directly via cell lysis, while Lv was shown to act as an opsonin, facilitating phagocytosis of bacteria by macrophages. Serum Vtg has been additionally reported to bind and neutralize infectious pancreatic necrosis virus. The antimicrobial properties of certain conserved domains present in zebrafish VtgAo2 were verified by producing them as recombinant proteins and testing their bioactivity. These domains encompassed the carboxy-terminal third of LvH, most of LvL, and all of b0-c. All three recombinant domains functioned as pattern-recognition receptors, binding both Gram-positive and Gram-negative bacteria and their isolated signature PAMPs, and those including portions of LvH or LvL also functioned as opsonins, promoting phagocytosis of *Escherichia coli* (—) and *Staphylococcus aureus* (þ) by carp, *Cyprinus carpio*, macrophages. Native Lv was associated with immune defense of rosy barb, *Pethia conchonius*, embryos and larvae, and Pv was shown to possess antimicrobial activity in zebrafish embryos and larvae. The protective immune functions of Vtg-derived YPs may extend through oocyte maturation and ovulation to embryonic and larval development, when the YPs have been degraded into smaller polypeptides. For example, three small polypeptides derived from the C-terminal 55 residues of a zebrafish Pv have been shown to have individual and enhanced collective activity against growth of *S. aureus* (þ) and *Aeromonas hydrophila* (–) bacteria and, thus, may have protective activity in developing embryos. Vtg may also contribute to immune priming by carrying immunological memory from mother to offspring, as it does in insects.

Vitellogenin's are also known to have antioxidant activity and the capability to suppress free-radical reactions in fish oocytes and to protect the host from oxidative stress. It has been demonstrated that hen egg yolk Pv exhibits strong antioxidant activity owing to its high serine and phosphorus content, which makes it a particularly strong iron-chelating agent. It has also been shown that zebrafish recombinant Pv (rPv) is an antioxidant capable of protecting against radical-mediated oxidation of cellular biomolecules (Sun and Zhang 2015). Several studies have demonstrated that Vtg regulates honey bee aging by boosting antioxidant balance of the body via its antioxidative properties, thus slowing down physiological aging and inhibiting inflammation by reducing oxidative stress. It is intriguing to speculate that Vtg may have similar activities in fishes.

9.9 Applications and Perspectives of Vitellogenesis

The past 20 years have seen great strides in our understanding of vitellogenesis in fishes. However, much remains to be learned. Our first glimpses into molecular mechanisms for control of Vtg gene transcription in zebrafish have revealed much complexity and the likelihood that Vtg gene promoters are sites of considerable crosstalk between numerous types of transcription factors, just as physiology studies revealed that several hormones besides estrogen modulate vitellogenesis. Almost nothing is known about the ways in which different types of Vtg are regulated by these processes. This will be a rich and rewarding area for future research. It is now understood that the multiplicity of Vtgs is the norm in fishes and that parallel multiplicity of Vtg receptors may exist, particularly in higher (acanthomorph) teleosts. These fishes exhibit different ratios of abundance of VtgAa, VtgAb, and VtgC in their bloodstream, and these ratios change during the reproductive cycle and differ between blood plasma and oocytes. These differences likely reflect variation in hepatic secretion of the different types of Vtg and also differences in their rates of sequestration by oocytes via receptor-mediated endocytosis, with both processes modulated to optimize the mixture of Vtg-specific YP products deposited in growing oocytes. These hypotheses remain to be experimentally verified.

Maturational proteolysis of YPs in acanthomorph fishes spawning pelagic eggs in seawater appears to be regulated to optimize oocyte hydration, egg buoyancy, and delivery of the preferred types of nutrients to early embryos and late-stage larvae. In such species whose eggs lack prominent oil droplets, VtgAa is neofunctionalized for susceptibility proteolysis, and it is hypothesized that the newly discovered receptor that preferentially binds VtgAa (Lrp13) may deliver it into a special compartment where it is fated for digestion during maturation, whereas the VtgAb is targeted by a different receptor (Vtgr) to a different compartment. The identity and functional details of these compartments need resolution.

The basic biology of the C-type Vtg is poorly understood. In higher teleosts, VtgC usually escapes significant proteolysis during final maturation, but it is uncertain how this is achieved. There is presently no definitive evidence that VtgC binds to a specific receptor that delivers it into a "protected" compartment nor is there evidence that VtgC even enters yolk vesicles where it would be subject to proteolysis by CatB/L during final maturation. It is known that dephosphorylated Vtg cannot bind to Vtg receptor(s), and VtgC, lacking a Pv domain, is constitutively "dephosphorylated." VtgC also lacks the vWfd domain (b′–c, Ct) thought to be important for dimerization and perhaps cell adhesion associated with receptor-mediated endocytosis. In short, the molecular biology of VtgC is a mystery and "low-hanging fruit" for a scientist wishing to make an important contribution to the study of fish vitellogenesis. Circulating Vtgs have served as markers for onset of puberty and progression of gonad maturation in female fishes, especially in aquaculture. The presence of Vtg in blood, mucus, and muscle has also been used to identify the gender of fishes that do not exhibit sexual dimorphism (see also Social and Reproductive Behaviors: Sexual Behavior in Fish). Assessment of the proper maturational proteolysis of Vtg-derived YPs, or of expression of cathepsin transcripts and proteins, has also been used as a marker of egg quality. In addition, Vtgs are frequently used to assess exposure of animals in aquatic environments to endocrine-disrupting chemicals (EDCs), specifically to EDCs that mimic the action of estrogens. Since Vtgs are produced in response to endogenous E2, they are potentially ideal markers for assessment of estrogenic EDCs. However, as disparities exist between the different types of Vtgs (or vtg transcripts) with regard to their sensitivity to induction by estrogen(s), the specific type of Vtg being measured to detect EDCs should be taken into account.

References

Adeel M, Song X, Wang Y, Francis D, Yang Y (2017) Environmental impact of estrogens on human, animal and plant life: a critical review. Environ Int 99:107–119

Bjornsson B (1997) The biology of salmon growth hormone: from day light to dominance. Fish Physiol Biochem 17:9–24

Bjornsson B, Johnsson J, Benedet S, Einarsdottir I, Hildahl J, Agustsson T, Johnsson E (2002) Growth hormone endocrinology of salmonids: regulatory mechanisms and mode of action. Fish Physiol Biochem 27:227–242

Canosa LF, Chang JP, Peter RE (2007) Neuroendocrine control of growth hormone in fish. Gen Comp Endocrinol 151:1–26

Carnevali O, Cionna C, Tosti L, Lubzens E, Maradonna F (2006) Role of cathepsins in ovarian follicle growth and maturation. Gen Comp Endocrinol 146:195–203

Dang Z (2016) Interpretation of fish biomarker data for identification, classification, risk assessment and testing of endocrine disrupting chemicals. Environ Int 92–93:422–441

Davis LK, NaoshiHiramatsu KH, Reading BJ, Matsubara T, Hara A, Sullivan CV, Pierce AL, Tetsuya Hirano E, Grau G (2007) Induction of three vitellogenins by 17beta-estradiol with concurrent inhibition of the growth hormone-insulin-like growth factor 1 axis in a Euryhaline teleost, the tilapia (Oreochromis mossambicus). Biol Reprod 77(4):614–625

Duan C (1997) The insulin-like growth factor system and its biological actions in fish. Am Zool 37(6):491–503

Hara A, Hirai H (1978) Comparative studies on immunochemical properties of female-specific serum protein and egg yolk proteins in rainbow trout (*Salmo gairdneri*). Comp Biochem Physiol 59B:339–343

Hara A, Hiramatsu N, Fujita T (2016) Vitellogenesis and choriogenesis in fishes. Fish Sci 82:187–202. https://doi.org/10.1007/s12562-015-0957-5

Hiramatsu N, Ichikawa N, Fukada H, Fujita T, Sullivan CV, Hara A (2002) Identification and characterization of proteases involved in specific proteolysis of vitellogenin and yolk proteins in salmonids. J Exp Zool 292:11–25

Hiramatsu N, Matsubara T, Fujita T, Sullivan CV, Hara A (2006) Multiple piscine vitellogenins: biomarkers of fish exposure to estrogenic endocrine disruptors in aquatic environments. Mar Biol 149:35–47

Jiao B, Huang X, Chan CB, Zhang L, Wang D, Cheng CH (2006) The co-existence of two growth hormone receptors in teleost fish and their diVerential signal transduction, tissue distribution and hormonal regulation of expression in seabream. J Mol Endocrinol 36:23–40

Kajimura S, Kawaguchi N, Kaneko T, Kawazoe I, Hirano T, Visitacion N, Grau EG, Aida K (2004) Identification of the growth hormone receptor in an advanced teleost, the tilapia (Oreochromis mossambicus) with special reference to its distinct expression pattern in the ovary. J Endocrinol 181:65–76

Li H, Zhang S (2017) Functions of vitellogenin in eggs. Results Probl Cell Differ 63:389–401

Li WS, Chen D, Wong AO, Lin HR (2005) Molecular cloning, tissue distribution, and ontogeny of mRNA expression of growth hormone in orange-spotted grouper (Epinephelus coioides). Gen Comp Endocrinol 144:78–89

Mahapatra S, Kabita S, Bhattacharya D, Sarkar S, Juin SK, Maitra S, Nath P (2017) Purification and development of ELISAs for two forms of vitellogenin in Indian walking catfish, Clarias batrachus (L.). Fish Physiol Biochem 43(2):477–491. https://doi.org/10.1007/s10695-016-0304-5. Epub 2017 Feb 28

Maitra S, Sahu R, Trehan N, Garg SK, Nath P (2007) Circannual variation in plasma vitellogenin and gonadotropin II levels in relation to annual ovarian cycle in female mrigal, Cirrhinus mrigala. Aquaculture 265:370–384

Makino K, Onuma TA, Kitahashi T, Ando H, Ban M, Urano A (2007) Expression of hormone genes and osmoregulation in homing chum salmon: a minireview. Gen Comp Endocrinol 152:304–309

Matsubara T, Nagae M, Ohkubo N, Andoh T, Sawaguchi S, Hiramatsu N, Sullivan CV, Hara A (2003) Multiple vitellogenins and their unique roles in marine teleosts. Fish Physiol Biochem 28:295–299

Meinhardt UJ, Ho KK (2006) Modulation of growth hormone action by sex steroids. Clin Endocrinol 65(4):413–422. https://doi.org/10.1111/j.1365-2265.2006.02676.x. PMID: 16984231

Melamed P, Rosenfeld H, Elizur A, Yaron Z (1998) Endocrine regulation of gonado- tropin and growth hormone gene transcription in fish. Comp Biochem Physiol C Pharmacol Toxicol Endocrinol 119:325–338

Mommsen TP, Walsh PJ (1988) Vitellogenesis and oocyte assembly. In: Hoar WS, Randall DJ (eds) Fish physiology, vol XI. Academic Press, New York, pp 347–406

Mosconi G, Carnevali O, Habibi HR, Sanyal R, Polzonetti-Magni AM (2002) Hormonal mechanisms regulating hepatic vitellogenin synthesis in the gilthead seabream, Sparus aurata. Am J Physiol Cell Physiol 283:673–678

Nath P, Maitra S (2001) Role of two plasma vitellogenins from Indian major carp (Cirrhinusmrigala) in catfish (Clarias batrachus) Vitellogenesis. Gen Comp Endocrinol 124:30–44

Nath P, Bhakta M, Maitra S, Sarkar S (1997) Vitellogenin induces vitellogenesis in the catfish, Clarias batrachus. In: Kawashima S, Kikuyama S (eds) Proceedings of the XIIIth international congress of comparative endocrinology, Yokohama, Japan, pp 1475–1479

Nath P, Sahu R, Sk K, Bhattacharya D (2007) Vitellogenesis with special emphasis on Indian fishes. Fish Physiol Biochem 33:359–366

Opresko LK, Wiley HS (1987) Receptor-mediated endocytosis in *Xenopus* oocytes. I. Characterization of the vitellogenin receptor system. J Biol Chem 262:4109–4115

Pan MJ, Bell WJ, Telfer WH (1969) Vitellogenic blood protein synthesis by insect fat body. Science 165:393–394

Patino R, Sullivan CV (2002) Ovarian follicle growth, maturation, and ovulation in teleost fish. Fish Physiol Biochem 26:57–70

Pelis RM, McCormick SD (2001) Effects of growth hormone and cortisol on Na þ-K þ-2Cl cotransporter localization and abundance in the gills of Atlantic salmon. Gen Comp Endocrinol 124:134–143

Pousis C, De Giorgi C, Mylonas CC, Bridges CR, Zupa R, Vassallo-Agius R et al (2011) Comparative study of liver vitellogenin gene expression and oocyte yolk accumulation in wild and captive Atlantic bluefin tuna (Thunnus thynnus L.). Anim Reprod Sci 123:98105

Reis-Henriques MA, Ferreira M, Silva L, Dias A (2000) Evidence for an involvement of vitellogenin in the steroidogenic activity of rainbow trout (Oncorhynchus mykiss) vitellogenic oocytes. Gen Comp Endocrinol 117:260–267

Riley LG, Hirano T, Grau EG (2002) Disparate effects of gonadal steroid hormones on plasma and liver mRNA levels of insulin-like growth factor-I and vitellogenin in the tilapia, Oreochromis mossambicus. Fish Physiol Biochem 26:223–230

Romano M, Rosanova P, Anteo C, Limatola E (2004) Vertebrate yolk protein: a review. Mol Reprod Dev 69:109

Sakai M, Kajita Y, Kobayashi M, Kawauchi H (1997) Immunostimulating effect of growth hormone: in-vivo administration of growth hormone in rainbow trout enhances resistance to Vibrio anguillarum infection. Vet Immunol Immunopathol 57:147–152

Sakamoto T, McCormick SD (2006) Prolactin and growth hormone in fish osmoregulation. Gen Comp Endocrinol 147:24–30

Sawaguchi S, Ohkubo N, Amano H, Hiramatsu N, Hara A, Sullivan CV, Matsubara T (2008) Controlled accumulation of multiple vitellogenins into oocytes during vitellogenesis in the barfin flounder, Verasper moseri. Cybium 32:262

Selman K, Wallace RA (1989) Cellular aspects of oocyte growth in teleosts. Zool Sci 6:211–231

Shen X, Steyrer E, Retzek H, Sanders EJ, Schneider WJ (1993) Chicken oocyte growth: receptor-mediated yolk deposition. Cell Tissue Res 272:459–471

Stifani S, Nimpf J, Schneider WJ (1990) Vitellogenesis in Xenopus laevis and chicken: cognate ligands and oocyte receptors. The binding site for vitellogenin is located on lipovitellin I. J Biol Chem 265(2):882–888

Sullivan CV, Hiramatsu N, Kennedy AM, Clark RW, Weber GM, Matsubara T, Hara A (2003) Induced maturation and spawning: opportunities and applications for research on oogenesis. Fish Physiol Biochem 28:481–486

Sun C, Zhang S (2015) Immune-relevant and antioxidant activities of vitellogenin and yolk proteins in fish. Nutrients 7:8818–8829

Sundstrom LF, Lohmus M, Johnsson JI, Devlin RH (2004) Growth hormone transgenic salmon pay for growth potential with increased predation mortality. Proc Biol Sci 271(suppl 5):S350–S352

Tyler CR, Santos EM, Prat F (1999) Unscrambling the egg – cellular, biochemical, 23 molecular and endocrine advances in oogenesis. In: Norberg B, Kjesbu OS, Taranger GL, Anderson E, Stefansson SO (eds) Proceedings of sixth international symposium on the reproductive physiology of fish. Institute of Marine Research and University of Bergen, Bergen, pp 237–280

Wallace RA, Begovac PC (1985) Phosvitins in *Fundulus* oocytes and eggs: preliminary chromatographic and electrophoretic analyses together with biological considerations. J Biol Chem 260:11268–11274

Yada T (2007) Growth hormone and fish immune system. Gen Comp Endocrinol 152:353–358

Zbikowska HM (2003) Fish can be first – advances in fish transgenesis for commercial applications. Transgenic Res 12:379–389

Vitellogenin Is a Biomarker

10

R. Thirumalaivasn, M. Devaprakash, and N. Sivakumar

Abstract

Primary understanding about vitellogenin (Vtg) is a female-specific reproductive protein, a precursor for egg yolk proteins including phosvitin (Pv) and lipovitellin (Lv), which provide nourishment for early embryos. Vtg biosynthesis is induced by a female reproductive hormone estrogen (E2, estrone (E1), estriol (E3)) or can be induced by estrogen-like compounds' presence in the environment. Environmentally available endocrine disruptor chemicals (EDCs) also possess estrogen-like activities. Continuous exposure of animals to these EDCs stimulates variable levels of Vtg biosynthesis in female, male, and juvenile fishes. Over-induced Vtg in male and juvenile fish causes serious effects on sex differentiation, infertility, abnormal gonad induction, and also reduces viable embryo development. In some cases, it promotes physiological defects in animals including kidney failure and cancer. The aberrant expression of Vtg is mainly by the estrogen induction or anti-androgenic activities of EDCs. Hence, Vtg can be a biomarker to diagnose and also identify the estrogenic chemicals there in water bodies. In this chapter, discuss the process of vitellogenesis, its properties, types and their importance to consider Vtg as a biomarker for fish growth in various environments. Further, methods used for detecting Vtg and its expression are also discussed.

Keywords

Biomarker · Endocrine disruptor · Environmental estrogen · Vitellogenin · Vitellogenesis

R. Thirumalaivasn · M. Devaprakash · N. Sivakumar (✉)
Department of Molecular Microbiology, School of Biotechnology, Madurai Kamaraj University, Madurai, Tamil Nadu, India

V. Baskaralingam, R. Vanichviriyakit (eds.), *Vitellogenin in Fishes-Diversification, Biological Properties, and Future Perspectives*, https://doi.org/10.1007/978-981-99-5340-0_10

Abbreviations

17βE2	17β-estradiol
AhR	Aryl hydrocarbon receptor
APEOs	Alkylphenol ethoxylates
BPA	Bisphenol A
Ct	C-terminal
DDT	Dichlorodiphenyltrichloroethane
DES	Diethylstilbestrol
E1	Estrone
E2	Estrogen
E3	Estriol
EDCs	Endocrine-disrupting chemicals
EDs	Endocrine disruptors
EEDCs	Estrogenic endocrine-disrupting chemicals
ELISA	Enzyme-linked immune sorbent assay
ERα and ERβ	Estrogen receptors
FSH	Follicle-stimulating hormone
GSI	Gonadosomatic index
kDa	kilo Dalton
Lv	Lipovitellin
NP	Nonylphenols
OP	Octyl phenols
PBB	Polybrominated biphenyls
PCB	Polychlorinated biphenyls
PCR	Polymerase chain reaction
Pv	Phosvitin
Q-PCR	Quantitative polymerase chain reaction
VSH	Vitellogenesis-stimulating hormone
Vtg	Vitellogenin
Yps	Yolk proteins
β′-c	β′-component
B-BHC	Benzene hexachloride
δ-BHC	Delta-hexachlorocyclohexane

10.1 Introduction

Vitellogenin (Vtg) is a yolk precursor glycolipophosphoprotein produced by both oviparous and ovoviviparous species. Vtg is naturally present in females, but trace amounts are present in males as well. It is one of the essential nutrient reserves for the early larval stages development of animals including embryonic development. Vtg is a high-molecular-weight protein (300–640 kDa), synthesized in liver tissues due to the exogenous activation of estrogen, especially in mature females (Sun and Zhang 2015). It requires oocyte development. The secreted Vtg reaches the oocytes through the serum and enters the oocytes by receptor-mediated endocytosis. Generally, males do not synthesize Vtg; upon exposure to estrogen or compounds that mimic estrogen, they can be induced to synthesize Vtg. In males, it has no functions, and in quite a short period, it can be degraded by plasma proteases. The elevated concentration of Vtg presence in plasma increases the protein content (Ferreira et al. 2009), and Vtg affects blood dynamics, function, and survival in juvenile fish due to kidney failure (Zhang et al. 2008). Similarly, finding Vtg in males indicates exogenic exposure to estrogenic compounds. Besides these, elevated Vtg is observed in disease conditions; hence, Vtg is considered a marker for the detection and determination of pollutants in the aquatic environment (Kadim and Risjani 2022), the maturity of fish for reproduction, and the diagnosis of certain diseases. This chapter discusses the biomarker properties of Vtg, the availability of Vtg in marine and freshwater fishes, a marker for the detection of environmental endocrine disruptors, sexing and maturity of fishes, and the health status of fishes.

10.2 Vitellogenesis in Fish

The process of vitellogenesis in fish which commence with the action of the follicle-stimulating hormone (FSH) released by the pituitary glands into the bloodstream stimulates the synthesis of the sex steroid hormone estrogen (estradiol-17), which triggers vitellogenesis in the liver. The Vtg gene in the nucleus is affected by estrogen via estrogen receptors after it binds to sex steroid hormone-binding globulin in the blood and is transported to hepatocytes (Lubzens et al. 2010). Hepatocytes' combination of estrogen and the estrogen receptor attach to the Vtg gene's promoter region, activating the gene to start and speed up transcription. Following the start of translation of the Vtg transcript products, the Vtg proteins undergo lipidation, phosphorylation, and glycosylation steps before being released into circulation. Blood Vtg enters the cell after binding to the Vtg receptor on the oocyte plasma membrane (Hiramatsu et al. 2005).

Circulated Vtg enters the oocyte by a receptor (VgR)-mediated endocytosis. Usually, VgR is bound with clathrin-coated pits and invaginated into the vesicle and enters into the peripheral ooplasm to form multivesicular bodies. The oocyte lysosome containing cathepsin D or B, an oocyte serine protease enzyme cleaves Vtg into yolk proteins such as phosvitin (Pv) and lipovitellin (Lv), is stored in the egg approximately 80–90% (Picchietti et al. 2004). The ovulated eggs containing Pv and Lv are utilized by the growing embryo as a source of nutrition. In teleosts, Vtg is the precursor for yolk proteins such as lipovitellin (Lv), phosvitin (Pv), β′-component (β′-c), and C-terminal (Ct) peptide (Sun and Zhang 2015). Lipovitellin (Lv) is the largest Vtg-derived lipoprotein and consists of two peptide chains named light chain (LvL) and heavy chain (LvH), as a source of amino acid and lipids for embryo development. Phosvitin (Pv), a serine-rich protein bound with phosphate and calcium, is essential for skeletal development. Other two small cysteine-rich (14 residues) proteins (β′-c and Ct) devoid of lipid and phosphate are involved in disulfide linkage essential for complex folding of Vtg peptide dimerization. The peptide β′-c is identified to release from the C-terminal end of Vtg in many fish species (Reading et al. 2018).

Naturally, male and female fishes, as well as immature juveniles, have hepatic estrogen receptors. The liver cells of female fish are typically exposed to estrogens and induced for vitellogenesis. Unusually, the production of vitellogenin by males, juveniles, or non-vitellogenic females can serve as a bioindicator of exposure to environmental estrogens (Kime et al. 1999; Caldwell et al. 2008). Vtg genes act in a dosage-dependent manner, which correlates the gene copy number with the speed of yoking. For example, fishes with multiple Vtg gene copies will produce a larger amount of eggs in a shorter time than other oviparous (Buisine et al. 2002; Hara et al. 2016).

Generally, the brain and thoracic ganglia release a putative hormone called vitellogenesis-stimulating hormone (VSH) that promotes vitellogenesis and ovarian development (Huberman 2000). Pituitary, granulosa, and theca cells of maturing and developing oocytes secrete gonadotropins and steroid hormones, respectively (Yaron and Levavi-Sivan 2011; Lee et al. 2021). The synthesis of steroids in specific ovarian cells is connected to various stages of the oocyte's development. In many species, the concentration of estradiol during vitellogenesis, which is related to the development of vitellogenic oocytes, increases the level of estrogen in plasma where the production of vitellogenin and chorionic by the liver (Anderson et al. 2017) is controlled by estradiol (Fig. 10.1).

The production of Vtg is a gonadotropin-dependent process (Thomas and Rahman 2009). An estradiol binding to Vtg receptors on hepatocytes leads to Vtg transcription. Vtg is posttranslational phosphorylated, glycosylated protein, and lipid groups are added before it is released into the bloodstream. As it reaches the developing oocytes, it is secreted in the blood as a homomeric complex that binds to a specific plasma membrane receptor and then is taken up by the oocytes through clathrin-mediated endocytosis (Carducci et al. 2019); further processes of proteolytic cleavage occur inside the oocytes to produce their derivative YPs. Some species' ovulated eggs contains up to 80–90% of the dry mass as yolk generated from Vtg (Reading et al. 2017).

10.3 Variations in Vtg

Until the mid-1990s, single Vtg had been known from the teleosts species. After the synthesis and characterization of the cDNA of Vtg from *Fundulus heteroclitus* by LaFleur et al. (1995), two different Vtg transcripts and their translated products there in teleost fishes, which gives new pattern to understand the vitellogenesis in fishes. There are multiple Vtg proteins, and their corresponding genes have been identified from different fishes, which give some confusion in understanding the physiology of vitellogenesis in fishes. Therefore, a clear interpretation is required for the identification of either duel or multiple types of Vtg in a single fish species. The complete Vtg protein structure consists of (NH2-LvH-Pv-LvL-β′-c-C-terminal

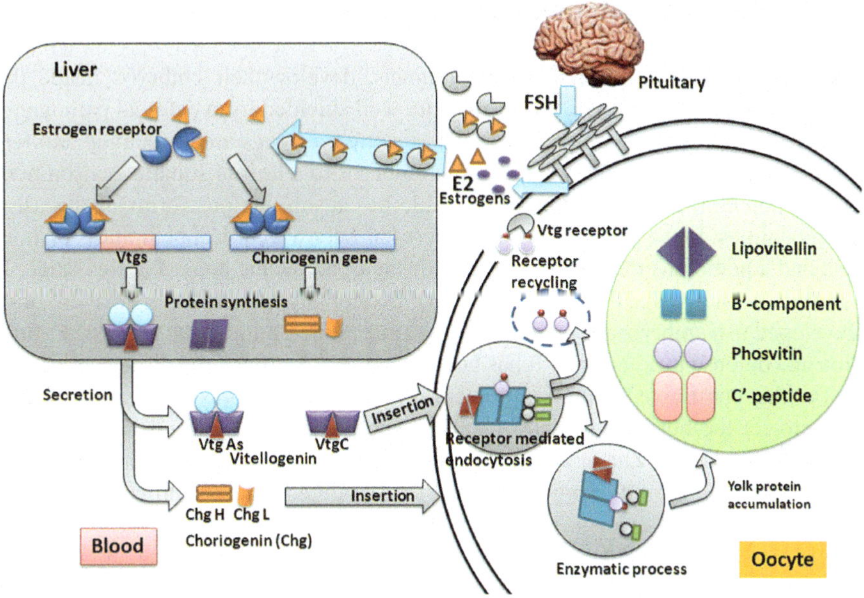

Fig. 10.1 Activation of Vtg genes by estrogenic compounds. The fish vitellogenin process starts with the secretion of follicle-stimulating hormone (FSH) from the pituitary gland, which mediated follicle cells developed in their surrounding that induces the sex hormone (estrogen, E2), which induces vitellogenesis in the liver. Through the blood streams it reaches the oocyte by receptor-mediated endocytosis and is degraded specifically by a cathepsin D-like enzyme and undergoes molecular cleavage to produce Lipovitellin, Phosvitin, and β′-C peptides in the cell. (Adapted and modified from Hiramatsu et al. 2005)

peptide-COOH) two types, namely, VtgA and VtgB. Type A and B are complete Vtg with variations of the Lv and Pv (Hara et al. 2016). A third one is completely lacking the Pv domain (NH2-LvH-LvL-COOH), or having a greatly shortened Pv domain, called VtgC (incomplete). C type has a lower content of phosphorus and serine residues, and considerably lowers molecular mass than either VgA or VgB. Some fish species contain all three forms of Vtg (A–C) (Sawaguchi et al. 2005; Koua et al. 2022), for example, mosquitofish (*Gambusia affinis*), white perch (*Morone americana*), red seabream (*Pagrus major*), white-edged rockfish (*Sebastes taczanowskii*), mummichog, and striped mullet (*Mugil cephalus*). Both types of complete (VtgA and VtgB) and incomplete Vtgs (VtgC) are widely reported in various teleosts, which gives an idea to pattern the fish species according to their genetic makeup as well as the induction process (Matsubara et al. 2003; Ye et al. 2022). There are seven distinct functional Vtg genes reported in zebrafish (Wang et al. 2020); besides, rainbow trout genome consists of 20 complete Vtg genes and ten pseudogenes with variations of less than 3% nucleotides, which express indistinguishable protein products (Trichet et al. 2000).

10.4 Role of Vitellogenin in Immune Factors

Most mammalian species, including humans, develop their embryos inside the mother's body, in the uterus, where they are well shielded from external pathogenic assaults. Contrarily, most fish and aquatic invertebrate eggs are discharged and fertilized externally, exposing the emerging embryos to a hostile aquatic environment full of potential pathogens that are capable of causing various forms of diseases, even fatalities. In the past two decades, there has been a significant emphasis on how fish and aquatic invertebrate embryos withstand pathogenic attacks at this stage. It is common knowledge that fish and aquatic invertebrates generate eggs that are fully developed fish embryos in an aquatic environment, including all the necessary nutrition and defense mechanisms. It has been shown that Vtgs and the proteins give rise to Yps that contribute to embryo protection. Vtg is an immunocompetent factor that can protect the host from microorganisms such as bacteria and viruses (Sun and Zhang 2015; Li and Zhang 2017). The Vtg purified from the protochordate amphioxus' (*Branchiostoma japonicum*) ovaries exhibited hemagglutinating activity against chick, toad, and grass carp erythrocytes; as well as antibacterial activity against the Gram-negative bacterium *Escherichia coli* concludes that Vtg plays an immune-relevant role. Vtg has been demonstrated to work as an acute phase protein in zebrafish, enabling the eradication of invasive microorganisms like *E. coli* and *Staphylococcus aureus* (Tong et al. 2010). The intraperitoneal injection of *E. coli* increases the serum level of Vtg in male rosy barbs *Puntius conchonius* (Shi et al. 2006), and the zebrafish skin was challenged with the Gram-negative bacterium *Citrobacter freundii*, which upregulated the expression of Vtg (Lu et al. 2013). Interestingly, Vtgs from protostomes appear to have antibacterial activities as Vtg from the scallop (*Patinopecten yessoensis*) has antibacterial properties against both Gram-positive and Gram-negative bacteria (Wu et al. 2015). Even in the nematode *C. elegans'* Vtg gene also plays a role in the nematode's ability to fight germs. After infection by the pathogen, a decreased survival was seen in the Vtg-knockdown *C. elegans* (Fischer et al. 2013). The improvement in nematode resistance to the pathogen *Photorhabdus luminescens* when Vtg synthesis was induced by estrogen 17-estradiol and phytoestrogen daidzein correlates with the association between invertebrate Vtg and its resistance to bacteria (Fischer et al. 2012). Vtg plays multifaceted immune-related roles in the marine fish *Hexagrammos otakii* where Vtg binds to both Gram-positive and Gram-negative bacteria, including *S. aureus* and *Pichia pastoris* (Li et al. 2008). Carp and zebrafish Vtgs also bind to *E. coli* and *S. aureus* (Tong et al. 2010). The ability to aggregate pathogens and detect invasive microbes is provided by the binding of Vtgs to bacteria. Notably, Vtgs remain uncleaved in the oocytes of amphioxus and nematode (Sharrock 1983; Sun and Zhang 2001). Thus, Vtgs in these animals may protect their oocytes and embryos against pathogenic attack. The Vtg of Atlantic salmon could neutralize the infectious pancreatic necrosis virus, indicating that Vtg is also engaged in the host's antiviral immunity. This implies that Vtg actively defend the host against infection. Additionally, Vtg play a role in the host's antimicrobial defense against broad-spectrum bacteria and viruses. Vtg functions as a multivalent pattern recognition receptor capable of binding to lipopolysaccharide, lipoteichoic acid, peptidoglycan, glucan, and a bactericidal molecule capable of damaging bacterial cell walls, and an opsonin capable of enhancing the phagocytosis of bacteria by macrophages (Garcia et al. 2010; Li et al. 2008; Sun and Zhang 2015; Zhang et al. 2011, 2015).

10.5 The Effect of Endocrine Disruptors on Fish Reproduction

Hazardous chemicals have many negative effects on aquatic animals' sex inversion, decrease the fertility rate, and troubles in embryonic development and also cause cancer and kidney disorder. When such chemicals bind with estrogen receptors, they can affect the secretion, transport, and physiology of hormones in the animal endocrine system, which are usually called endocrine disruptors (EDs). Agrochemicals such as atrazine, dieldrin, and toxaphene; natural hormones such as 17β-estradiol (17βE2), estrone (E1), and estriol (E3); surfactants such as alkylphenol ethoxylates (APEOs), phytoestrogens, diethylstilbestrol (DES), dioxin (2,3,7,8-TCDD), bisphenol A (BPA), nonylphenols (NP), phthalates, styrene, polybrominated biphenyls (PBB), octyl phenols (OP), and polychlorinated biphenyls (PCB); and also heavy metals such as cadmium, lithium, barium, chromium, arsenic, and antimony are some examples of estrogenic compounds (Rasier et al. 2006; Diamanti-Kandarakis et al. 2010; Yang et al. 2015; Pamplona-Silva et al. 2018) that alter the endocrine system. These EDs are classified into three types based on their functions such as (a) mimic or block natural testosterone, (b) androgenic compounds, and (c) thyroidal compounds. Compounds mimic natural estrogenic hormone-like 17βE2 have endocrine disruptor function and affect the fertility and reproductive properties of fishes (Schug et al. 2011). In general, the EDs interfering the metabolism of fish by two different mechanisms, such as alteration in target gene expression and also modify the membrane receptor functions (Schug et al. 2011). Some agrochemicals inhibit cellular signaling pathways and enzyme hormones resulting in variations in the level of hormones in the plasma (Kiyama and Wada-Kiyama 2015). The compounds such as α-endosulfan, aldrin, cyhalothrin, dicofol, cypermethrin, deltamethrin, fenvalerate, glyphosate, DDT, β-BHC, δ-BHC, permethrin, diazinon, prothiofos pyriproxyfen, methoxychlor, CNP-amino, prothiofos, tolclofos-methyl, thiabendazole, and cyanofenphos exhibit elevated estrogenic activity (Barber et al. 2015; Kiyama and Wada-Kiyama 2015; Pamplona-Silva et al. 2018).

In marine and freshwater habitats, these ED substances' contact with females, effects can even worse affect embryogenesis (Snyder et al. 1999). For example, contact with juvenile females induces early vitellogenesis and increases plasma Vtgs. Naturally, both female and male fishes have Vtg receptors; however, mature females are normally exposed to estrogens (17αEE2; E1 and E3); Vtg gene can be induced to synthesizing Vtg followed by the development of oocytes. But in male fishes, it is related to a decreased level of testosterone. Exogenous induction of environmental estrogens in male fishes affects renal and gonadal pathology and develops follicles in testicles (Andersen et al. 2003) resulting in reproductive defects (reduced sperm count and nonmotile sperms) and increasing intersex animals (Niemuth and Klaper 2015). These intersex dioic fishes behaved in both male and female properties with less potential to generate active sperms and low fertilization (Osman et al. 2015). Due to the EDCs, the intersex fishes showed low body weight and low gonadosomatic index (GSI). For example, fishes exposed to NP and OP increase gonadal alteration specifically fibrosis, necrosis, hypertrophy of connective tissue, sex interchange, and delayed gamete development (Martin-Skilton et al. 2006). Various fish species such as *Cyprinus carpio, Micropterus* spp., and *Ictalurus punctatus* from the Colorado River (EUA) showed various levels of reproductive disorders due to a high level of plasma Vtg in males, especially increasing intersex species, i.e., males with testicular oocytes and females with sperm in the ovaries (Hinck et al. 2007).

10.6 Mechanisms of Endocrine Disruption

Endocrine disruptors could use several possible mechanisms to disrupt the endocrine system as well as direct interaction with hormone receptors to elevate or quench its functions. When fishes are exposed to stressful environments or toxic chemicals, they immediately responded to changes in the concentration of hormones and their related functions, which leads to cause adverse effects including cellular damage, metabolic defects, and organ failure (Shanle and Xu 2011). Certain chemicals generally mimic estrogen activities, for example, alkylphenol ethoxylates, bisphenol A, phthalates, and phytoestrogens (e.g., flavonoids, genistein, sitosterol), OP-DDT, and PCBs which have either increasing or decreasing natural estrogen activities (Safe and Gaido 1998; Snyder et al. 1999). The possible mechanism of EDCs is given in Fig. 10.2.

Endocrine-disrupting chemicals (EDCs) interact with estrogen receptors (ERα and ERβ); they are DNA-binding proteins which activate the genomic and nongenomic estrogen receptor activity through either direct or indirect mechanisms. In indirect mechanism, EDCs interact with aryl hydrocarbon receptor (AhR) and modulate metabolic enzyme function-mediated activation of estrogen receptors. Bisphenolics and organochlorine pesticides and phytoestrogens target ER through

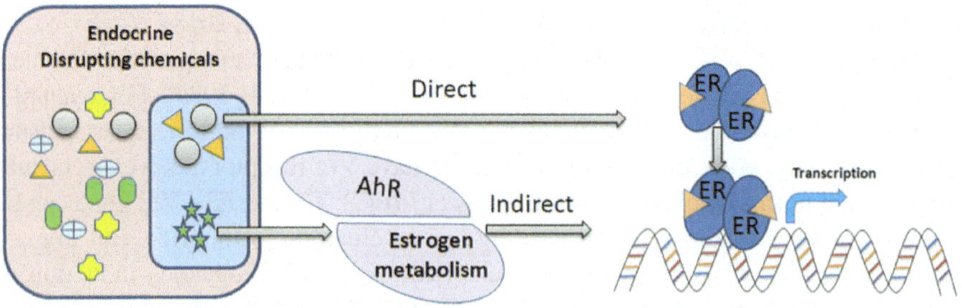

Fig. 10.2 Mechanism of endocrine disruption

multiple modes. Some can exhibit tissue-specific activity. Therefore, finding the effects of EDCs in aquatic animals may explore more stringent way through complex tissue-selective alleviation of ERs and other cellular signals (Shanle and Xu 2011).

10.7 Biomarkers for EDC Assessments

Our environment is highly polluted with various hazardous chemicals which have several toxicological effects on biology including endocrine disruptions named endocrine disruptor chemicals (EDCs). There is a growing concern about EDCs and their effect on wildlife and humans (Kumar et al. 2020), specifically in male and female reproductive disorders (Marlatt et al. 2022), cancer, obesity, and other disorders. An individual is continuously exposed to EDCs, which could affect the physiology of the system and elevate certain bio-chemicals end to disorder (e.g., growth, development and reproduction). The environmental stimulation by EDCs affects system biomarkers to elevate their levels including secondary sexual characteristics, gonadosomatic index, plasma steroids, Vtg, and gonad histology (Ankley and Johnson 2004; Marlatt et al. 2022).

10.8 Biomarker for Environmental Estrogens

Due to anthropogenic activities, our environment is highly polluted with several toxic chemicals which include endocrine-disrupting chemicals (EDCs) that have a considerable effect on androgen and thyroid receptors (US-EPA 1998). Several of such EDCs resemble the functional properties of 17β-estradiol (E2), for example, estrogenic chemicals such as organochlorine pesticides, nonylphenol (NP), bisphenol A, octylphenol (OP), and the derivatives of polychlorinated biphenyls (PCBs) (Jones et al. 2000; Kumar et al. 2020). They are structurally different chemicals and mimic the functions of endogenous steroid estrogens, bind on the estrogen receptor (ER), but less powerful than E2 (Anderson et al. 2017). Usually, they are surface active in water and affect aquatic biota, most specifically fishes, increasing the concentration of egg yolk precursor protein (Vtg) in the blood. EDCs stimulate the elevated secretion of Vtg in the blood; increased concentration affects the body functions. Continuous exposure to such EDCs causes kidney failure. Generally, Vtg is under E2 control and is also involved in body functions such as reproductive development and change of behavior. Hence, Vtg is considered a biomarker for the detection of environmental EDCs. Several studies reported the increasing concentration of Vtg in the blood when exposed to different estrogenic chemicals in their habitat (Tran et al. 2019).

10.9 Biomarker Properties

Biomarkers are defined as biological measures of an elevated response of indicators under normal biological processes and or pathogenic processes to a therapeutic intervention. Biomarkers can be used for the assessment of the physiological status of the organisms to illustrate the health status of the individuals for the therapeutic plan. An ideal biomarker consists of several properties such as consistent across genders, easy to measure, safe, cost-effective, and modifiable with treatment. Various cellular biochemicals, blood components, and system parameters are considered biomarkers for biological assay. Biological macromolecules such as DNA, RNA, protein, carbohydrates, enzymes, and hormones are generally used as biomarkers for the diseased status of individuals as well as to identify pathogens (van der Oost et al. 2003). The egg yolk precursor protein vitellogenin (Vtg) is also considered a biomarker for estrogenic activities. The integrated approach such as plasma Vtg and gonad histology is used to understand the toxicological effect of fish reproduction.

10.10 Vitellogenin (Vtg) Is a Biomarker for Estrogenic Activity Determination

The aquatic environment is tremendously affected due to various human-made activities, using chemical pesticides, fertilizers, and hazardous chemicals for agricultural purposes. Usually, pesticide compounds are low to strong endocrine-disrupting functions; hence they are classified as endocrine-disrupting chemicals (EDCs), and they also contain estrogenic activities. Mostly, estrogenic activities induce the secretions of Vtg in the blood, especially in male and juvenile fishes. The Vtg production is mainly controlled with E2 in the liver of female fishes at the time of maturation for the development of ovaries and oocytes (Kumar et al. 2020; Lee et al. 2021). However, a traceable amount of Vtg is also found in the blood of male fishes due to exogenous estrogenic chemicals. Several reports revealed the presence of Vtg in male fish when exposed to estrogenic endocrine-disrupting chemicals (EEDCs). It is common in the majority of Asian countries, the United States, Europe, and Africa (Ortiz-Zarragoitia et al. 2014; Marlatt et al. 2022).

To detect the presence of EEDCs in the aquatic environment, the plasma Vtg level in fishes could be analyzed through various immunoassays. Various subtypes of Vtg (VtgC (incomplete), VtgAa and VtgAb) have been detected through this method (Amano et al. 2008; Williams et al. 2014). The Electrophoresis method is generally used to detect both complete and incomplete types of Vtg. However, real-time PCR methods have been used to measure various subtypes used based on gene-specific primers. The Vtg gene expression profile and the Vtg protein expression can be compared and assessed for the quantitative determination of EEDC induction. Blood plasma Vtg protein and Vtg mRNA profile in liver tissues represent estrogenic activity (Hiramatsu et al. 2006). For example, transcriptional fingerprint analysis of experimental Rainbow trout liver tissues was compared with blood plasma

E2 level (Benninghoff and Williams 2008) due to EEDC (Jung et al. 2012) in the environment. RT-PCR assay of Vtg mRNA in Mediterranean tuna fish (*Thunnus thynnus*) exposed to estrogenic compounds revealed two different Vtgs (Barucca et al. 2006).

10.11 Detection of Estrogenic Endocrine-Disrupting Chemicals

Agricultural and aquaculture farmers use various chemicals for increasing the productivity of food substances. However, the residues of toxic chemicals such as organochlorine pesticides and the derivatives of PCBs (Jones et al. 2000) have mild to high estrogenic activity (EEDC) which disrupts the endocrine system of the fishes. In field studies, biomarker induction (Vtgs) in male fish is the best method for tracking a single compound from the mixture. In addition, a biomarker can provide reliable information about the real laboratory data, compared to field data with abnormalities. In general, continuous exposure to estrogenic chemicals in fishes has variably induced secretions of Vtg in blood plasma. If we need to know the level of such EEDC in the water system, detect the quantity and level of Vtg in fish blood. The level of fish Vtg index based on the environmental EEDC activity in the field was first reported in 1995 (Sumpter and Jobling 1995). Several experiments were conducted to investigate the effects of EEDC on fish. For example, caged male rainbow trout in a drainage channel increases the level of blood Vtg compared with the control environment due to the presence of nonylphenol in the drainage water, demonstrating the effect of environmental EEDCs (Hara et al. 2016). Usually, the fishes exposed to EEDCs, could directly affect the physiology of Vtg-mediated abnormal reproduction, especially in fin fishes (Jung et al. 2012) under laboratory studies. Fujita et al. (2004) observed the serum level variation of VtgAs (complete form) and VtgC (incomplete form) in male masu salmon exposed to estrogen. During maturation female fish showed a higher concentration of Vtgs in their blood that was similar to the estrogenic induction of Choriogenins (Chgs). The level of complete and incomplete Vtgs is at 1:2 ratios. However, in male masu salmon, trace amounts of incomplete Vtg (VtgC) and Chg have been detected throughout the year (Fujita et al. 2008) which may be due to the presence of environmental estrogens (EEDCs). This type of assessment helps to detect the effects of EEDCs also in marine environments. For example, red lip mullet and grey mullet showed high estrogen activity in urban coastal waters and also showed abnormal gonads, including ovotestes (Aoki et al. 2010; Kumar et al. 2020). The principle ODID method is explained in Fig. 10.3.

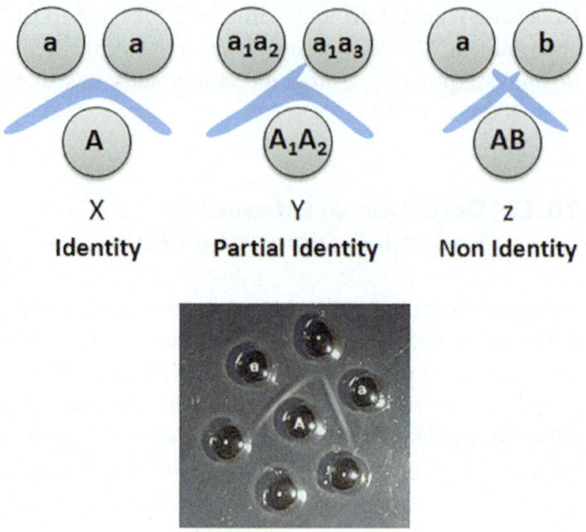

Fig. 10.3 Principle ODID method

10.12 Methods of Vitellogenin Detection

10.12.1 Indirect Method

Instead of measuring Vtg, other related components acid-labile phosphorous (ALP) in the blood by HPLC could determine the level of Vtg. Vtg is highly phosphorylated in the blood. The level of plasma APL is determined by the amount of phosphate attached to Vtg using simple acid digestion followed by HPLC (Kramer et al. 1998; Mugiya and Tanahashi 1998). Acid-labile phosphorus in blood serum can also be measured to estimate Vtg indirectly. The Electrophoresis method is also used, first, to separate plasma proteins by electrophoresis based on the molecular weight of Vtg protein and quantified by densitometry (phosphorus dye staining). This method is comparatively easy than complex immunochemical methods. But it is less sensitive than immunoblotting. These methods are therefore more applicable when the degree of Vtg induction is high and where differences between exposure groups are relatively great. The specificity of the electrophoretic procedures can be improved by transferring the separated proteins to a blotting membrane and performing Western blotting procedures with an antibody specific for Vtg (Table 10.1).

10.12.2 Immunological Methods

The Vtg protein(s) that are present in the bloodstream of fish or cultured hepatocytes have been identified and quantified using a variety of techniques. Due to its large molecular weight, it can be easily detected by observing high-molecular-weight peaks that occur during gel filtration chromatography (Hara et al. 2016). However, quantitative and qualitative analyses based on immunological techniques using

Table 10.1 Methods to analyze Vtg and estrogen signals in fishes

Fish studied	Induction chemical	Methods used for detection of Vtg	References
Rainbow trout	E2, 4-NP(4-*tert* nonylphenol)	ELISA	Tyler et al. (2002)
Oryzias latipes and *Oncorhynchus mykiss*	E2	ELISA	Iguchi et al. (2006)
Japanese medaka	EE2, 17beta-trenbolone	HPG-PCR array	Zhang et al. (2008)
Zebrafish	Letrozole (LET), tamoxifen	qPCR for Vtg1, Vtg2, and eralpha	Sun et al. (2010)
Oryzias latipes	EE2, letrozole, tamoxifen	Real-time PCR (qPCR)	Sun et al. (2011a, b)
Rainbow trout	4-Nonylphenol	ELISA	Naderi et al. (2017)
Zacco platypus	Estradiol-17β	ELISA	Lim et al. (2013)
Clarias batrachus	17β-estradiol (E_2)	ELISA, SDS-PAGE, Western blot, immune histochemistry	Garnayak et al. (2013)
Rainbow trout hepatocytes	17β-estradiol, estrone, diethylstilbestrol, hexestrol, genistein	Quantitative real-time PCR (qPCR); ELISA	Markell et al. (2014)
Pimephales promelas	17β-estradiol (β-E2) and 17α-ethinylestradiol	QPCR, LC-MS/MS	Ankley et al. (2017)
Oryzias melastigma	Estradiol (E_2)	ELISA, SDS-PAGE, LC-MS/MS	Yi et al. (2018)
Gambusia yucatana	17β-estradiol	Q-PCR	Rendon Von Osten et al. (2019)
Acanthopagrus latus	Bisphenol-A, 17β-estradiol (E2)	RT-PCR	Negintaji et al. (2019)
Sea bass	EE2, Benzo[a]pyrene and $CdCl_2$	ELISA	Prasatkaew et al. (2019)
Primary hepatocytes	17β-E2, DES, and HES	ELISA	Li et al. (2021)
Oncorhynchus mykiss	17β-E2	ELISA, qRT-PCR	Chen et al. (2021)

antibodies are typically superior and highly sensitive. Antisera raised against Vtg of female fish and egg yolk are considered as Ab for detection of Vtg in animals exposed to estrogens.

10.12.3 Ouchterlony Double Immunodiffusion

This is the simplest detection technique where blood serum and antibodies are placed in several wells on an agar gel plate and allowed to grow for several hours up

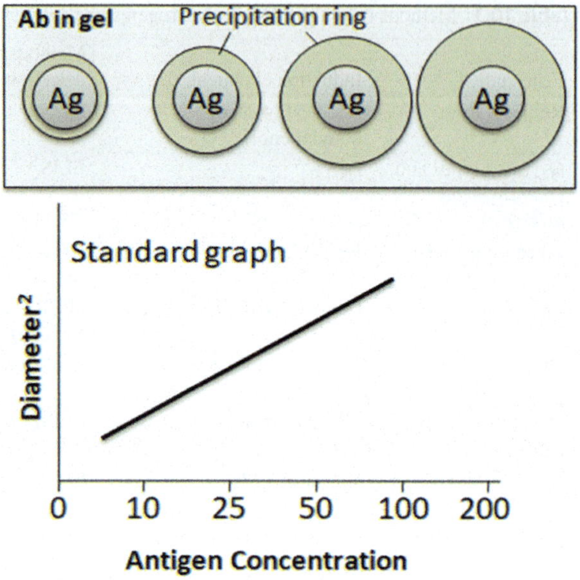

Fig. 10.4 Mancini single radial immunodiffusion assay

to overnight. After that, an antibody-antigen precipitation line (precipitation reaction) develops. This approach is semiquantitative and has a detection sensitivity of roughly 10 g/ml; compared to other approaches, it takes relatively large numbers of antibodies and has a low sensitivity for Vtg detection. Nonetheless, there are benefits to its ease of use in terms of both manipulation and apparatus required (Hara 1987).

10.12.4 Mancini Method (Single Radial Immunodiffusion)

Similar to the Ouchterlony double immunodiffusion, this quantitative approach is sensitive Ag-Ab reaction. Antibody-containing agar gel plates are made beforehand, and the concentrations of Vtg are determined by quantifying the size of the precipitation ring that forms around the sample well. Under normal circumstances, this method of detection and quantification is sufficient for *Salmonidae* species since Vtg levels in the blood at the height of vitellogenesis are fairly high (about 40 mg/ml) (Hara 1987; Hiramatsu et al. 1997) (Fig. 10.4).

10.12.5 Sensitive High-Throughput Immunoassays

Other than Mancini and Ouchterlony immune precipitation assay, the methods of radioimmunoassay (RIA), enzyme immunoassay (EIA), fluorescent immunoassay (FIA), and chemiluminescent immunoassay (CLIA) are all quantitative assays with high sensitivity (>10,000-fold better sensitivity than agar plate precipitation

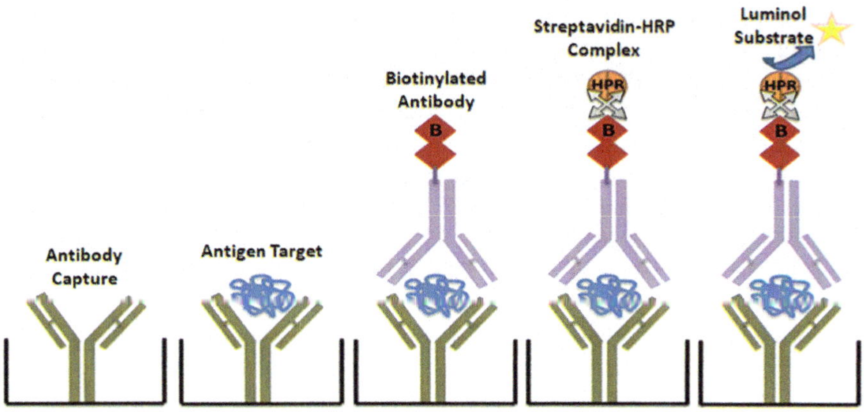

Fig. 10.5 Sandwich-based CLIA for detection of Vtg

methods). These techniques employ several tagging strategies using tracers such as radioactive material, enzymes, fluorescent material, and chemiluminescent material and also share a common detection procedure (pg/ml) using 96-well microtiter plates pre-coated with Vtg-specific antibody. Sandwich-based CLIA method was effective for measuring Vtg in *Salmonidae* without using radiolabeled material (Hiramatsu et al. 2005; Hara et al. 2016). Afterward, standards and samples were put on the plate along with known amounts of Vtgs. The Vtgs were then sandwiched between antibodies (antiLv antibodies) that were dyed with chemiluminescent acridinium-tagged antibodies (Fukada et al. 2001; Hiramatsu et al. 2005; Haruna et al. 2018) (Fig. 10.5).

10.12.6 Nucleic Acid-Based Methods

Northern blotting: The concentration of Vtg protein in the blood of the fishes is directly related to the expression of the Vtg gene. Tissue-specific expression of Vtg can be detected through quantification of mRNA by Northern blot (Le Guellec et al. 1988). Northern blotting is the preliminary quantification method with great specificity of mRNAs due to the use of radiolabeled oligonucleotide probes (Mellanen et al. 1999; Miller et al. 1999). For example, Vtg level has been detected through this method in tilapia (*Oreochromis* sp.) (Lazier et al. 1996), rockfish (*Sebastes schlegelii*) (Jung et al. 2006), and *Anguilla japonica* (Wang and Lou 2006). The differences in mRNA splicing are also detected through this method. One of the disadvantages of this method is the toxicants that alter the mRNA concentration, making it hard to establish the concentration-dependent response of estrogen induction.

Reverse transcriptase-polymerase chain reaction (RT-PCR): This method has been used to measure Vtg gene expression in different fish species either by natural or artificial induction. This method is highly sensitive than Northern blotting, due to the real-time quantification of Vtg-specific primers. Several RT-PCR methods have been designed based on gene-specific primers. One limitation is the need for "housekeeping" genes as internal control, and it can be quantitatively determined through semiquantitative RT-PCR. For example, Vtg mRNA expression level was estimated in fishes such as *Oncorhynchus mykiss* (Arukwe et al. 2000), *Mugil soiuy* (An et al. 2006), and *Scophthalmus maximus* (Dang and Sun 2011). Kim et al. (2020) used the RT-PCR method for the effect of water temperatures on walleye Pollock for sox9a expression in males and cyp19a and vitellogenin (Vtg) expression in females. Forner-Piquer et al. (2020) studied the effect of EDCs, bisphenol A (BPA) on zebrafish gonad development, and Vtg. Q-PCR has been used to determine changes in Vtg gene expression in black molly fish (*Poecilia sphenops*) exposed to pyrogenic hydrocarbon and petroleum from Campeche Sound (Maurilio and Rendon von 2020). Sa-an et al. (2022) detected Vtg mRNA in the liver in estuarine eyebrow goby, *Oxyurichthys ophthalmonema* by qPCR assay (Fig. 10.6).

All the above methods have been widely used to detect the natural induction of estrogens and the level of Vtg in plasma, during the adverse condition like environmental estrogenic chemicals, xenotoxins, etc., under laboratory as well as field exposure studies that provide valuable information regarding the use of Vtg as a potential biomarker for detection of sexual maturity of animals as well as the environmental toxicants. The following section highlights the importance of the Vtg biomarker to determine the dosage level response of various agrochemicals and other industrial chemicals which possess endocrine disruptor functions.

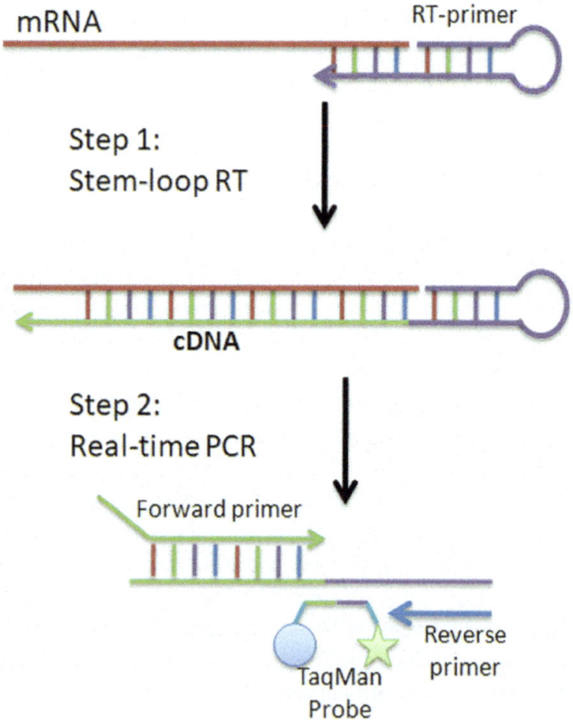

Fig. 10.6 TaqMan/Stem-loop primer for RT-PCR

10.13 Environmental Impact Analysis by Vtg Biomarker

Fishes under laboratory exposure is different from field exposures. Under laboratory conditions, animals are exposed to either single or multiple pure compounds; hence, the induction of Vtg is quantifiable, but animals in their natural habitat could exposed to an unknown concentration of multiple compounds along with various environmental factors that can affect the host and differentially express the Vtg and other related genes responsible for gonad development and reproduction. In animals under controlled environmental studies, there is a possibility that a single chemical with inducible concentration alters the genetic regulation of vitellogenesis without interfering with any differential environmental factors. Through laboratory studies a variety of chemicals such as DDT residues, nonylphenol, octyl phenol and PCB compounds, and their estrogenic effects were analyzed with various fishes (Table 10.2), which can provide a clear understanding of the toxicological effects and Vtg induction in experimental fishes. It is more environmentally relevant (Giesy et al. 2000; Hara et al. 2016). However, the ecological relevance of Vtg induction studies is limited. The majority of the studies were executed at the laboratory level to find the estrogenic effects of environmental chemicals, and how they related to the natural estrogen of the specific fish species (Ahmadpanah et al. 2019).

In recent years, there is a growing interest in environmental protection, how to protect our environment from toxic chemicals, and their impact on the health of living organisms. Most agriculture chemicals and the waste of refineries and pharmaceutical industries possess estrogenic compounds. They have a direct impact on our environment; specifically, it has an adverse effect on wildlife in aquatic habitat and also on humans. Several research outcomes elaborate the impacts of environmental estrogens, includs natural hormones such as estrone (E1), estradiol (E2), estriol (E3), estrol (E4), synthetic hormones ethinyloestradiol (EE2), and compounds showing estrogen-like activities (EDCs) on differentiation of vitellogenesis in animals. Naturally, estrogens are female hormones; during sexual maturity their concentration increases in the ovary for the development of ovarian follicles and synthesis of egg yolk protein, and also for the proper functioning of male organisms (Wojnarowski et al. 2021).

The fishes and mammals' endocrine systems have some degree of similarity with significant impact by the EECs. The EDCs affect living organisms not only on sexual behaviors and reproduction; it affects their immune status and also body functions (Szwejser et al. 2016; Chaves-Pozo et al. 2018). The risk for fish is due to the accumulation of such EDCs in the environmental sediments, and its periodic release in the aquatic environment affects the natural reproductive physiology of both female and male fishes (Thrupp et al. 2018). Some EDCs possess anti-androgen properties. Androgens are important hormones secreted by the hypothalamic-pituitary-gonadal (HPG) axis in vertebrates essential for the sexual maturity of adult males. In fishes, two androgen hormones such as testosterone and 11-ketotestosterone play an essential function to control sexual differentiation and reproductive behavior in adults (Wojnarowski et al. 2021). For example, flutamide, vinclozolin (VZ), *p,p′*-DDE, 4-*tert*-octylphenol, and bisphenol exhibits anti-androgen properties that cause feminization of male fish which results in the induction of plasma Vtgs,

Table 10.2 List of studies used to analyze endocrine disrupting effects of various chemicals

Fish	EDC	Duration	Method	Observation	References
Oncorhynchus mykiss	EE2, NP	3 weeks	*In vivo*, sub-adult/adult, waterborne	Plasma Vtg, testicular growth	Jobling et al. (1996)
Salmo salar	NP	30 days	*In vivo*, early life, i.p. injection	HSI, plasma Vtg	Madsen et al. (1997)
Xiphophorus helleri	NP, BPA	3–60 days	*In vivo*, sub-adult/adult, early life, waterborne	Growth, hepatic *vtg*, reproductive damage	Kwak et al. (2001)
Pimephales promelas	BPA	164 days	*In vivo*, transgenerational, waterborne	Plasma Vtg; F1 egg production and hatchability	Sohoni et al. (2001)
Oncorhynchus mykiss	EE2, NP, sewage treatment effluent	48 h	*In vitro*, binding assay	Estrogenic activity, hepatic Vtg, ZRP	Ackermann et al. (2002)
Oncorhynchus mykiss	NP	3 months	*In vivo*, transgenerational	Hatching rate; intersex; plasma E2, T, and Vtg	Schwaiger et al. (2002)
Oryzias latipes	NP	3 weeks	*In vivo*, sub-adult/adult	Egg production, fertility, GSI, plasma Vtg,spermatogenesis, testis-ova	Kang et al. (2003)
Salmo salar	NP	5 days	*In vivo*, early life, waterborne	Body weight, plasma IGF1	Arsenault et al. (2004)
Danio rerio	EE2, NP	3 weeks (*in vivo*)	*In vitro*, binding assay; *in vivo*, sub-adult/adult, waterborne	Plasma Vtg, GSI, estrogenicity	Tsai and Liao (2006)
Danio rerio	EE2	3 weeks	*In vivo*, sub-adult/adult, waterborne	Hepatic *vtg*, *er*, and *igfbp1*	Martyniuk et al. (2007)
Oncorhynchus mykiss	BPA	3 h	*In vivo*, early life, ovarian fluid	Growth, hatching, yolk absorption, whole embryo GH levels, *ghr, igfs*, and *igfr*s	Aluru et al. (2010)
Danio rerio	Phthalate	3 weeks	*In vivo*, sub-adult/adult, waterborne	GVBD, fecundity, ovulation; ovarian *lhr, mPRb, ptgs2*, BMP15; plasma Vtg	Carnevali et al. (2010)

Fish	EDC	Duration	Method	Observation	References
Pimephales promelas	EE2	48 h	*In vivo*, sub-adult/adult, waterborne	Hepatic *erα* and *vtg*, testicular *ar* and *er*, testicular *cyp17*	Garcia-Reyero et al. (2011)
Oryzias latipes	EE2	1–6 weeks	*In vivo*, sub-adult/adult, waterborne	Hepatic *vtg* and *chg*, oocyte marker 42Sp50, testis-ova, zona pellucida related genes	Hirakawa et al. (2012)
Micropterus salmoides	EE2	60 days	*In vivo*, sub-adult/adult, dietary	Gonadal *cyp19a*, gonadal and hepatic *er*, plasma Vtg, HSI, GSI	Colli-Dula et al. (2014)
Oryzias latipes	MPs	2 months	*In vivo*, sub-adult/adult, dietary	Hepatic *chg, erα, vtg*	Rochman et al. (2014)
Oncorhynchus mykiss	EE2	48 h	*In vitro*, cell/organ culture	*er*, *vtg*, *ghr1*, *igfbp1*	Hultman et al. (2015)
Labeo bata	Bisphenol A	96-h	*In vivo*, sub-adult/adult, dietary	Plasma Vtg, hematology, blood biochemicals, hepatic ERα/ERβ	Mukherjee et al. (2020)
Oncorhynchus mykiss	Tonalide (AHTN)	6 weeks	*In vivo*, sub-adult/adult, dietary	Lipid peroxidation in caudal kidney tissue, Vtg	Hodkovicova et al. (2020)
Danio rerio	Metals cadmium (Cd) and zinc (Zn)	21 days	*In vivo*/adult males	Vitellogenin (Vtg) detection, embryo-larval development	de Alkimin and Fracácio (2020)
Danio rerio	Acetochlor/estradiol (E_2)	21 days	*In vivo*, sub-adult/adult, waterborne	Fish ovarian development, ovarian vitellogenin (Vtg), *bmp15*, *gdf9*	Zhang et al. (2020)
Cyprinus carpio	Carbamazepine (CBZ)	28 days	*In vivo,* waterborne	Blood biochemicals, vtg, oxidative stress, and damage to liver tissues	Liang et al. (2022)
Danio rerio	Triclosan (TCS)	150 days	*In vivo,* waterborne	Vtg quantitative	Qiao et al. (2022)
Danio rerio	EE2	95 days/64 days post-fertilization	*In vivo,* waterborne	*esr1* and *esr2a,* sexual development, Vtg	Kernen et al. (2022)
Danio rerio	PPCPs	120 days	*In vivo,* waterborne	*vtg, cyp17*, and *17βhsd*	Hamid et al. (2022)
Danio rerio	Carbamazepine (CBZ), and progesterone (P4)	28 days	*In vivo,* adult exposure tests	Vtg, biotransformation enzymes (EROD, GST), and oxidative stress marker (DNAsb)	Ács et al. (2022)

reduction of gonadosomatic index, and reduction of secondary sex characteristics (Kinnberg and Toft 2003; Golshan and Alavi 2019). This effect on fish in the aquatic environment is an indication of the presence of anti-androgen pollutants which may affect humans either by direct exposure or through food chains. Further, it causes sex change in fishes, for example, EDCs cause excess production of Vtg in males to become intersex in fathead minnow, *Pimephales promelas* (Green et al. 2015). Feminization and intersex fish resulting from the EDCs may change in the gender ratio of the population (Hill Jr and Janz 2003). The elevated level of estrogens in the plasma affects the quality of egg production in fish species, for example, *Pimephales promelas I (*Miller et al. 2007), *Danio rerio (*Hill Jr and Janz 2003; Schäfers et al. 2007), *Oncorhynchus mykiss (*Ahmadpanah et al. 2019), and *Cyprinus carpio (*Barse et al. 2006). Another interesting effect of EDCs combined with environmental temperature is to increase morphological behavioral and developmental changes and also cause reproductive disorders in fishes. This phenomenon may correlate with global climate change and its effect on the reproductive physiology of animals including humans (Cox et al. 2018; Jackson 2020; Wojnarowski et al. 2021), the resulting increase in fertility issues. Moreover, EDCs increase defective Vtg synthesis (ovovitellin), circulated in plasma causing a toxic effect on fish. Long-term exposure to EDCs differentiates the expression of natural estrogens (E1, E2, and E3) in fishes, which can affect the Vtg gene regulation; therefore, its level may be differentiated in the blood of fishes, considered as a biomarker to investigate the concentration of environmental estrogens. Table 10.2 describes the use of Vtg as a biomarker for the detection of various EDCs.

10.14 Conclusions

Vitellogenin (Vtg) is a suitable biomarker for the detection of estrogenic and anti-androgenic potentials of various compounds used in environmental applications. Quantification of Vtg is relatively simple by a range of susceptible techniques. Living organisms like oviparous and viviparous have responded with several endocrine signals. EDCs also have a similar functional role to interact with the endocrine signals and activate the estrogenic stimulatory hormones followed by subsequent Vtg gene induction. Vtg gene regulation is tightly controlled with estrogen receptor (ER); it has a direct effect on the response of xenoestrogens and it increases the availability of E2. The level of Vtg in the blood is the direct response of ER receptiveness or hormones that regulate E2. Hence, Vtg proteins and Vtg mRNA are suitable biomarkers for the detection of environmental estrogens and toxic estrogen-like compounds that could easily differentiate the level of Vtg induction between the fishes exposed with compounds in the field and control groups of fishes in laboratory exposure.

However, any biological assays use commonly for various systems should have certain limitations. One such limitation is poor responsiveness of Vtg production under several adverse conditions, even reproductive effects that are the responses most related to Vtg production. In addition, Vtg production in females is a cyclic process; therefore, it can be assessed only in males to interpret the responses. Further, the ecological importance of Vtg in males is still fewer than in females. As with most biological responses, Vtg production is affected by a variety of other environmental factors. Several field exposure studies revealed some controversial responses of estrogenic signals received by the fishes exposed under multiple EDCs contained in water bodies. Therefore, the use of Vtg, E2 signals, and transcript of its respective genes are considered detection marker with some limitations.

Acknowledgments The authors would like to thank the DST-PURSE Phase-III facility, Madurai Kamaraj University, for providing Internet facility and RUSA (MHRD) programs.

Author Contributions All authors listed have made a considerable intellectual contribution to this chapter and approved it for publication.

Conflict of Interest Authors declare that there is no conflict of interest.

References

Ackermann GE, Brombacher E, Fent K (2002) Development of a fish reporter gene system for the assessment of estrogenic compounds and sewage treatment plant effluents. Environ Toxicol Chem 21(9):1864–1875

Ács A, Liang X, Bock I, Griffitts J, Ivánovics B, Vásárhelyi E et al (2022) Chronic effects of carbamazepine, progesterone and their mixtures at environmentally relevant concentrations on biochemical markers of zebrafish (Danio rerio). Antioxidants 11(9):1776

Ahmadpanah K, Soltani M, Rajabi Islami H, Shamsaie M (2019) Effects of nonylphenol on hematological parameters and immune responses in immature rainbow trout (Oncorhynchus mykiss). Mar Freshw Behav Physiol 52(4):151–165

de Alkimin GD, Fracácio R (2020) Analysis of vitellogenin by histochemical method as an indicator of estrogenic effect in male Danio rerio exposed to metals. Environ Sci Pollut Res 27:17789–17793

Aluru N, Leatherland JF, Vijayan MM (2010) Bisphenol a in oocytes leads to growth suppression and altered stress performance in juvenile rainbow trout. PLoS One 5(5):e10741

Amano H, Fujita T, Hiramatsu N, Kagawa H, Matsubara T, Sullivan CV, Hara A (2008) Multiple vitellogenin-derived yolk proteins in gray mullet (*Mugil cephalus*): disparate proteolytic patterns associated with ovarian follicle maturation. Mol Reprod Dev 75:1307–1317

An L, Hu J, Zhang Z, Yang M (2006) Quantitative real-time RT-PCR for determination of vitellogenin mRNA in so-iuy mullet (Mugil soiuy). Anal Bioanal Chem 386:1995–2001

Andersen H, Siegrist H, Halling-Sørensen B, Ternes TA (2003) Fate of estrogens in a municipal sewage treatment plant. Environ Sci Technol 37(18):4021–4026

Anderson K, Pankhurst N, King H, Elizur A (2017) Effects of GnRHa treatment during vitellogenesis on the reproductive physiology of thermally challenged female Atlantic salmon (Salmo salar). PeerJ 5:e3898

Ankley GT, Johnson RD (2004) Small fish models for identifying and assessing the effects of endocrine-disrupting chemicals. Int Lab Anim Res J 45:467–481

Ankley GT, Feifarek D, Blackwell B, Cavallin JE, Jensen KM, Kahl PS, Randolph E, Saari T, Villeneuve DL (2017) Re-evaluating the significance of estrone as an environmental estrogen. Environ Sci Technol 51(8):4705–4713

Aoki JY, Nagae M, Takao Y, Hara A, Lee YD, Yeo IK et al (2010) Survey of contamination of estrogenic chemicals in Japanese and Korean coastal waters using the wild grey mullet (Mugil cephalus). Sci Total Environ 408(3):660–665

Arsenault JTM, Fairchild WL, MacLatchy DL, Burridge L, Haya K, Brown SB (2004) Effects of water-borne 4-nonylphenol and 17β-estradiol exposures during parr-smolt transformation on growth and plasma IGF-I of Atlantic salmon (Salmo salar L.). Aquat Toxicol 66(3):255–265

Arukwe A, Celius T, Walther BT, Goksøyr A (2000) Effects of xenoestrogen treatment on zona radiata protein and vitellogenin expression in Atlantic salmon (Salmo salar). Aquat Toxicol 49(3):159–170

Barber LB, Loyo-Rosales JE, Rice CP, Minarik TA, Oskouie AK (2015) Endocrine disrupting alkylphenolic chemicals and other contaminants in wastewater treatment plant effluents, urban streams, and fish in the Great Lakes and Upper Mississippi River regions. Sci Total Environ 517:195–206

Barucca M, Canapa A, Olmo E, Regoli F (2006) Analysis of vitellogenin gene induction as a valuable biomarker of estrogenic exposure in various Mediterranean fish species. Environ Res 101(1):68–73

Barse AV, Chakrabarti T, Ghosh TK, Pal AK, Jadhao SB (2006) One-tenth dose of LC50 of 4-tert-butylphenol causes endocrine disruption and metabolic changes in Cyprinus carpio. Pestic Biochem Physiol 86(3):172–179

Benninghoff AD, Williams DE (2008) Identification of a transcriptional fingerprint of estrogen exposure in rainbow trout liver. Toxicol Sci 101(1):65–80

Buisine N, Trichet V, Wolff J (2002) Complex evolution of vitellogenin genes in salmonid fishes. Mol Gen Genomics 268(4):535–542

Caldwell DJ, Mastrocco F, Hutchinson TH, Länge R, Heijerick D, Janssen C et al (2008) Derivation of an aquatic predicted no-effect concentration for the synthetic hormone, 17α-ethinyl estradiol. Environ Sci Technol 42(19):7046–7054

Carducci F, Biscotti MA, Canapa A (2019) Vitellogenin gene family in vertebrates: evolution and functions. Eur Zool J 86(1):233–240

Carnevali O, Tosti L, Speciale C, Peng C, Zhu Y, Maradonna F (2010) DEHP impairs zebrafish reproduction by affecting critical factors in oogenesis. PLoS One 5(4):e10201

Chaves-Pozo E, García-Ayala A, Cabas I (2018) Effects of sex steroids on fish leukocytes. Biology 7(1):9

Chen H, Bi B, Kong L, Rong H, Su Y, Hu Q (2021) Seasonal changes in plasma hormones, sex-related genes transcription in brain, liver and ovary during gonadal development in female rainbow trout (*Oncorhynchus mykiss*). Fishes 6:62. https://doi.org/10.3390/fishes6040062

Colli-Dula RC, Martyniuk CJ, Kroll KJ, Prucha MS, Kozuch M, Barber DS, Denslow ND (2014) Dietary exposure of 17-alpha ethinylestradiol modulates physiological endpoints and gene signaling pathways in female largemouth bass (Micropterus salmoides). Aquat Toxicol 156:148–160

Cox MK, Peterson KN, Tan D, Novak PJ, Schoenfuss HL, Ward JL (2018) Temperature modulates estrone degradation and biological effects of exposure in fathead minnows. Sci Total Environ 621:1591–1600

Dang W, Sun L (2011) Determination of internal controls for quantitative real time RT-PCR analysis of the effect of Edwardsiella tarda infection on gene expression in turbot (Scophthalmus maximus). Fish Shellfish Immunol 30(2):720–728

Diamanti-Kandarakis E, Economou F, Palimeri S, Christakou C (2010) Metformin in polycystic ovary syndrome. Ann N Y Acad Sci 1205(1):192–198

Ferreira F, Santos MM, Castro LFC, Reis-Henriques MA, Lima D, Vieira MN, Monteiro NM (2009) Vitellogenin gene expression in the intertidal blenny Lipophrys pholis: a new sentinel species for estrogenic chemical pollution monitoring in the European Atlantic coast? Comparative Biochemistry and Physiology Part C: Toxicology & Pharmacology 149(1):58–64

Fischer M, Regitz C, Kahl M, Werthebach M, Boll M, Wenzel U (2012) Phytoestrogens genistein and daidzein affect immunity in the nematode Caenorhabditis elegans via alterations of vitellogenin expression. Mol Nutr Food Res 56(6):957–965

Fischer M, Regitz C, Kull R, Boll M, Wenzel U (2013) Vitellogenins increase stress resistance of Caenorhabditis elegans after Photorhabdus luminescens infection depending on the steroid-signaling pathway. Microbes Infect 15(8–9):569–578

Forner-Piquer I, Beato S, Piscitelli F, Santangeli S, Di Marzo V, Habibi HR et al (2020) Effects of BPA on zebrafish gonads: focus on the endocannabinoid system. Environ Pollut 264:114710

Fujita T, Fukada H, Shimizu M, Hiramatsu N, Hara A (2004) Quantification of serum levels of precursors to vitelline envelope proteins (choriogenins) and vitellogenin in estrogen treated masu salmon, Oncorhynchus masou. Gen Comp Endocrinol 136(1):49–57

Fujita T, Fukada H, Shimizu M, Hiramatsu N, Hara A (2008) Molecular cloning and characterization of three distinct choriogenins in masu salmon, Oncorhynchus masou. Molecular Reproduction and Development: Incorporating Gamete Research 75(7):1217–1228

Fukada H, Haga A, Fujita T, Hiramatsu N, Sullivan CV, Hara A (2001) Development and validation of chemiluminescent immunoassay for vitellogenin in five salmonid species. Comp Biochem Physiol A Mol Integr Physiol 130(1):163–170

Garcia J, Munro ES, Monte MM, Fourrier MC, Whitelaw J, Smail DA, Ellis AE (2010) Atlantic salmon (Salmo salar L.) serum vitellogenin neutralises infectivity of infectious pancreatic necrosis virus (IPNV). Fish Shellfish Immunol 29:293–297

Garcia-Reyero N, Lavelle CM, Escalon BL, Martinović D, Kroll KJ, Sorensen PW, Denslow ND (2011) Behavioral and genomic impacts of a wastewater effluent on the fathead minnow. Aquat Toxicol 101(1):38–48

Garnayak SK, Mohanty J, Rao TV, Sahoo SK, Sahoo PK (2013) Vitellogenin in Asian catfish, Clarias batrachus: purification, partial characterization and quantification during the reproductive cycle by ELISA. Aquaculture 392–395:148–155

Giesy JP, Pierens SL, Snyder EM, Miles-Richardson SR, Kramer VJ, Snyder SA, Nichols KM, Villeneuve DL (2000) Effects of 4-nonyl phenol on fecundity and biomarkers of estrogenicity in fathead minnows (Pimephales promelas). Environ Toxicol Chem 19(5):1368–1377

Golshan M, Alavi SMH (2019) Androgen signaling in male fishes: examples of anti-androgenic chemicals that cause reproductive disorders. Theriogenology 139:58–71

Green C, Brian J, Kanda R, Scholze M, Williams R, Jobling S (2015) Environmental concentrations of anti-androgenic pharmaceuticals do not impact sexual disruption in fish alone or in combination with steroid oestrogens. Aquat Toxicol 160:117–127

Hamid N, Junaid M, Manzoor R, Duan JJ, Lv M, Xu N, Pei DS (2022) Tissue distribution and endocrine disruption effects of chronic exposure to pharmaceuticals and personal care products mixture at environmentally relevant concentrations in zebrafish. Aquat Toxicol 242:106040

Hara A (1987) Studies on female-specific serum proteins (vitellogenin) and egg yolk proteins in teleosts: immunochemical, physicochemical and structural studies. Mem Fac Fish Hokkaido Univ 34:1–59

Hara A, Hiramatsu N, Fujita T (2016) Vitellogenesis and choriogenesis in fishes. Fish Sci 82:187–202. https://doi.org/10.1007/s12562-015-0957-5

Haruna A, Akihiro K, Naoshi H, Toshiaki F, Takashi T, Jun-ya A, Kiyoshi S, Hirohiko K, Akihiko H (2018) Development of specific chemiluminescent immunoassays for three subtypes of vitellogenin in grey mullet (Mugil cephalus). Gen Comp Endocrinol 271:30. https://doi.org/10.1016/j.ygcen.2018.10.020

Hill RL Jr, Janz DM (2003) Developmental estrogenic exposure in zebrafish (Danio rerio): I. Effects on sex ratio and breeding success. Aquat Toxicol 63(4):417–429

Hinck JE, Blazer VS, Denslow ND, Echols KR, Gross TS, May TW et al (2007) Chemical contaminants, health indicators, and reproductive biomarker responses in fish from the Colorado River and its tributaries. Sci Total Environ 378(3):376–402

Hirakawa I, Miyagawa S, Katsu Y, Kagami Y, Tatarazako N, Kobayashi T et al (2012) Gene expression profiles in the testis associated with testis–ova in adult Japanese medaka (Oryzias latipes) exposed to 17α-ethinylestradiol. Chemosphere 87(7):668–674

Hiramatsu N, Shimizu M, Fukada H, Kitamura M, Ura K, Fuda H, Hara A (1997) Transition of serum vitellogenin cycle in Sakhalin taimen (Huchoperryi). Comp Biochem Physiol C: Pharmacol Toxicol Endocrinol 118(2):149–157

Hiramatsu N, Cheek AO, Sullivan CV, Matsubara T, Hara A (2005) Vitellogenesis and endocrine disruption. In: Mommsen TP, Moon TW (eds) Biochemistry and molecular biology of fishes, vol 6: Environmental toxicology. Elsevier, Amsterdam, pp 431–471

Hiramatsu N, Matsubara T, Fujita T, Sullivan CV, Hara A (2006) Multiple piscine vitellogenins: biomarkers of fish exposure to estrogenic endocrine disruptors in aquatic environments. Mar Biol 149:35–47

Hodkovicova N, Enevova V, Cahova J, Blahova J, Siroka Z, Plhalova L et al (2020) Could the musk compound tonalide affect physiological functions and act as an endocrine disruptor in rainbow trout? Physiol Res 69(suppl 4):S595

Huberman A (2000) Shrimp endocrinology. A review. Aquaculture 191(1–3):191–208

Hultman MT, Song Y, Tollefsen KE (2015) 17α-Ethinylestradiol (EE2) effect on global gene expression in primary rainbow trout (Oncorhynchus mykiss) hepatocytes. Aquat Toxicol 169:90–104

Iguchi T, Irie F, Urushitani H, Tooi O, Kawashima Y, Roberts M et al (2006) Availability of in vitro vitellogenin assay for screening of estrogenic and anti-estrogenic activities of environmental chemicals. Environ Sci 13(3):161–183

Jackson T (2020) Climate change. QED Publishing, London

Jobling S, Sumpter JP, Sheahan D, Osborne JA, Matthiessen P (1996) Inhibition of testicular growth in rainbow trout (Oncorhynchus mykiss) exposed to estrogenic alkylphenolic chemicals. Environ Toxicol Chem 15(2):194–202

Jones PD, De Coen WM, Tremblay L, Giesy JP (2000) Vitellogenin as a biomarker for environmental estrogens. Water Sci Technol 42(7–8):1–14

Jung JH, Shim WJ, Addison RF, Baek JM, Han CH (2006) Protein and gene expression of VTG in response to 4-nonylphenol in rockfish (Sebastes schlegelii). Comp Biochem Physiol C Toxicol Pharmacol 143(2):162–170

Jung JH, Bae SH, Baeck KL, Shim WJ, Kim DJ, Han CH (2012) Effects of 4-nonlyphenol exposure on P. olivaceus and S. schlegeli Vitellogenesis. Dev Reprod 16(1):9–18

Kadim MK, Risjani Y (2022) Biomarker for monitoring heavy metal pollution in aquatic environment: an overview toward molecular perspectives. Emerg Contam 8:195–205

Kang IJ, Yokota H, Oshima Y, Tsuruda Y, Hano T, Maeda M et al (2003) Effects of 4-nonylphenol on reproduction of Japanese medaka, Oryziaslatipes. Environ Toxicol Chem 22(10):2438–2445

Kernen L, Phan A, Bo J, Herzog EL, Huynh J, Segner H, Baumann L (2022) Estrogens as immunotoxicants: 17α-ethinylestradiol exposure retards thymus development in zebrafish (Danio Rerio). Aquat Toxicol 242:106025

Kim S, Ji K, Shin H, Park S, Kho Y, Park K et al (2020) Occurrences of benzalkonium chloride in streams near a pharmaceutical manufacturing complex in Korea and associated ecological risk. Chemosphere 256:127084

Kime D, Nash J, Scott A (1999) Vitellogenesis as a biomarker of reproductive disruption by xenobiotics. Aquaculture 177(1–4):345–352. https://doi.org/10.1016/S0044-8486(99)00097-6

Kinnberg K, Toft G (2003) Effects of estrogenic and antiandrogenic compounds on the testis structure of the adult guppy (Poecilia reticulata). Ecotoxicol Environ Saf 54(1):16–24

Kiyama R, Wada-Kiyama Y (2015) Estrogenic endocrine disruptors: molecular mechanisms of action. Environ Int 83:11–40

Koua NZD, Henry J, Corre E, Pontin J, Bernay B, Nunez J (2022) Immuno-enzymatic and proteomic approaches for sexing the African Bonytongue (Heterotisniloticus Cuvier, 1829). Aust Fish 7(3):106

Kramer VJ, Miles-Richardson S, Pierens SL, Giesy JP (1998) Reproductive impairment and induction of alkaline-liable phosphate, a biomarker of estrogen exposure, in fathead minnows exposed to waterborne estradiol. Aquat Toxicol 40:335–360

Kumar Singh S (2020) Understanding the assessment of vitellogenin and yolk protein immune-relevant and antioxidant activities in teleost. Journal of Entomology and Zoology Studies 8(5):762–768

Kwak HI, Bae MO, Lee MH, Lee YS, Lee BJ, Kang KS et al (2001) Effects of nonylphenol, bisphenol A, and their mixture on the viviparous swordtail fish (Xiphophorus helleri). Environ Toxicol Chem 20(4):787–795

LaFleur GJ, Byrne BM, Kanungo J, Nelson LD, Greenberg RM, Wallace RA (1995) Fundulus heteroclitus vitellogenin: the deduced primary structure of a piscine precursor to noncrystalline, liquid-phase yolk protein. J Mol Evol 41:505–521

Lazier CB, Langley S, Ramsey NB, Wright JM (1996) Androgen inhibition of vitellogenin gene expression in tilapia (Oreochromis niloticus). Gen Comp Endocrinol 104:321–329

Le Guellec K, Lawless K, Valotaire Y, Kress M, Tenniswood M (1988) Vitellogenin gene expression in male rainbow trout (Salmo gairdneri). Gen Comp Endocrinol 71:359–371

Lee J, Moon KW, Ji K (2021) Systematic review of exposure to bisphenol A alternatives and its effects on reproduction and thyroid endocrine system in zebrafish. Appl Sci 11(4):1837

Li X, Liu X, Jia Z, Wang T, Zhang H (2021) Screening of estrogenic endocrine-disrupting chemicals in meat products based on the detection of vitellogenin by enzyme-linked immunosorbent assay. Chemosphere 263:128251

Li H and Zhang S (2017) Chapter 17: functions of Vitellogenin in eggs, In Oocytes, results and problems in cell differentiation, Kloc (ed.), Springer. pp. 389–401

Li Z, Zhang S, Liu Q (2008) Vitellogenin functions as a multivalent pattern recognition receptor with an opsonic activity. PLoS One 3(4):e1940

Liang X, Csenki Z, Ivánovics B, Bock I, Csorbai B, Molnár J et al (2022) Biochemical marker assessment of chronic carbamazepine exposure at environmentally relevant concentrations in juvenile common carp (Cyprinus carpio). Antioxidants 11(6):1136

Lim ES, Lee EH, Kim MH, Han CH, Lee SK, Kim J (2013) Development of an enzyme-linked immunosorbent assay using vitellin for vitellogenin measurement in the Pale Chub, Zacco platypus. Environ Health Toxicol 28:e2013016

Lu A, Hu X, Wang Y, Shen X, Zhu A, Shen L, Ming Q, Feng Z (2013) Comparative analysis of the acute response of zebrafish Danio rerio skin to two different bacterial infections. J Aquat Anim Health 25:243–251

Lubzens E, Young G, Bobe J, Cerdà J (2010) Oogenesis in teleosts: how fish eggs are formed. Gen Comp Endocrinol 165:367–489

Madsen SS, Mathiesen AB, Korsgaard B (1997) Effects of 17β-estradiol and 4-nonylphenol on smoltification and vitellogenesis in Atlantic salmon (Salmo salar). Fish Physiol Biochem 17:303–312

Markell LK, Mingoia RT, Peterson HM, Yao J, Waters SM, Finn JP et al (2014) Endocrine disruption screening by protein and gene expression of vitellogenin in freshly isolated and cryopreserved rainbow trout hepatocytes. Chem Res Toxicol 27(8):1450–1457

Marlatt VL, Bayen S, Castaneda-Cortès D, Delbès G, Grigorova P, Langlois VS et al (2022) Impacts of endocrine disrupting chemicals on reproduction in wildlife and humans. Environ Res 208:112584

Martin-Skilton R, Lavado R, Thibaut R, Minier C, Porte C (2006) Evidence of endocrine alteration in the red mullet, Mullus barbatus from the NW Mediterranean. Environ Pollut 141(1):60–68

Martyniuk CJ, Gerrie ER, Popesku JT, Ekker M, Trudeau VL (2007) Microarray analysis in the zebrafish (Danio rerio) liver and telencephalon after exposure to low concentration of 17alpha-ethinylestradiol. Aquat Toxicol 84(1):38–49

Matsubara T, Nagae M, Ohkubo N, Andoh T, Sawaguchi S, Hiramatsu N et al (2003) Multiple vitellogenins and their unique roles in marine teleosts. Fish Physiol Biochem 28:295–299

Maurilio L-F, Rendon von J O 2020. Expression of estrogenic response genes in black mollies (Poecilia sphenops) exposed to pyrogenic hydrocarbon and petroleum from Campeche sound. Agric Sci Dig 10.18805/ag.D-142

Mellanen P, Soimasuo M, Holmbom B, Oikari A, Santti R (1999) Expression of the vitellogenin gene in the liver of juvenile whitefish (Coregonus lavaretus L. s.l.) exposed to effluents from pulp and paper mills. Ecotoxicol Environ Saf 43:133–137

Miller MR, Wentz E, Ong S (1999) Acetaminophen alters estrogenic responses in vitro: inhibition of estrogen-dependent vitellogenin production in trout liver cells. Toxicol Sci 48(1):30–37

Miller DH, Jensen KM, Villeneuve DL, Kahl MD, Makynen EA, Durhan EJ, Ankley GT (2007) Linkage of biochemical responses to population-level effects: a case study with vitellogenin in the fathead minnow (Pimephales promelas). Environ Toxicol Chem 26(3):521–527

Mugiya Y, Tanahashi A (1998) Inhibitory effects of aluminium on vitellogenin induction by estradiol-17β in the primary culture of hepatocytes in the rainbow trout Oncorhynchus mykiss. Gen Comp Endocrinol 109(1):37–43

Mukherjee U, Samanta A, Biswas S, Das S, Ghosh S, Mandal DK, Maitra S (2020) Bisphenol A-induced oxidative stress, hepatotoxicity and altered estrogen receptor expression in Labeo bata: impact on metabolic homeostasis and inflammatory response. Ecotoxicol Environ Saf 202:110944

Naderi M, Keyvanshokooh S, Salati AP, Ghaedi A (2017) Combined or individual effects of dietary vitamin E and selenium nanoparticles on humoral immune status and serum parameters of rainbow trout (Oncorhynchus mykiss) under high stocking density. Aquaculture 474:40–47

Negintaji A, Safahieh A, Zolgharnein H, Matroodi S (2019) Vitellogenin gene expression and sex steroid levels as biomarkers in yellowfin seabream (Acanthopagrus latus) exposed to bisphenol-A. Iran J Toxicol 1:27–33

Niemuth NJ, Klaper RD (2015) Emerging wastewater contaminant metformin causes intersex and reduced fecundity in fish. Chemosphere 135:38–45

Ortiz-Zarragoitia M, Bizarro C, Rojo-Bartolomé I, Diaz de Cerio O, Cajaraville MP, Cancio I (2014) Mugilid fish are sentinels of exposure to endocrine disrupting compounds in coastal and estuarine environments. Mar Drugs 12(9):4756–4782

Osman A, Alsomait H, Seshadri S, El-Toukhy T, Khalaf Y (2015) The effect of sperm DNA fragmentation on live birth rate after IVF or ICSI: a systematic review and meta-analysis. Reprod Biomed Online 30(2):120–127

Pamplona-Silva MT, Mazzeo DEC, Bianchi J, Marin-Morales MA (2018) Estrogenic compounds: chemical characteristics, detection methods, biological and environmental effects. Water Air Soil Pollut 229:1–27

Picchietti S, Taddei AR, Scapigliati G, Buonocore F, Fausto AM, Romano N et al (2004) Immunoglobulin protein and gene transcripts in ovarian follicles throughout oogenesis in the teleost Dicentrarchuslabrax. Cell Tissue Res 315:259–270

Prasatkaew W, Nanthanawat P, Thanomsit C (2019) Assessment of endocrine disrupting chemicals exposure in sea bass (Lates calcarifer) and wild fishes using vitellogenin as a biomarker. Environ Asia 12(2):69

Qiao Y, He J, Han P, Qu J, Wang X, Wang J (2022) Long-term exposure to environmental relevant triclosan induces reproductive toxicity on adult zebrafish and its potential mechanism. Sci Total Environ 826:154026

Rasier G, Toppari J, Parent AS, Bourguignon JP (2006) Female sexual maturation and reproduction after prepubertal exposure to estrogens and endocrine disrupting chemicals: a review of rodent and human data. Mol Cell Endocrinol 254:187–201

Reading BJ, Andersen LK, Ryu YW, Mushirobira Y, Todo T, Hiramatsu N (2018) Oogenesis and egg quality in finfish: yolk formation and other factors influencing female fertility. Aust Fish 3(4):45

Reading BJ, Sullivan CV, Schilling J (2017) Vitellogenesis in fishes. Reference module in life sciences. Elsevier BV, Amsterdam. https://doi.org/10,1016,03076-4

Rendon Von Osten J, Aguayo-Dione G, Dzul-Caamal R, Lara-Flores M (2019) Expression of estrogenic response genes to different concentration of 17ß-estradiol in male mosquitofish (Gambusia yucatana). Iran J Fish Sci 18(2):272–282

Rochman CM, Kurobe T, Flores I, Teh SJ (2014) Early warning signs of endocrine disruption in adult fish from the ingestion of polyethylene with and without sorbed chemical pollutants from the marine environment. Sci Total Environ 493:656–661

Sa-an AC, Quinitio GF, Anasco NC, Traifalgar RFM, Nillos MGG (2022) Identification and quantification of vitellogenin gene in eyebrow goby (Oxyurichthys ophthalmonema). Biodiversitas 23(11)

Safe SH, Gaido K (1998) Phytoestrogens and anthropogenic estrogenic compounds. Environ Toxicol Chem 17(1):119–126

Sawaguchi S, Ohkubo N, Koya Y, Matsubara T (2005) Incorporation and utilization of multiple forms of vitellogenin and their derivative yolk proteins during vitellogenesis and embryonic development in the mosquitofish, Gambusia affinis. Zool Sci 22(6):701–710

Schäfers C, Teigeler M, Wenzel A, Maack G, Fenske M, Segner H (2007) Concentration-and time-dependent effects of the synthetic estrogen, 17α-ethinylestradiol, on reproductive capabilities of the zebrafish, Danio rerio. J Toxic Environ Health A 70(9):768–779

Schug TT, Janesick A, Blumberg B, Heindel JJ (2011) Endocrine disrupting chemicals and disease susceptibility. J Steroid Biochem Mol Biol 127(3–5):204–215

Schwaiger J, Mallow U, Ferling H, Knoerr S, Braunbeck T, Kalbfus W, Negele RD (2002) How estrogenic is nonylphenol?: a transgenerational study using rainbow trout (Oncorhynchus mykiss) as a test organism. Aquat Toxicol 59(3–4):177–189

Shanle EK, Xu W (2011) Endocrine disrupting chemicals targeting estrogen receptor signaling: identification and mechanisms of action. Chem Res Toxicol 24(1):6–19

Sharrock WJ (1983) Yolk proteins of Caenorhabditis elegans. Dev Biol 96:182–188

Shi X, Zhang S, Pang Q (2006) Vitellogenin is a novel player in defense reactions. Fish Shellfish Immunol 20:769

Snyder SA, Keith TL, Verbrugge DA, Snyder EM, Gross TS, Kannan K, Giesy JP (1999) Analytical methods for detection of selected estrogenic compounds in aqueous mixtures. Environ Sci Technol 33(16):2814–2820

Sohoni PCRT, Tyler CR, Hurd K, Caunter J, Hetheridge M, Williams T et al (2001) Reproductive effects of long-term exposure to bisphenol A in the fathead minnow (Pimephales promelas). Environ Sci Technol 35(14):2917–2925

Sumpter JP, Jobling S (1995) Vitellogenesis as a biomarker for estrogenic contamination of the aquatic environment. Environ Health Perspect 103(suppl 7):173–178

Sun X, Zhang S (2001) Purification and characterization of a putative vitellogenin from the ovary of amphioxus (Branchiostoma belcheri tsingtaunese). Comp Biochem Physiol B: Biochem Mol Biol 129(1):121–127

Sun C, Zhang S (2015) Immune-relevant and antioxidant activities of vitellogenin and yolk proteins in fish. Nutrients 7(10):8818–8829

Sun L, Wen L, Shao X, Qian H, Jin Y, Liu W, Fu Z (2010) Screening of chemicals with anti-estrogenic activity using in vitro and in vivo vitellogenin induction responses in zebrafish (Danio rerio). Chemosphere 78(7):793–799

Sun L, Shao X, Chi J, Hu X, Jin Y, Fu Z (2011a) Transcriptional responses in the brain, liver and gonad of Japanese rice fish (Oryzias latipes) exposed to two anti-estrogens. Comp Biochem Physiol C Toxicol Pharmacol 153(4):392–401

Sun L, Shao X, Hu X, Chi J, Jin Y, Ye W, Fu Z (2011b) Transcriptional responses in Japanese medaka (Oryzias latipes) exposed to binary mixtures of an estrogen and anti-estrogens. Aquat Toxicol 105(3–4):629–639

Szwejser E, Verburg-van Kemenade BM, Maciuszek M, Chadzinska M (2017) Estrogen-dependent seasonal adaptations in the immune response of fish. Horm Behav 88:15–24. https://doi.org/10.1016/j.yhbeh.2016.10.007

Thomas P, Rahman MS (2009) Biomarkers of hypoxia exposure and reproductive function in Atlantic croaker: a review with some preliminary findings from the northern Gulf of Mexico hypoxic zone. J Exp Mar Biol Ecol 381:S38–S50

Thrupp TJ, Runnalls TJ, Scholze M, Kugathas S, Kortenkamp A, Sumpter JP (2018) The consequences of exposure to mixtures of chemicals: something from 'nothing' and 'a lot from a little' when fish are exposed to steroid hormones. Sci Total Environ 619:1482–1492

Tong Z, Li LR, Zhang S (2010) Vitellogenin is an acute phase protein with bacterial-binding and inhibiting activities. Immunobiology 215:898–902

Tran TKA, Yu RMK, Islam R, Nguyen THT, Bui TLH, Kong RYC et al (2019) The utility of vitellogenin as a biomarker of estrogenic endocrine disrupting chemicals in molluscs. Environ Pollut 248:1067–1078

Trichet V, Buisine N, Mouchel N, Moran P, Pendas AM, Le Pennec JP, Wolff J (2000) Genomic analysis of the vitellogenin locus in rainbow trout (Oncorhynchus mykiss) reveals a complex history of gene amplification and retroposon activity. Mol Gen Genet MGG 263:828–837

Tsai JW, Liao CM (2006) Mode of action and growth toxicity of arsenic to tilapia Oreochromis mossambicus can be determined bioenergetically. Arch Environ Contam Toxicol 50:144–152

Tyler CR, van Aerle R, Nilsen MV, Blackwell R, Maddix S, Nilsen BM et al (2002) Monoclonal antibody enzyme-linked immunosorbent assay to quantify vitellogenin for studies on environmental estrogens in the rainbow trout (Oncorhynchus mykiss). Environ Toxicol Chem 21(1):47–54

U.S. Environmental Protection Agency, 1998. Guidelines for Ecological Risk Assessment. Washington, DC EPA/630/R-95/002F

Van der Oost R, Beyer J, Vermeulen NP (2003) Fish bioaccumulation and biomarkers in environmental risk assessment: a review. Environ Toxicol Pharmacol 13(2):57–149

Wang YS, Lou SW (2006) Structural and expression analysis of hepatic vitellogenin gene during ovarian maturation in Anguilla japonica. J Steroid Biochem Mol Biol 100(4–5):193–201

Wang L, Yan R, Yang Q, Li H, Zhang J, Shimoda Y et al (2020) Role of GH/IGF axis in arsenite-induced developmental toxicity in zebrafish embryos. Ecotoxicol Environ Saf 201:110820

Williams VN, Reading BJ, Amano H, Hiramatsu N, Schilling J, Salger SA, Williams TI, Gross K, Sullivan CV (2014) Proportional accumulation of yolk proteins derived from multiple vitellogenins is precisely regulated during vitellogenesis in striped bass (*Morone saxatilis*). J Exp Zool 321A:301–315

Wojnarowski K, Podobiński P, Cholewińska P, Smoliński J, Dorobisz K (2021) Impact of estrogens present in environment on health and welfare of animals. Animals 11(7):2152

Wu B, Liu Z, Zhou L, Ji G, Yang A (2015) Molecular cloning, expression, purification and characterization of vitellogenin in scallop Patinopecten yessoensis with special emphasis on its antibacterial activity. Dev Comp Immunol 49:249–258

Yang O, Kim HL, Weon JI, Seo YR (2015) Endocrine-disrupting chemicals: review of toxicological mechanisms using molecular pathway analysis. J Cancer Prev 20(1):12

Yaron Z, Levavi-Sivan B (2011) Endocrine regulation of fish reproduction. In: Encyclopedia of fish physiology: from genome to environment, vol 2, no 2. Academic, San Diego, pp 1500–1508

Ye Z, Zhao T, Wei Q, Lin H, Zhang Y, Li S (2022) Distinct roles of estrogen receptors in the regulation of vitellogenin expression in orange-spotted grouper (Epinephelus coioides). Int J Mol Sci 23(15):8632

Yi X, Li C, Zhong X, Gong Y (2018) Development of a lipovitellin-based sandwich ELISA for determination of vitellogenin in the marine medaka Oryzias melastigma. Chemosphere 197:477–484

Zhang X, Hecker M, Park JW, Tompsett AR, Newsted J, Nakayama K et al (2008) Real-time PCR array to study effects of chemicals on the hypothalamic–pituitary–gonadal axis of the Japanese medaka. Aquat Toxicol 88(3):173–182

Zhang S, Wang S, Li H, Li L (2011) Vitellogenin, a multivalent sensor and an antimicrobial effector. Int J Biochem Cell Biol 43(3):303–305

Zhang S, Dong Y, Cui P (2015) Vitellogenin is an immunocompetent molecule for mother and offspring in fish. Fish Shellfish Immunol 46(2):710–715

Zhang Y, Xue W, Long R, Yang H, Wei W (2020) Acetochlor affects zebrafish ovarian development by producing estrogen effects and inducing oxidative stress. Environ Sci Pollut Res 27:27688–27696

Fish Vitellogenin Induction and Its Related Egg Yolk Protein

11

Maharajan Athisuyambulingam
and Vaseeharan Baskaralingam

Abstract

Vitellogenin (Vtg) is secreted by liver, which is then consumed by the maturing oocyte by receptor-mediated endocytosis. There, it is split into the two main yolk proteins in the developing embryo. The ability of marine teleosts to osmoregulate their eggs is significantly influenced by the yolk proteins. Vitellogenin is a useful biomarker for endocrine disruptors to screen chemical substances in an aquatic ecosystem. Fish oogenesis requires extremely precise regulatory processes and directly affects the ultimate egg quality and seed integrity, especially when it comes to the production of the yolk globules, egg envelope, and oil globules. One of the most fascinating biological phenomena is the process by which a primary oocyte grows by several orders of magnitude while gathering or synthesising everything required to finally be fertilised and support the full development of a new life.

Keywords

Teleosts · Endocytosis · Oil globules · Osmoregulation · Oocyte

M. Athisuyambulingam (✉)
PG & Research Department of Zoology, Khadir Mohideen College (Affiliated to Bharathidasan University, Tiruchirappalli), Adirampattinam, Tamil Nadu, India

V. Baskaralingam
Department of Animal Health and Management, Alagappa University, Karaikudi, Tamil Nadu, India

V. Baskaralingam, R. Vanichviriyakit (eds.), *Vitellogenin in Fishes-Diversification, Biological Properties, and Future Perspectives*,
https://doi.org/10.1007/978-981-99-5340-0_11

11.1 Introduction

The majority of fish egg yolk is made of vitellogenins, which are produced in the liver and then taken up from the maternal circulation by developing oocytes via receptor-mediated endocytosis. These vitellogenins are then processed enzymatically to create yolk proteins, which are then stored in the ooplasm. A good biomarker for endocrine disruptors is vitellogenin (Vtg), which can be used to screen chemicals in aquatic ecosystems. In order to assess the oestrogenic action, it is helpful to measure the serum levels of Vtg in fish organ homogenate. Although adult male fish and young fish do not create Vtg, the precursor to egg yolk is formed in the female liver (Heppell et al. 1995). Male fish create higher VTG when they are in contact with exogenous oestrogen. Antibodies against purified rainbow trout Vtg have been found in a wide range of animals, including birds, amphibians, and reptiles in addition to fish. There have been reports of Vtg measurements in a variety of fish species from rivers all over the world (Sugawara 2011).

Fish contain vtg protein, which can be assessed through repeated sampling. Fish are harmful in several ways when Vtg protein levels are high, especially in the kidney. Kidney failure from Vtg disrupts blood dynamics and function and decreases young fish survival (Hiramatsu et al. 2002a, b, c, d). Endogenous steroid hormones appear to be reduced even while endocrine disruptors increase Vtg synthesis. Reproduction and fertility may suffer if endogenous steroid levels are decreased (Kanetoshi et al. 2004). *Pleronectusyohohamae* fish have been shown to have intersex gonads and high Vtg concentrations (Hashimoto et al. 2000). Oocytes were present in the testes of adult male medaka subjected to octylphenol, which caused a suppression of spermatogenesis and the emergence of intersex. Even though the lack of Vtg expression in an aquatic ecosystem cannot be construed as a reproductive consequence, Vtg expression is thought to be an endocrine disruptor. After 8 weeks of exposure, medaka were exposed to lower concentrations of the environmental oestrogen o,p′-DDT, which had a bigger impact since vitellogenin expression is more vulnerable to estrogenic effects than fertility and hatching success (Cheek et al. 2001). Since the oestrogen receptor and AhR pathways interact, DLCs significantly reduced the production of the Vtg protein in fish (Bemanian et al. 2004). The Vtg screening approach makes it possible to identify endocrine disruptors and evaluate risk for aquatic species.

11.2 Egg Yolk Protein

In oviparous species, the deposit of yolk components into oocytes during oogenesis and their mobilisation throughout embryogenesis are important strategies for productive reproduction. The majority of the proteins and lipids in oocyte yolks are made by the enzymatic cleavage of complex precursors, principally Vtg and very low-density lipoprotein, as was already demonstrated (Schneider 1996; Kwon et al. 2001). The primary component of the yolk of vertebrate eggs is vitellogenin (Sullivan and Yilmaz 2018). Yolk is then stored until the last stages of oogenesis, at which point it is released and utilised by the embryo to hydrate buoyant eggs and provide nutrients for embryogenesis (Sire et al. 1994). The precursor of the egg yolk protein (Vtg), which is produced by E 2 in the liver, is secreted and transported by blood to the ovary where it is then taken up by mature oocytes (Pawlowski et al. 2000). Vtg is a complex phospholipoglycoprotein with a large molecular weight (MW; 250–600 kDa) that binds calcium (ibid.). The molecule's protein backbone is connected to significant functional groups, including phosphate, lipid, and specific carbohydrates, according to Vtg's as a phospholipoglycoprotein (Silversand and Haux 1995). The ability of Vtg to bind ions also serves as a substantial source of minerals for the oocytes. The transformation of circulating Vtg into yolk and depositing it in the oocyte is what causes oocyte development in fish (Wallace 1985). These specific oocyte Vtg receptors are clustered in clathrin-coated pits. Coated vesicles from the oocytes combine with golgian lysosomes to produce multivesicular formations (Le Menn et al. 2000). The growth of both eggs and larvae depends on the process of vitellogenesis because it provides an essential source of nourishment. Furthermore, cortisol, other lipophilic hormones like thyroxin, and maternal sex steroids are all present in teleost eggs and can all enter the egg through Vtg (Mommsen et al. 1999). The biological purpose(s) of the hormones found in eggs is/are unknown. Nonetheless, it has been hypothesised that they may act as metabolites or as pharmacological synergists in the early stages of development.

11.3 Egg Shell Protein

The envelope of the animal egg plays crucial roles in the reproductive and developmental processes. At the beginning, it acts as an interface between the egg and the sperm, and according to Grierson and Neville (1981), it also acts as an interface between the embryo and its surroundings. The egg envelope, often referred to as the zona radiate, is a crucial structural element of the fish egg shell and may be distinguished by its striated appearance under a light microscope (Plate 1). Three to four well conserved glycoproteins make up the zona proteins in eutherian mammals and fish, though comparing them can be challenging due to terminological and nomenclatural differences.

In-depth research has been done on the genes responsible for producing zona proteins. For instance, the zona pellucida, or Zp2, homologues' coding sequence and exon-intron mappings in mice, pigs, and humans (Taya et al. 1995). Consequently, it has become increasingly clear that eutherian mammals and teleost fish share the same zona pellucida and egg envelope proteins, respectively. Recent research has revealed the peculiar connection between the proteins from the teleostane and mammalian egg membranes (Oppen-Berntsen et al. 1999) (Fig. 11.1). It was found that the majority of thermophilic animals produce Zr-protein in the liver. For instance, in addition to Atlantic salmon, cod, and rainbow trout (Murata et al. 1995), medaka, *Oryzias latipes* (Lee et al. 2002), winter flounder, *Pseudopleuronectes americanus* (Lyons et al. 1993), and gilthead seabream, *Sparus aurata* (Del-Giacco et al. 1998),

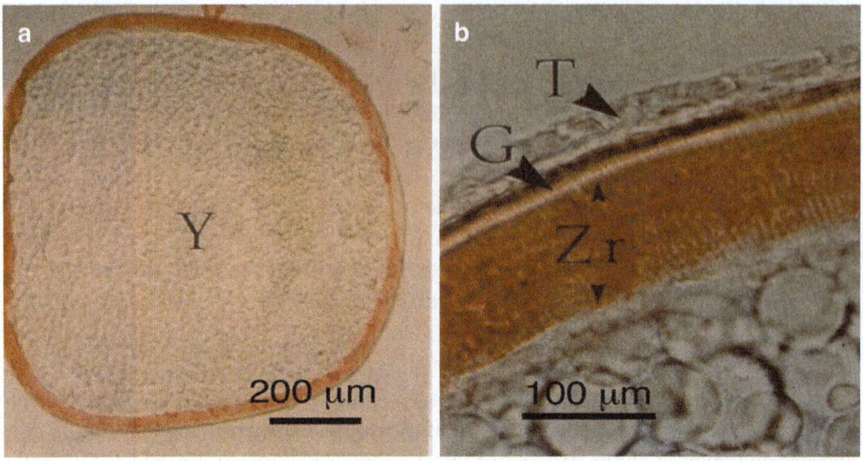

Fig. 11.1 An ovarian follicle from a cod (*Gadus morhua*) that had been immunohistochemical stained with an anti-cod rabbit serum for zona radiata proteins. In the liver of most fish species, vitellogenin (Y) and the zona radiata proteins (Zr) are both produced and delivered to the ovary. (**a**) Whole-oocyte section showing zona radiata staining that is specific and does not react with yolk material (Y). (**b**) A cod follicle that has been magnified more. Zr stands for zone radiata (positively stained). Arrowheads denote the follicle cells (theca, T, and granulosa, G). Unstained yolk granules are represented by spherical entities. (With permission from the author and the University of Basque Country Press (UBC Press), Oppen-Berntsen et al. (1999) have been reproduced)

synthesise Zr-protein in the liver. Other species seem to produce Zr-protein in the ovary, including carp *Cyprinus carpio* (Chang et al. 1997) and pipefish *Syngnathus scovelli* (Begovac and Wallace 1989).

11.4 Fish Vitellogenin

Chicken oogenesis has been studied ever since the early 1900s, when it was thought that the synthesis of egg yolk proteins and low-molecular-weight substances like amino acids was what caused oocyte growth. Based on a biochemical examination using African clawed frogs, *Xenopus laevis*, and the unique notion that egg yolk proteins are first produced in the maternal liver and then transferred to mature oocytes via the blood, it was determined in 1974 that Vtg was the precursor of egg yolk proteins (Bergink and Wallace 1974). Akihiko Hara et al. 2015 have identified and purified Vtg and three egg yolk proteins, Lv, Pv, and β′-component (β′-c), during the research of different fish species. These fish included rainbow trout (Hara and Hirai 1978) and Sakhalin taimen *Huchoperryi* (Hiramatsu and Hara 1996). He observed the molecular cleavage of Vtg to yield Lv, Pv, and β′-c using a variety of antibodies against the purified proteins, immunodiffusion, and Western blotting techniques.

In the Sakhalin taimen, Vtg is a 540 kDa dimer made up of two 240 kDa subunits that each contain 23 kDaPv, a 34 kDa β′-c protein made up of a 17 kDa peptide, and

330 kDaLv, which is formed of two heavy chains (92 kDa) and two light chains (29 kDa). Bovine cathepsin D was used to degrade biotin-labeled Vtg into egg yolk proteins, which suggests that Vtg is specifically degraded in eggs by a cathepsin D-like enzyme (Hiramatsu et al. 2002a, b, c, d). However, a Vtg receptor was found, which led to the conclusion that the Lv domain within the Vtgmolecule serves as a receptor-binding site with reference to the particular uptake of Vtg into eggs (Hiramatsu et al. 2001). The main component of egg yolks is Lv, which accounts for 20% of its mass in terms of lipid content. Lv is a significant breakdown product of Vtg. Lv is a dimer composed of a heavy chain (LvH) and a light chain, two distinct polypeptides (LvL). Various kinds of amino acids and lipids, which are crucial food supplies for embryogenesis, are rich in Lv. Pv, on the other hand, is a phosphoprotein with a phosphorus concentration of 10% and serine (Ser) accounting for 50% of the amino acid content. As a result, Pv is difficult to identify using standard staining methods and has a very low antigenicity due to its high phosphorus content. It is believed that minerals linked to the Ser residues in this phosphoprotein are crucial for osteogenesis and metabolic processes.

11.5 Fish Vitellogenin Induction in Egg Yolk Protein

Also found the third component of the egg yolk protein, β′-c, in rainbow trout. They classified it as a type of serum protein since it lacks phosphate and fat. A similar protein was identified from coho salmon eggs and given the name β′-c by Markert and Vanstone (1971). White perch *Morone americana* (Hiramatsu et al. 2002a, b, c, d), grey mullet *Mugil cephalus* (Amano et al. 2007), and barfin flounder *Veraspermoseri* (Matsubara and Koya 1997) are a few examples of fish from which it has been isolated outside of the salmonid family. Thus, β′-c is regarded as a common egg yolk protein in teleosts. For the first time in the Elasmobranch, we reported that the eggs of the clouded cat shark *Scyliorhinus torazame* contain β′-c as well (Yamane et al. 2013). Shimizu et al. (2009) recently found that β′-c is one of the allergens that contribute to the symptoms of fish egg allergy. Similar to fish β′-c, a 40 kDa glycoprotein (YG40) with many Cys residues was discovered in chicken eggs (Yamamura et al. 1995). The fourth egg yolk protein component, thought to have come from the furthest C-terminus of the Vtg polypeptide (C-terminal coding domain), was recently discovered from a study of the barfin flounder (Matsubara et al. 2003). Based on these findings, it is thought that the molecular structure of the egg yolk proteins in Vtg can be expressed as NH2-(LvH)-(Pv)-(LvL)-(β′-c)-(C-terminal coding domain)-COOH (Hiramatsu et al. 2002a, 2006).

During vitellogenesis, Vtg undergoes a minimal amount of degradation to form at least three different types of egg yolk proteins (Lv, Pv, and β′-c), which are then stored in eggs. The first proteolysis refers to the initial degradation that occurs during vitellogenesis, while the second and third proteolysis refer to the subsequent degradations that occur during final maturation and embryogenesis, respectively (Hiramatsu et al. 2002a). Salmonids that deposit their eggs in fresh water, on the other hand, do not exhibit this second proteolysis. Matsubara and Sawano (1995) described three distinct egg yolk protein types (Lv, Pv, and -c) in the barfin flounder that underwent first proteolysis and further degraded during the final maturation (second proteolysis), with the majority of Pv and -c egg yolk proteins degrading into free amino acids (FAA). On the other hand, nothing is understood about the third, concurrent with embryogenesis, proteolysis of egg yolk proteins. Using salmonids, we discovered that three different egg yolk proteins (Lv, Pv, and -c) underwent different degradations after fertilisation, with LvH being broken down into smaller products and Pv being dephosphorylated after the stage of the eyed embryo, whereas -c was preserved throughout embryo development and no second proteolysis was noticed (Hiramatsu et al. 2002a).

Three unique Vtg types were successfully identified for the first time from the blood plasma of white perch in 2002 (Hiramatsu et al. 2002b). Fish Vtgs can be classified as either complete or incomplete (Hiramatsu et al. 2006). The basic amino acid sequence of the whole Vtg is made up of the previously characterised LvH, Pv, LvL, ′-c, and C-terminal coding domains of the five egg yolk protein portions. As a result of the homology analysis, the complete Vtg was further divided into the type A (VtgA) and type B subgroups (VtgB).

11.6 Conclusion

Fish oogenesis requires extremely precise regulatory processes and directly affects the ultimate egg quality and seed integrity, especially when it comes to the production of the yolk globules, egg envelope, and oil globules. One of the most fascinating biological phenomena is the process by which a primary oocyte grows by several orders of magnitude while gathering or synthesising everything required to finally be fertilised and support the full development of a new life.

References

Amano H, Fujita T, Hiramatsu N, Shimizu M, Sawaguchi S, Matsubara T, Kagawa H, Nagae M, Sullivan CV, Hara A (2007) Egg yolk proteins in gray mullet (*Mugil cephalus*): purification and classification of multiple lipovitellins and other vitellogenin-derived yolk proteins and molecular cloning of the parent vitellogenin genes. J Exp Zool A Ecol Genet Physiol 307:324–341

Begovac PC, Wallace RA (1989) Major vitelline envelope proteins in pipefish oocytes originate within the follicle and are associated with the Z3 layer. J Exp Zool 251:56–73

Bemanian V, Male R, Goksøyr A (2004) The aryl hydrocarbon receptor-mediated disruption of vitellogenin synthesis in the fish liver: cross-talk between AHR- and ERα-signaling pathways. Comp Hepatol 3:2

Bergink EW, Wallace RW (1974) Precursor-product relationship between amphibian vitellogenin and yolk proteins, lipovitellin and phosvitin. J Biol Chem 249:2897–2903

Chang YS, Hsu CC, Wang SC, Tsao CC, Huang FL (1997) Molecular cloning, structural analysis, and expression of carp ZP2 gene. Mol Reprod Dev 46:258–267

Cheek AO, Brouwer TH, Carroll S, Manning S, McLachlan JA, Brouwer M (2001) Experimental evaluation of vitellogenin as a predictive biomarker for reproductive disruption. Environ Health Perspect 109(7):681–690

Del-Giacco L, Vanoni C, Bonsignorio D, Duga S, Mosconi G, Santucci A, Cotelli F (1998) Identification and spatial distribution of the mRNA encoding the gp49 component of the gilthead sea bream, *Sparus aurata*, egg envelope. Mol Reprod Dev 49:58–69

Grierson JP, Neville AC (1981) Helicoidal architecture of fish eggshell. Tissue Cell 13:819–830

Hara A, Hirai H (1978) Comparative studies on immunochemical properties of female-specific serum protein and egg yolk proteins in rainbow trout (*Salmo gairdneri*). Comp Biochem Physiol 59B:339–343

Hara A, Hiramatsu N, Fujita T (2015) Vitellogenesis and choriogenesis in fishes. Fish Sci 82:187. https://doi.org/10.1007/s12562-015-0957-5

Hashimoto S, Bessho H, Hara A, Nakamura M, Iguchi T, Fujita K (2000) Elevated serum vitellogenin levels and gonadal abnormalities in wild male flounder (*Pleuronectes yokohamae*) from Tokyo Bay, Japan. Mar Environ Res 49:37–53

Heppell SA, Denslow ND, Folmar LC, Sullivan CV (1995) Universal assay of vitellogenin as a biomarker for environmental estrogens. Environ Health Perspect 103(suppl 7):9–15

Hiramatsu N, Fukada H, Sullivan CV, Hara A (2001) Simple and sensitive detection of vitellogenin receptor(s) in Sakhalin taimen (*Huchoperryi*). Bull Fish Sci Hokkaido Univ 52:5–9

Hiramatsu N, Hara A (1996) Relationship between vitellogenin and its related egg yolk proteins in Sakhalin taimen (Huchoperryi). Comp Biochem Physiol A Physiol 115:243–251

Hiramatsu N, Matsubara T, Hara A, Donato DM, Hiramatsu K, Denslow ND, Sullivan CV (2002a) Identification, purification and classification of multiple forms of vitellogenin from white perch (*Morone americana*). Fish Physiol Biochem 26:355–370

Hiramatsu N, Matsubara T, Weber GM, Sullivan CV, Hara A (2002b) Vitellogenesis in aquatic animals. Fish Sci 68(suppl I):694–699

Hiramatsu N, Ichikawa N, Fukada H, Fujita T, Sullivan CV, Hara A (2002c) Identification and characterization of proteases involved in specific proteolysis of vitellogenin and yolk proteins in salmonids. J Exp Zool 292:11–25

Hiramatsu N, Hara A, Hiramatsu K, Fukada H, Weber GM, Denslow ND, Sullivan CV (2002d) Vitellogenin-derived yolk proteins of white perch, Morone americana: purification, characterization and vitellogenin-receptor binding. Biol Reprod 67:655–667

Hiramatsu N, Matsubara T, Fujita T, Sullivan CV, Hara A (2006) Multiple piscine vitellogenins: biomarkers of fish exposure to estrogenic endocrine disruptors in aquatic environments. Mar Biol 149:35–47

Kanetoshi A, Katsura E, Fujimoto T, Kojima H, Hori Y, Fukada H, Takahara S, Hara A (2004) Study on the screening test of endocrine disrupting chemicals using carp (*Cyprinus carpio*) hepatocyte culture: vitellogenin induction test by estrogens. Rep Hokkaido Inst Public Health 54:1–6

Kwon JY, Prat F, Randal C, Tyler CR (2001) Molecular characterization of putative yolk processing enzymes and their expression during oogenesis and embryogenesis in rainbow trout (*Oncorhynchus mykiss*). Biol Reprod 65:1701–1709

Le Menn F, Davali B, Pelissero C, Ndiaye P, Bon E, Perazzolo L, Rodriguez JN (2000) A new approach to fish vitellogenesis. In: Norberg B, Kjesbu OS, Taranger GL, Andersson E, Stefansson SO (eds) Proceedings of the 6th international symposium on the reproductive physiology of fish, 4–9 July 1999, Bergen. John Grieg A/S, Bergen, pp 281–284

Lee C, Na J, Lee K, Park K (2002) Choriogenin mRNA induction in male medaka, *Oryziaslatipes* as a biomarker of endocrine disruption. Aquat Toxicol 61:233–241

Lyons CE, Payette KL, Price JL, RCC H (1993) Expression and structural-analysis of a teleost homolog of a mammalian zona-pellucida gene. J Biol Chem 268:21351–21358

Markert JP, Vanstone WE (1971) Egg proteins of coho salmon (Oncorhynchus kisutch). J Fish Res Board Can 28(1853–1856):57

Matsubara T, Koya Y (1997) Course of proteolytic cleavage in three classes of yolk proteins during oocyte maturation in barfin flounder, *Veraspermoseri*, a marine teleost spawning pelagic eggs. J Exp Zool 272(34–45):58

Matsubara T, Sawano K (1995) Proteolytic cleavage of vitellogenin and yolk proteins during vitellogenin uptake and oocyte maturation in barfin flounder (*Veraspermoseri*). J Exp Zool 272:34–45

Matsubara T, Nagae M, Ohkubo N, Andoh T, Sawaguchi S, Hiramatsu N, Sullivan CV, Hara A (2003) Multiple vitellogenins and their unique roles in marine teleosts. Fish Physiol Biochem 28:295–299

Mommsen TP, Vijayan MM, Moon TW (1999) Cortisol in teleosts: dynamics. Mechanism of action and metabolic regulation. Rev Fish Biol Fish 9:211–268

Murata K, Sasaki T, Yasumasu S, Iuchi I, Enami J, Yasumasu I, Yamagami K (1995) Cloning of cDNAs for the precursor protein of a low-molecular-weight subunit of the inner layer of the egg envelope (chorion) of the fish *Oryzias-latipes*. Dev Biol 167:9–17

Oppen-Berntsen DO, Arukwe A, Yadetie F, Lorens JB, Male R (1999) Salmon eggshell protein expression: a marker for environ- mental estrogens. Mar Biotechnol 1:252–260

Pawlowski S, Islinger M, Völkl A, Braunbeck T (2000) Temperature-dependent vitellogenin-mRNA expression in primary cultures of rainbow trout (*Oncorhynchus mykiss*) hepatocytes at 14 and 18°C. Toxicol In Vitro 14:531–540

Schneider WJ (1996) Vitellogenin receptors: oocyte-specific members of the low-density lipoprotein receptor supergene family. Int Rev Cytol 166:103–137

Shimizu Y, Nakamura A, Kishimura H, Hara A, Watanabe K, Saeki H (2009) Major allergen and its IgE cross-reactivity among salmonid fish roe allergy. J Agric Food Chem 57:2314–2319

Silversand C, Haux C (1995) Fatty acid composition of vitellogenin from four teleost species. J Comp Physiol 164:593–599

Sire MF, Babin PJ, Vernier JM (1994) Involvement of the lysosomal system in yolk protein deposit and degradation during vitellogenesis and embryonic development in trout. J Exp Zool 269:69–83

Sugawara T (2011) Screening systems for endocrine disruptors. In: Reproductive and developmental toxicology. Academic, London, pp 893–902

Sullivan CV, Yilmaz O (2018) Vitellogenesis and yolk proteins, fish. In: Encyclopedia of reproduction, vol 6, 2nd edn. Elsevier, Amsterdam, pp 266–277

Taya T, Yamasaki N, Tsubamoto H, Hasegawa A, Koyama K (1995) Cloning of a cDNA coding for porcine zona-pellucida glycolprotein ZP1 and its genomic organization. Biochem Biophys Res Comm 207:790–799

Wallace RA (1985) Vitellogenesis and oocyte growth in non-mammalian vertebrates. In: Browder LW (ed) Developmental biology, vol 1. Plenum Press, New York, pp 127–177

Yamamura JI, Adachi T, Aoki N, Nakajima H, Nakamura R, Matsuda T (1995) Precursor-product relationship between chicken vitellogenin and the yolk proteins: the 40 kDa yolk plasma glycoprotein is derived from the C-terminal cysteine rich domain of vitellogenin II. Biochim Biophys Acta 1244:384–394

Yamane K, Yagai T, Nishimiya O, Sugawara R, Amano H, Fujita T, Hiramatsu N, Todo T, Matsubara T, Hara A (2013) Characterization of vitellogenin and its derived yolk proteins in cloudy catshark (*Scyliorhinustorazame*). Fish Physiol Biochem 39:373–390

Vitellogenin Receptor in Fishes

12

Maharajan Athisuyambulingam
and Ganapiriya Viswambaran

Abstract

The main component of the yolk of vertebrate eggs is vitellogenin. As essential nutrients for developing embryos, all fishes produce vitellogenins. The majority of the components required to create and support a new life are delivered to the ooplasm of oocytes by vitellogenin when they develop by orders of magnitude during oogenesis. Vitellogenin is a complex glycophospholipoprotein in the blood of mature female fishes produced by the liver in response to circulating estrogen, released into the bloodstream and taken up by growing oocytes, and chemically modified to form a suite of egg yolk proteins. Vtg is unique to mature females, hence measuring vitellogenin expression or plasma levels is thought to be a helpful method for assessing female maturity in relation to changes in peripheral gonadal steroid levels. Nevertheless, yolk precursor proteins can be found in males or juveniles exposed to estrogens. This protein is typically not detectable in males or juveniles.

Keywords

Vertebrate · Steroid · Oogenesis · Ooplasm · Receptor

12.1 Introduction

Hepatic production and bloodstream secretion of vitellogenin are key components of vertebrate vitellogenesis. Pan et al. (1969) suggested the word vitellogenin to describe the female-specific blood-borne yolk precursors seen in insects. The term

M. Athisuyambulingam (✉) · G. Viswambaran
PG & Research Department of Zoology, Khadir Mohideen College (Affiliated to Bharathidasan University, Tiruchirappalli), Adirampattinam, Tamil Nadu, India

V. Baskaralingam, R. Vanichviriyakit (eds.), *Vitellogenin in Fishes-Diversification, Biological Properties, and Future Perspectives*,
https://doi.org/10.1007/978-981-99-5340-0_12

vitellogenin has mainly been accepted to denote yolk precursor proteins found in the blood of oviparous vertebrates. Subsequently, the same sort of yolk precursors have been reported in many different phyla. The majority of taxa, including teleost fishes, amphibians, and reptiles (Morini et al. 2020), have a single protein called vitellogenin that is present in the blood. Oocyte growth in oviparous species is dependent on the uptake of nutrients and their storage as yolk, whose constituents are subsequently used by the embryo during early stages of its development. One of the most varied groups of vertebrates is the teleosts. Their capacity to use different reproductive systems to fill a variety of ecological niches has been a key factor in their success. These reproductive systems are behavioral as well as biological. Different gender differentiation, fertilization techniques, and the quantity of spawning cycles are all features of biological systems. Mating systems and parental care are examples of behavioral systems.

12.2 Vitellogenin in Fishes

A complex glycol-phospho-lipoprotein called vitellogenin is created by the liver in response to circulating estrogen in the blood of mature female fishes. It is then released into the bloodstream, taken up by developing oocytes, and chemically changed to form a group of egg yolk proteins (Specker and Sullivan 1994). Vitellogenin makes its way through multiple cellular and extracellular layers surrounding the oocyte from the capillary network displacing blood at the periphery of the follicle. After penetrating the vitellogenin reaches the oolemma by passing along the oocyte microvilli and internalized through specific cell surface receptors. The synthesis of vitellogenin in hepatocytes has been vastly studied and is well demonstrated to be under the control of estradiol through esterogen receptors.

Receptors that transport vitellogenin into oocytes are of pivotal significance to egg-laying animals since they mediate a key step in oogenesis. In order to satisfy the avid demand of the developing oocytes for circulating vitellogenin, precise mechanisms must have evolved to assure efficient incorporation of vitellogenin. The knowledge of the vitellogenin receptor system in fish is very scarce compared to birds, amphibians, and invertebrates. The piscine receptor for vitellogenin was first characterized in the ovary of coho salmon, *Oncorhyncuskysutch* (Stifani et al. 1990). Later specific receptors for vitellogenin were identified in rainbow trout *Oncorhynchus mykiss* (Rodriguez et al. 1996). Though the vitellogenin-specific receptors are highly conserved between species, specific characteristics are there for different species that contain species with different reproductive strategies. The information on vitellogenin receptor system will be necessary to understand the mechanisms regulating oocyte growth in fish.

In fish, the vitellogenin is specifically incorporated into oocytes by receptor-mediated endocytosis at specific areas known as coated pits (Barber et al. 1991). In fishes, vitellogenin (Vg) binds to a specific receptor (VgR) on the oocyte surface and is then sequestered via receptor-mediated endocytosis. A membrane receptor on the oocyte surface with a high affinity for Vg, called the Vg receptor (VgR), mediates the endocytotic process. Ooytes sequester vitellogenin through a process of receptor-mediated endocytosis. There is specific binding of vitellogenin to fish ovarian membrane preparations using a salmonid species to confirm the existence of a receptor-mediated system for vitellogenin internalization in fishes. The salmon vitellogenin receptor was found to resemble the vitellogenin receptor of chicken and Xenopus with regard to its estimated mass, binding kinetics, ligand specificity, and localization to the ovary (Tyler and Lancaster 1993; Rodriguez et al. 1996).

12.3 Vitellogenin Receptor in Fishes

The vitellogenin receptor (VTGR), which is also found in non-oviparous vertebrates like humans, is essential for oocyte development in egg-laying mammals. The primary source of nutritional reserves for the growing ebryos is the integration and proteolytic cleavage of vitellogenin into oocytes. The VTGR, also known as very-low-density lipoprotein receptor (VLDLR), is a member of the low-density lipoprotein receptor superfamily (LDLR). The low-density lipoprotein receptor (LDLR) family includes the vitellogenin receptor (Hussain et al. 1999). Members of the LDLR family are involved in lipid metabolism in both vertebrates and invertebrates by binding a variety of ligands.

In teleosts, the vitellogenin receptors have been described including, for example, rainbow trout, *Oncorhynchus mykiss* (Prat et al. 1998), white perch, *Morone americana* (Hiramatsu et al. 2002), cutthroat trout, *Oncorhynchus clarkii* (Mizuta et al. 2017), or in eels. In fish, the molecular weight of the vitellogenin receptors for coho salmom (*Oncorhyncus kisutch*) 100 kDa (Stifani et al. 1990), common carp (*Cyprinus carpio*) 90 kDa (Le Menn and Núñez Rodriguez 1991), and rainbow trout (*Oncorhynchus mykiss*) 200 kDa (Tyler and Lancaster 1993).

Generally, circulating vitellogenin forms a complex with the vitellogenin receptor at the plasma membrane of oocytes (Wall and Patel 1987). Internalized as coated vesicles into the cytoplasm is the vitellogenin receptor complex. The mature yolk protein is created from vitellogenin in the end, and the receptor for vitellogenin is then attracted to the cell membrane via tubular vesicles. Hence, the vitellogenin receptor is essential for fish oogenesis, which is further demonstrated by data that genetic deletions or mutations of the receptor could impede or cause aberrant ovarian development as well as, occasionally, female sterility. Five highly conserved structural domains make up the vitellogenin receptor: the ligand-binding domain, the EGF precursor homology domain, the O-linked sugar domain, the transmembrane domain, and the cytosolic domain (Tufail and Takeda 2009).

The most dominant trigger of vitellogenin expression is the ovarian steroid hormone 17β-estradiol (E2) that is synthesized under the regulation of the hypothalamic–pituitary–gonad axis (Polzonetti-Magni et al. 2004). The evidence available thus far supports the hypothesis that particular nuclear estrogen receptors serve as the primary mediators of estrogen activity (ERs). Using estrogen receptors on the hepatocytes, the estrogen acts on the vitellogenin gene in the nucleus by attaching to sex steroid hormone-binding globulin in the blood. Hepatocytes' combination of estrogen and the oestrogen receptor attaches to the vitellogenin gene's promoter region and activates it to start and speed up transcription. Blood vitellogenin enters the cell after binding to the vitellogenin receptor on the oocyte plasma membrane. Circulating estradiol travels into the liver cells and attaches to an esteradiol receptor, causing a conformational shift and dimerization. The vitellogenin gene's protmoter region contains incomplete esterogen receptor sequences or esterogen response elements, which the dimerized complexes bind to to start the gene's expression and produce vitellogenin.

Estrogens diffuse into the cell during this process and bind to ERs, which are found in the cytosol or nucleus of the target cells. The ERs form homo- or hetero dimers after ligand interaction, and these dimers bind to particular palindromic estrogen response elements (ERE) sequences (Gruber et al. 2004) in the promoter region of estrogen-responsive genes, resulting in recruitment of coactivators or corepressors to the promoter. This results in altered quantities of mRNA and related protein synthesis, which triggers the physiological response. (Kinge et al. 2004). In teleosts, there are several forms of esterogen receptors. Three esterogen receptor ER subtypes were described so far for fish and include the estrogen receptor 1, estrogen receptor 2b, and estrogen receptor 2a [with the gene names of estrogen receptor 1 (esr1), estrogen receptor 2b (esr2b), and estrogen receptor 2a (esr2a), respectively) (Menuet et al. 2002; Hawkins et al. 2000)]. The term "nongenomic activities" refers to some of estrogens' effects that happen so quickly that they cannot rely on RNA and protein synthesis. They entail triggering protein-kinase cascades, which ultimately result in the phosphorylation and activation of transcription factors (TFs) in the nucleus, regulating gene expression. Vitellogenin is strictly regulated by E2-dependent up-regulation of esrl expression in the liver. Vtg is unique to mature females, hence measuring vitellogenin expression or plasma levels is thought to be a helpful method for assessing female maturity in relation to changes in peripheral gonadal steroid levels. Nevertheless, yolk precursor proteins can be found in males or juveniles exposed to estrogens. This protein is typically not detectable in males or juveniles.

Since E2 stimulation increases hepatic ER1 mRNA expression, esterogen receptor 1 appears to be crucial for the transcription of vitellogenin genes (Mushirobira et al. 2018). The complexes are endocytosed in clathrin-coated pits that invaginate to form coated vesicles after vitellogenin binds to their receptors. These endocytosed vesicles form multivesicular structures when they fusion with lysosomes in the peripheral ooplasm. The proteolytic enzyme cathepsin D is present in the lysosomes, and it is possible that other enzymes (like cathepsin B) colocalize with the imported vitellogenins. When the pH of multivesicular structures drops, catepsin D is activated and cleaves vitellogenin into yolk proteins, which are then stored as yolk granules, globules, platelets, or as liquid yolk in the ooplasm (Romano et al. 2004). The isolation of a cDNA encoding for the trout oocyte Vtg receptor led to the identification of the receptor's site of production. As shown by in situ hybridization experiments, transcripts are not found in ovarian somatic cells and are exclusively found in oocytes and the absence of transcripts from oocytes during their phase of rapid growth, when receptors are absorbing Vtg at the highest rates.

12.4 Conclusion

Regulation of Vtg receptor gene expression has potential commercial implications in addition to scientific ones, at least in fishes. Additionally, vitellogenesis has been suggested as a biomarker system for estrogenic contamination of the aquatic environment.

References

Barber DL, Sanders EJ, Aebersold R, Schneider WJ (1991) The receptor for yolk lipoprotein deposition in the chicken oocyte. J Biol Chem 266:18761–18770

Gruber CJ, Gruber DM, Gruber IM, Wieser F, Huber JC (2004) Anatomy of the estrogen response element. Trends Endocrinol Metab 15:73–78

Hawkins MB, Thornton JW, Crews D, Skipper JK, Dotte A, Thomas P (2000) Identification of a third distinct estrogen receptor and reclassification of estrogen receptors in teleosts. Proc Natl Acad Sci U S A 97:10751–10756

Hiramatsu N, Hara A, Hiramatsu K, Fukada H, Weber GM, Denslow ND, Sullivan CV (2002) Vitellogenin-derived yolk proteins of white perch, Morone americana: purification, characterization, and vitellogenin-receptor binding. Biol Reprod 67:665–667

Klinge CM, Jernigan SC, Mattingly KA, Risinger KE, Zhang J (2004) Estrogen response element-dependent regulation of transcriptional activation of estrogen receptors alpha and beta by coactivators and corepressors. J Mol Endocrinol 33:387–410

Le Menn F, Núñez Rodriguez J (1991) Receptors mediate endocytosis of VTG in fish follicle. In: Scott AP, Sumpter JP, Kime DE, Rolfe MS (eds) Proceeding of the 4th international symposium on the reproductive physiology of fish. Sheffield University Press, Sheffield, pp 300–302

Mahmood Hussain M, Bakillah A, Nayak N, Gregory S. (1998) Shelness, Amino Acids 430–570 in Apolipoprotein B Are Critical for Its Binding to Microsomal Triglyceride Transfer Protein Jour of Biol Chem. 273(40):25612–25615

Menuet A, Pellegrini E, Anglade I, Blaise O, Laudet V, Kah O, Pakdel F (2002) Molecular characterization of three estrogen receptor forms in zebrafish: binding characteristics, transactivation properties, and tissue distributions. Biol Reprod 66:1881–1892

Mizuta H, Mushirobira Y, Nagata J, Todo T, Hara A, Reading BJ, Sullivan CV, Hiramatsu N. (2017) Ovarian expression and localization of clathrin (Cltc) components in cut throat trout, Oncorhynchus clarki: evidence for Cltc involvement in endocytosis of vitellogenin during oocyte growth. Compar Bioche and Physio, Part A: Molecul and Integ Physiol. 212:24–34.

Morini M, Lafont AG, Maugars G, Baloche S, Dufour S, Asturiano JF, Pérez L. (2020) Identification and stable expression of vitellogenin receptor through vitellogenesis in the European eel, Animal. 14(6):1213–1222.

Mushirobira Y, Nishimiya O, Nagata J, Todo T, Hara A, Reading BJ, Hiramatsu N. (2018) Molecular cloning of vitellogenin gene promoters and in vitro and in vivo transcription profiles following estradiol-17β administration in the cutthroat trout. Gen Comp Endocrino. 267: 157–166.

Pan ML, Bell WJ, Telfer WH. (1969) Vitellogenic blood protein synthesis by insect fat body. Science. 165:393–394.

Polzonetti-Magni AM, Mosconi G, Soverchia L, Kikuyama S, Carnevali O (2004) Multihormonal control of vitellogenesis in lower vertebrates. Int Rev Cytol 239:1–46

Prat F, Coward K, Sumpter J, Tyler C. (1998) Molecular characterization and expression of two ovarian lipoprotein receptor in the rainbow trout, Oncorhynchus mykiss Biol of Reprod. 58:1146–1153.

Rodriguez J-N, Bon E, Le Menn F (1996) Vitellogenin receptors during vitellogenesis in the rainbow trout Oncorhynchus mykiss. J Exp Zool 274:163–170
Romano M, Rosanova P, Anteo C, Limatola E (2004) Vertebrate yolk proteins: a review. Mol Reprod Dev 69:109–118
Specker JL, Sullivan CV (1994) Vitellogenesis in fishes: status and perspectives. In: Davey KG, Peter RE, Tobe SS (eds) Perspectives in comparative endocrinology. National Research Council, Ottawa, pp 304–315
Stifani S, Le Menn F, Rodriguez J-N, Schneider WJ (1990) Regulation of oogenesis: the piscine receptor for vitellogenin. Biochem Biophys Acta 1045:271–279
Tufail M, Takeda M (2009) Insect vitellogenin/lipophorin receptors: molecular structures, role in oogenesis, and regulatory mechanisms. J Insect Physiol 55:87–103. https://doi.org/10.1016/j.jinsphys.2008.11.007
Tyler CR, Lancaster P (1993) Isolation and characterization of the receptor for vitellogenin from follicles of the rainbow trout, Oncorhynchus mykiss. J Comp Physiol B 163:225–233
Wall DA, Patel S (1987) The intracellular fate of vitellogenin in Xenopus oocytes is determined by its extracellular concentration during endocytosis. Jour of Biol Chemist. 262 (30):14779–14789

Molecular Cloning and Induction of Vitellogenesis

13

Muthukumar Abinaya and Vaseeharan Baskaralingam

Abstract

Fish hepatocytes produce vitellogenin, a calcium-binding glycolipophosphoprotein, which can be induced in fishes to synthesize it by either estradiol or xenoestrogens. Regarding estrogenic compounds, the lowest effective doses of vitellogenin were reported, and it can be triggered via xenoestrogens. In aquatic ecosystems, estrogenic molecules such as natural steroidal estrogens and synthetic pollutants are extensively abundant. In recent times, the induction of Vtg has been suggested as a potential biomarker to detect estrogenic pollutants. Indeed, the estrogenicity of distinct compounds and mixtures was intensively explored in laboratories and field research, primarily in fish. However, due to our understanding of the endocrinology of aquatic invertebrates, little focus has been placed on studying the effects of xenoestrogen on these organisms. This chapter describes about the induction of Vtg in aquatic invertebrates which response to exposure of estrogenic compounds in both experimental and natural environments. At last, this chapter points out the molecular cloning on how these estrogenic compounds can modulate Vtg in fishes for a better understanding of the environmental behavior.

Keywords

M. Abinaya
Crustacean Molecular Biology and Genomics Division, Department of Animal Health and Management, Biomaterials and Biotechnology in Animal Health Lab, Alagappa University, Karaikudi, Tamil Nadu, India

V. Baskaralingam (✉)
Department of Animal Health and Management, Alagappa University, Karaikudi, Tamil Nadu, India

V. Baskaralingam, R. Vanichviriyakit (eds.), *Vitellogenin in Fishes-Diversification, Biological Properties, and Future Perspectives*,
https://doi.org/10.1007/978-981-99-5340-0_13

Vitellogenin · Xenoestrogens · Induction · Molecular cloning

Abbreviations

AR	Androgen receptors
BP	Benzophenone
BPA	Bisphenol A
Ca	Calcium
DDT	Dichloro-diphenyl-trichloroethane
E1	Estrone
E2	Estradiol
E3	Estriol
EDCs	Endocrine-disrupting chemicals
EE2	Ethinylestradiol
ELISA	Enzyme-linked immuno sorbent assay
ER	Estrogen receptor
EREs	Estrogen response elements
GtH-I	Gonadotropin I
NP	Nonylphenol
NPCs	Nonylphenol carboxylates
NPEs	Nonylphenol ethoxylates
OP	Octylphenol
OPEs	Octylphenol ethoxylates
PCBs	Polychlorinated biphenyls
Vn	Vitellins
Vtg/Vg/VTG	Vitellogenin
VtgR	Vitellogenin receptor
Zn	Zinc

13.1 Introduction

Vitellins (Vn) are egg yolk proteins that provide energy reserves for embryonic development in oviparous species and are mostly derived from vitellogenins (Vg), which are their main precursors. They are high-density glycolipophosphoproteins (300 to 700 kDa) with Ca and Zn ligands that depending on the species and vertebrates like fish show similar traits to these proteins (Nagler et al. 1987). Fish are one of the first target organisms for these actions of the EDCs in aquatic environments, and EDCs may promote toxic effects on fish reproduction by acting in the hypothalamic–pituitary–gonad–liver (HPGL) axis, which is responsible for the maintenance of the reproduction in these animals (Ankley et al. 2009; Hachfi et al. 2012; Kar et al. 2021). An important process promoted by HPGL axis is vitellogenin (Vtg) synthesis (Sumpter and Jobling 1995). In mature females, vg synthesis occurs in reaction to endogenous estrogens, and it is often lower in juveniles. Despite being present in males, the Vg gene is typically not expressed. It may, however, be activated by xenoestrogens (Wahli et al. 1981; Flouriot et al. 1995). These substances are referred to as "endocrine-disrupting chemicals" because they are part of a sizable, diverse collection of environmental pollutants known to be capable of altering endocrine processes. Furthermore, several of these contaminants are endocrine disruptor chemicals (EDCs), which evidence for the occurrence of these chemicals in the aquatic environment. In aquatic ecosystems, estrogenic molecules such as natural steroidal estrogens and synthetic pollutants are extensively abundant. Estrogenic compounds act by preventing hormonal binding to hormone receptors or by binding specific hormone receptors, mimicking the action of endogenous estrogens (Falconer et al. 2006).

In accordance with the most recent studies (Carducci et al. 2019; Hiramatsu et al. 2013; Yilmaz et al. 2018), teleost and other vertebrates exhibit multiple types of Vtg, which are associated with the expression of one to three Vtgs. These variants have similarities but can also vary on the basis of molecular structure as well as function while the developmental process. As a result, the evaluation of Vtg, both in terms of gene and protein expression in male fish, is regarded as a well-established biomarker in studies examining the effects of estrogenic drugs. However, there is still little knowledge about possible biomarkers related to its endocrine system (Barcellos et al. 2001; Costa et al. 2010; Fernandes et al. 2021). In this way, expanding knowledge about Vtg and its expansion as a biomarker for the exposure of EDCs in fishes. Henceforth, the present chapter focused to briefly summarize the induction of vitellogenin in fishes through estrogenic compounds and its molecular cloning were explored.

13.2 Endocrine-Disrupting Chemicals (EDCs) in Aquatic Environments

In the past decades, increasing attention has been given to evaluating adverse effects of EDCsin aquatic environments. Key research examined the impacts of EDCs on aquatic creatures and their methods of action (Depledge and Billinghurst 1999; Segner et al. 2003). According to Neubert (1997), EDCs are a diverse range of chemicals that might affect an organism's endocrine system. These chemicals can be either natural or manmade. According to Soto et al. (1995), EDCs can mimic the sex steroid hormones estrogens and androgens, by binding to hormone receptors or influencing cell signaling pathways; block, prevent, and alter hormonal binding to hormone receptors or influence cell signaling pathways; alter production and breakdown of natural hormones; and modify levels and function of hormone receptors.

Among EDCs, xenoestrogens received major attention owing to their capability to mimic natural estrogens (estrogen mimics) (WHO/IPCS 2002). The relatively low specificity of estrogen receptors facilitates estrogenic activity of xenoestrogens. The following endocrine distributions in aquatic environment are listed in Table 13.1.

These substances may activate estrogen receptors by direct binding or through other receptor and/or signal transduction pathways and may also prevent hormonal

Table 13.1 The chemicals implicated on endocrine disruption in aquatic environment

Endocrine Distributing Chemicals (EDCs) in aquatic environment		References
Steroidal estrogens	Both natural (Estradiol (E2), Estriol (E3) and Estrone (E1)), and Synthetic (Ethinylestradiol (EE2), Mestranol (MES))	Desbrow et al. (1998), Pojana et al. (2004)
Non-steroidal synthetic estrogenic compounds	Nonylphenol (NP), Nonylphenol ethoxylates (NPEs), Nonylphenol carboxylates (NPCs), Octylphenol (OP), Octylphenol ethoxylates (OPEs), Benzophenone (BP), Bisphenol A (BPA)	Pojana et al. (2004)
Phytoestrogens	Genistein	Safe and Gaido (1998)
Pesticides	Endosulfan, Dichloro-diphenyl-trichloroethane (DDT), Dieldrin, Alachlor, Atrazine, Nitrofen	DeRosa et al. (1998)
Polychlorinated biphenyls (PCBs)	NA	DeRosa et al. (1998)
Plasticisers	Phthalates	Harris et al. (1997)
Heavy metals	Mercury, Cadmium and Lead	DeRosa et al. (1998)

binding to hormone receptors (Soto et al. 1995; Gillesby and Zacharewski 1998; Kirk et al. 2003). It is also difficult to determine which chemicals, both natural and manufactured, may behave as hormone-like compounds in aquatic biota because of the variability in chemical structures of environmental estrogens. Risk assessment of EDC exposure is a challenge since these contaminants are present as complex mixtures that can trigger several mechanisms of estrogenic/antiestrogenic activity (Kiyama and Wada-Kiyama 2015). Biomarkers are often used in risk assessment and to assess effects of EDCs on wildlife. The evaluation of numerous endpoints, including steroid hormone levels and steroidogenesis as the functional and pathological parameters in target tissues, is necessary for the successful integration of biomarkers in EDC assessment (Carballo et al. 2005; Hecker et al. 2002; Hinck et al. 2007; Vajda et al. 2011).

13.3 Induction of Vtg

One of the most intriguing aspects of Vtg is that it can be made in systems where its synthesis would not typically take place by exposure to estrogen, such as male liver, in vitro liver tissue, and cultured hepatocytes. Vitellogenesis is initiated by the pituitary gland in response to environmental cues (Kime et al. 1999). The pituitary responds to hypothalamic stimulation (Kime et al. 1999; Nagahama 1983) by secreting gonadotropin I (GtH-I), a peptide hormone that regulates ovarian function and promotes meiotic maturation. GtH-I circulates in the blood plasma and, upon reaching the ovaries, stimulates follicle cells to synthesize estradiol, which is released into the serum, where it may bind to sex hormone-binding protein. However, it is the unbound portion of the hormone that is physiologically active (the free hormone hypothesis) as the protein hormone complex is too large to cross capillary walls (Mendel 1989). Estradiol diffuses freely across the membrane of liver cells and binds to the estrogen receptors, so initiating transcription and translation of Vtg. Vtg accumulates in the hepatocytes to be modified by the rough endoplasmic reticulum and the Golgi apparatus, and is then released to the blood plasma. On reaching the ovaries, it passes through the follicle cells along a capillary network to reach the oocyte surfaces. Vtg incorporation into the oocyte is also under GtH-I control (Kime et al. 1999) via receptor-mediated endocytosis into yolk platelets (Tyler et al. 1990) (Fig. 13.1).

In this regard, a number of potential biomarkers have been validated and used in the study of endocrine disruption in fish and other aquatic organisms (Arukwe et al. 2000; Jones et al. 2000; Kleinkauf et al. 2004; Mitchelmore and Rice 2006; Matozzo et al. 2008; Zaccaroni et al. 2009). Induction of vitellogenin (Vtg) synthesis in immature and male fish has been the most commonly used biomarker of exposure to estrogenic contaminants in the aquatic environments (Naderi et al. 2015; Wheeler et al. 2005).

Endocrine-disrupting chemicals (EDCs) appear to be a threat to the reproductive fitness of aquatic organisms. Because of structural similarity with endogenous

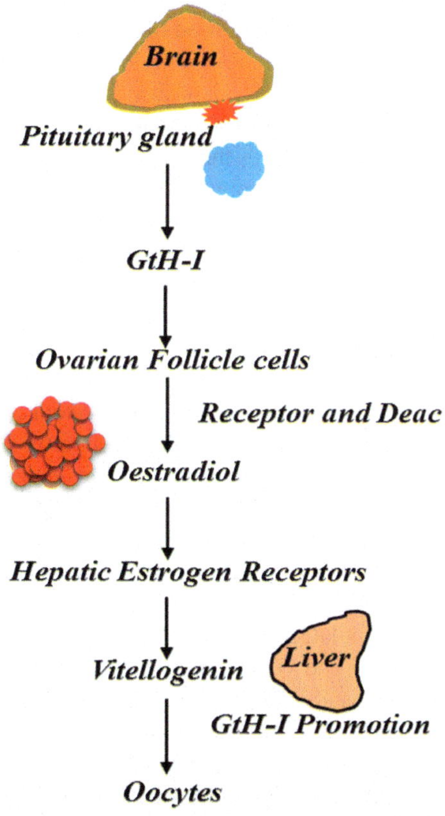

Fig. 13.1 Induction of vitellogenin

17-estradiol (E2), EDCs are able to interfere with normal hormone functions and induce estrogenic properties in fish (Goksøyr 2006). Although, these exogenous estrogens are less potent than steroid estrogens, they are more persistent in the environment, accumulate in fish, and may exhibit synergistic effects with other exogenous estrogens or steroid estrogens. In addition, one of these EDCs is bisphenol A (BPA) (Jobling et al. 2003).

13.4 Induction of Vtg by 17β-Estradiol

The hepatic induction of vitellogenin (Vtg) synthesis in male and juvenile oviparous fish, which normally only occurs in maturing females following stimulation by 17β-estradiol, has been proposed as a sensitive biomarker for estrogenic chemicals of exogenous origin (Sumpter and Jobling 1995). The induction of this biomarker protein has been utilized as a suitable testing method for chemical screening and environmental monitoring in vivo (Arukwe et al. 1997; Knudsen and Pottinger 1998; Andersen et al. 1999). Although in vivo testing of chemicals has become increasingly important in the risk assessment of estrogenic chemicals, in vitro bioassays may still act as a useful tool for rapid and cost-efficient screening of chemicals. Induction of vitellogenin (Vtg) synthesis in immature, and male fish has been the most commonly used biomarker of exposure to estrogenic contaminants in the aquatic environments (Wheeler et al. 2005).

It has been well established like in other oviparous vertebrates (Byrne et al. 1989; Selman et al. 1993). The studies described here demonstrate that the administration of estradiol induces the synthesis of vitellogenin. According to Kishida et al. (1992), the first time, the presence of vitellogenin in the surface mucus of a teleost fish is found out by detecting vitellogenin on female striped bass during their spawning migration. In addition to E2, several other steroid hormones are involved in up- or downregulation of Vg synthesis in vivo and in vitro conditions by the hepatocytes (Lethimonier et al. 2000). Additionally, 4-nonylphenol and 17β-estradiol exposure caused juvenile male yellowfin seabream (*A.s latus*) to produce Vtg (Naderi et al. 2014). Hence, its indirect quantification is achieved by the measurement of ALP, which has been widely used in different aquatic organisms like fish and bivalve mollusks (Gagnaire et al. 2009; Kramer et al. 1998; Ricciardi et al. 2008; Verslycke et al. 2002). According to Verslycke et al. (2002), there is a significant association between fish ALP levels and actual VTG levels as determined by particular immunotechnique assays. Similar to this, two protein bands were seen as a result of treating young male *A. latus* with E2. Prior to this, Naderi et al. (2015) reported the occurrence of two distinct bands of Vtg in *A. latus*. It demonstrates once again that the stimulated protein is Vtg. Furthermore, other investigations (Chu-Koo et al. 2009; Li and Wang 2005; Van den Belt et al. 2004) have reported the development of two bands of Vtg. These elevated levels of plasma ALP clearly indicate the VTtg induction in 4-NP-treated fish. Similarly, Christensen et al. (1999) observed a significant increase in plasma levels of ALP in male flounders (*Platichthysflesus*) treated with different doses of 4-NP. Induction in plasma ALP concentration was also found in adult fathead minnows (*Pimephalespromelas*) exposed to E2 through waterborne exposure (Kramer et al. 1998). Additionally, E2 therapy through diet led to a considerable rise in plasma ALP levels over the course of 85 days in comparison to controls in *Percafluviatilis* juvenile Eurasian perch (Mandiki et al. 2005). Another xenoestrogen chemical, 4-tert-octylphenol, was found to cause a significant increase in plasma ALP levels in the sand goby *Pomatoschistus minutus* (Robinson et al. 2004). Likewise, McCormick et al. (2005) found that the intraperitoneal administration of juvenile Atlantic salmon (*Salmo salar*) by 4-NP and E2 led to the induction of Vtg and total calcium in the plasma. This was consistent with the results of Christiansen et al. (1998), who reported on sexually immature salmonids exposed to 4-NP.

Verslycke et al. (2002) noted the considerable elevation of plasma protein in rainbow trout *Oncorhynchus mykiss* after intraperitoneal injection of EE2 and waterborne exposure. The induction of plasma protein was also reported in European eel (*Anguilla anguilla*) exposed to waterborne EE2 (Versonnen et al. 2004). Nevertheless, the protein induction (induced by 200 mg of 4-NP) in *A. latus* was much stronger than EE2-treated *O. mykiss* and *A. Anguilla* (45 and 75 mg/mL, respectively), despite the fact that the estrogenic substance used in those experiments had greater potential than NP (Folmar et al. 2002; Van den Belt et al. 2004).

13.5 Induction of Vtg by Bisphenol-A

BPA is utilized in the manufacture of polycarbonate plastics and epoxy resins and is found in food packaging, electronic instruments, fungicides, dental sealants, and paper coatings (Crain et al. 2007; Flint et al. 2012; Rubin 2011). It is produced worldwide, with an estimated production of 3.9 million tons in 2006 (Ballesteros-Gómez et al. 2009). Of this amount, about 100 tons is released into the environment (Rykowska and Wasiak 2006). BPA detection in aquatic environments and human tissues (Carwile et al. 2011; Geens et al. 2012; Sánchez-Avila et al. 2012) originated from increases in consumption of BPA products in recent decades.

The effects of BPA on aquatic organisms have been reported in previous studies. Based on the mentioned characteristics, Vtg could be considered as an appropriate biomarker for xenoestrogens in aquatic environments (Fujiwara et al. 2005; Scott et al. 2006; Hiramatsu et al. 2006; Johnson et al. 2008; Matozzo et al. 2008; Naderi et al. 2014; Teta and Naik 2017) For example, the induction of Vitellogenin production in fish exposed to BPA has been reported (Lindholst et al. 2003). Concentrations of BPA as low as 16 mg/L have been shown to interfere with spermatogenesis in mature male fathead minnows (*Pimephalespromelas*) (Sohoni et al. 2001).

The plasma Vtg levels were increased in treated immature fish in comparison with control groups after exposure to different doses of BPA, indicating an estrogenic effect of this compound. Several investigations have found that xenoestrogens induce total plasma protein in fish (Verslycke et al. 2002; Van den Belt et al. 2004; Versonnen et al. 2004; Naderi et al. 2015). For instance, *Cyprinus carpio* was exposed to BPA in water for 21 days at a concentration of 100 mg/L, and the elevated plasma Vtg was seen (Virk et al. 2014). Additionally, immature male yellowfin seabream (*A. latus*) treated to 10, 50, 100, 150, and 200 mg/g of 4-nonylphenol over a 14-day period had their plasma Vtg concentration (ALP) induced (Naderi et al. 2015). Elevated plasma VTG was seen in *Dicentrarchuslabrax* (Correia et al. 2007), *Carassius auratus* (Hatef et al. 2012), and *Sebastes schlegeli* (Keum et al. 2005) exposed to different doses of BPA. Tabata et al. (2003) reported that the estrogenic activity of BPA was decreased by chlorination treatment. However, several studies have shown that estrogenic potencies of chlorinated BPA were greater than BPA in vitro (Hu et al. 2002; Fukazawa et al. 2002). Bisphenol A (BPA) has been reported to behave as an endocrine disrupter below acute toxic levels (Milligan et al. 1998; Schafer et al. 1999; Lindholst et al. 2000; Yokota et al. 2000; Ishibashi et al. 2001; Sohoni et al. 2001; Kang et al. 2002).

13.6 Molecular Cloning Aspects of Vtg Induction

Nuclear receptors, one of the main targets of EDCs, are a superfamily of proteins that play a crucial role in the hormone system of vertebrates. Estrogen binds to and activates the estrogen receptor (ER), which initiates a protein conformational change. Activated estrogen/ER complexes dimerize and then bind to estrogen response elements (EREs) in a promoter region of the Vtg gene, which is followed by transcription (Nelson and Habibi 2013). ERα and ERβ and androgen receptors (AR) pertain to the nuclear receptor superfamily of ligand-activated transcription factors that modulate specific gene expression (Kuiper and Gustafsson 1997; Mosselman et al. 1996).

Three nuclear estrogens, such as ERs, ERα, ERβ, and ERγ, have been detected and characterized in fish, including largemouth bass *Micropterus salmoides* (Sabo-Attwood et al. 2004), zebrafish *Danio rerio* (Menuet et al. 2002), and Atlantic croaker *Micropogonias undulatus* (Hawkins et al. 2000). The vast majority of teleost have at least three distinct subtypes, like ERα, ERβ, and ERγ (Hawkins et al. 2000; Hawkins and Thomas 2004; Ma et al. 2000). According to studies conducted on the tilapia *Oreochromis mossambicus*, hepatic er is upregulated in response to estrogen (Davis et al. 2007) and this upregulation is correlated closely to the induction of VTG/Vtg synthesis in fishes (Nagler et al. 2012; Sabo-Attwood et al. 2004). Recently, a second form of hepatic erα (erα2), which was considered to be a minor er transcript, has been discovered in *Oncorhynchus mykiss* rainbow trout female vitellogenic (Nagler et al. 2007, 2012).

It is generally agreed that the induction of Vtg in hepatocytes is mediated through the binding of ligand (ERα & ERβ) complexes to activate the promoter region of the estrogen response element in DNA, resulting in increased mRNA transcription and subsequent translation followed by posttranslational modification to yield a mature Vtg protein that is detectable in plasma (Pakdel et al. 1991). For example, it was found that 17-estradiol caused the mud carp *Cirrhinusmolitorella* to express nuclear receptor and vitellogenin genes at the molecular level and in mRNA (Liang and Fang 2012). The mRNA expression of Vtg was undetectable in the liver of male zebrafish and medaka (Tong et al. 2004) and little discernible in the liver of male *Cyprinodon variegatus* sheepshead minnows (Bowman et al. 2000). However, E2 induced significant expression of Vtg in the liver of male fish, which suggested that the Vtg gene is a sensitive biomarker to monitor estrogenic effects in fish (Tong et al. 2004). The function of the Vtg gene expressed in tissues other than the liver is unknown (Mikawa et al. 2006).

13.7 Conclusion and Future Perspectives

In conclusion, vitellogenin is used as a biomarker for endocrine disruption in fish. Vtg are frequently used to assess exposure of animals in aquatic environments to EDCs, specifically to EDCs that mimic the action of estrogens. Because Vtgs are created in response to endogenous (E2), they could be appropriate indicators for assessing estrogenic EDCs. However, as disparities exist between the different types of Vtgs (or Vtg transcripts) with regard to their sensitivity to induction by estrogen(s), while it is important to consider the precise type of Vtg being evaluated in order to find EDCs. After the test has been properly optimized, Vtg synthesis occurs. In estrogenic substances, such 17-estradiol, nonylphenol, and bisphenol A, the lowest effective amounts of vitellogenin have been observed. This induction occurred at substantial higher concentrations than required for E2, and the estrogenic response for some of the environmental estrogens seemed to be limited by acute toxic stress on the cells. In future directions, the molecular aspects of vitellogenin induction and the impact of others chemicals on VTG genes in vivo and in vitro are necessary to investigate.

Acknowledgments The authors appreciatively acknowledge the financial support of DST-INSPIRE fellowship-IF160623, New Delhi, India.

Conflict of Interest The authors declare no conflict of interest.

References

Andersen HR, Andersson AM, Arnold SF, Autrup H, Barfoed M, Beresford NA, Grandjean P (1999) Comparison of short-term estrogenicity tests for identification of hormone-disrupting chemicals. Environ Health Perspect 107:89–108

Ankley GT, Bencic DC, Breen MS, Collette TW, Conolly RB, Denslow ND, Watanabe KH (2009) Endocrine disrupting chemicals in fish: developing exposure indicators and predictive models of effects based on mechanism of action. Aquat Toxicol 92(3):168–178

Arukwe A, Knudsen FR, Goksøyr A (1997) Fish zona radiata (eggshell) protein: a sensitive biomarker for environmental estrogens. Environ Health Perspect 105(4):418–422

Arukwe A, Celius T, Walther BT, Goksøyr A (2000) Effects of xenoestrogen treatment on zona radiata protein and vitellogenin expression in Atlantic salmon (*Salmo salar*). Aquat Toxicol 49(3):159–170

Ballesteros-Gómez A, Rubio S, Pérez-Bendito D (2009) Analytical methods for the determination of bisphenol A in food. J Chromatogr A 1216(3):449–469

Barcellos LJG, Woehl VM, Wassermann GF, Quevedo RM, Ittzés I, Krieger MH (2001) Plasma levels of cortisol and glucose in response to capture and tank transference in *Rhamdiaquelen* (Quoy & Gaimard), a South American catfish. Aquac Res 32(2):121–123

Bowman CJ, Kroll KJ, Hemmer MJ, Folmar LC, Denslow ND (2000) Estrogen-induced vitellogenin mRNA and protein in sheepshead minnow (*Cyprinodon variegatus*). Gen Comp Endocrinol 120(3):300–313

Byrne BM, Gruber MABG, Ab G (1989) The evolution of egg yolk proteins. Prog Biophys Mol Biol 53(1):33–69

Carballo M, Aguayo S, de la Torre A, Muñoz MJ (2005) Plasma vitellogenin levels and gonadal morphology of wild carp (*Cyprinus carpio* L.) in a receiving rivers downstream of sewage treatment plants. Sci Total Environ 341(1–3):71–79

Carducci F, Biscotti MA, Canapa A (2019) Vitellogenin gene family in vertebrates: evolution and functions. Eur Zool J 86(1):233–240

Carwile JL, Ye X, Zhou X, Calafat AM, Michels KB (2011) Canned soup consumption and urinary bisphenol A: a randomized crossover trial. JAMA 306(20):2218–2220

Christensen LJ, Korsgaard B, Bjerregaard P (1999) The effect of 4-nonylphenol on the synthesis of vitellogenin in the flounder *Platichthysflesus*. Aquat Toxicol 46(3–4):211–219

Christiansen PD, Brozek M, Hansen BW (1998) Energetic and behavioral responses by the common goby, *Pomatoschistusmicrops* (krøyer), exposed to linear alkylbenzene sulfonate. Environ Toxicol Chem Int J 17(10):2051–2057

Chu-Koo F, Dugué R, Alván Aguilar M, Casanova Daza A, Alcántara Bocanegra F, Chávez Veintemilla C, Nuñez J (2009) Gender determination in the Paiche or Pirarucu (*Arapaima gigas*) using plasma vitellogenin, 17β-estradiol, and 11-ketotestosterone levels. Fish Physiol Biochem 35:125–136

Correia AD, Freitas S, Scholze M, Gonçalves JF, Booij P, Lamoree MH, Reis-Henriques MA (2007) Mixtures of estrogenic chemicals enhance vitellogenic response in sea bass. Environ Health Perspect 115(Suppl 1):115–121

Costa DM, Neto FF, Costa MDM, Morais RN, Garcia JRE, Esquivel BM, Ribeiro CO (2010) Vitellogenesis and other physiological responses induced by 17-β-estradiol in males of freshwater fish *Rhamdiaquelen*. Comp Biochem Physiol C Toxicol Pharmacol 151(2):248–257

Crain DA, Eriksen M, Iguchi T, Jobling S, Laufer H, LeBlanc GA, Guillette LJ Jr (2007) An ecological assessment of bisphenol-A: evidence from comparative biology. Reprod Toxicol 24(2):225–239

Davis LK, Hiramatsu N, Hiramatsu K, Reading BJ, Matsubara T, Hara A, Grau EG (2007) Induction of three vitellogenins by 17beta-estradiol with concurrent inhibition of the growth hormone-insulin-like growth factor 1 axis in a euryhaline teleost, the tilapia (*Oreochromis mossambicus*). Biol Reprod 77(4):614–625

Depledge MH, Billinghurst Z (1999) Ecological significance of endocrine disruption in marine invertebrates. Mar Pollut Bull 39(1–12):32–38

DeRosa C, Richter P, Pohl H, Jones DE (1998) Environmental exposures that affect the endocrine system: public health implications. J Toxicol Environ Health B Crit Rev 1(1):3–26

Desbrow CEJR, Routledge EJ, Brighty GC, Sumpter JP, Waldock M (1998) Identification of estrogenic chemicals in STW effluent. 1. Chemical fractionation and in vitro biological screening. Environ Sci Technol 32(11):1549–1558

Falconer IR, Chapman HF, Moore MR, Ranmuthugala G (2006) Endocrine-disrupting compounds: a review of their challenge to sustainable and safe water supply and water reuse. Environ Toxicol Int J 21(2):181–191

Fernandes LDSP, Mathias FT, Richardi VS, Cardoso CC, Silva de Assis HC (2021) Cloning, partial sequencing and 17β-estradiol modulation of hepatic vitellogenin gene of the neotropical catfish *Rhamdiaquelen*. J Appl Ichthyol 37(4):545–552

Flint S, Markle T, Thompson S, Wallace E (2012) Bisphenol A exposure, effects, and policy: a wildlife perspective. J Environ Manag 104:19–34

Flouriot G, Pakdel F, Ducouret B, Valotaire Y (1995) Influence of xenobiotics on rainbow trout liver estrogen receptor and vitellogenin gene expression. J Mol Endocrinol 15(2):143–151

Folmar LC, Hemmer MJ, Denslow ND, Kroll K, Chen J, Cheek A, Grau EG (2002) A comparison of the estrogenic potencies of estradiol, ethynylestradiol, diethylstilbestrol, nonylphenol and methoxychlor in vivo and in vitro. Aquat Toxicol 60(1–2):101–110

Fujiwara Y, Fukada H, Shimizu M, Hara A (2005) Purification of two lipovitellins and development of immunoassays for two forms of their precursors (vitellogenins) in medaka (*Oryziaslatipes*). Gen Comp Endocrinol 143(3):267–277

Fukazawa H, Watanabe M, Shiraishi F, Shiraishi H, Shiozawa T, Matsushita H, Terao Y (2002) Formation of chlorinated derivatives of bisphenol A in waste paper recycling plants and their estrogenic activities. J Health Sci 48(3):242–249

Gagnaire B, Gagné F, André C, Blaise C, Abbaci K, Budzinski H, Garric J (2009) Development of biomarkers of stress related to endocrine disruption in gastropods: alkali-labile phosphates, protein-bound lipids and vitellogenin-like proteins. Aquat Toxicol 92(3):155–167

Geens T, Neels H, Covaci A (2012) Distribution of bisphenol-A, triclosan and N-nonylphenol in human adipose tissue, liver and brain. Chemosphere 87(7):796–802

Gillesby BE, Zacharewski TR (1998) Exoestrogens: mechanisms of action and strategies for identification and assessment. Environ Toxicol Chem Int J 17(1):3–14

Goksøyr A (2006) Endocrine disruptors in the marine environment: mechanisms of toxicity and their influence on reproductive processes in fish. J Toxic Environ Health A 69(1–2):175–184

Hachfi L, Couvray S, Simide R, Tarnowska K, Pierre S, Gaillard S, Prévot-D'Alvise N (2012) Impact of endocrine disrupting chemicals [EDCs] on hypothalamic-pituitary-gonad-liver [HPGL] axis in fish. World J Fish Mar Sci 4(1):14–30

Harris CA, Henttu P, Parker MG, Sumpter JP (1997) The estrogenic activity of phthalate esters in vitro. Environ Health Perspect 105(8):802–811

Hatef A, Alavi SMH, Abdulfatah A, Fontaine P, Rodina M, Linhart O (2012) Adverse effects of bisphenol A on reproductive physiology in male goldfish at environmentally relevant concentrations. Ecotoxicol Environ Saf 76:56–62

Hawkins MB, Thomas P (2004) The unusual binding properties of the third distinct teleost estrogen receptor subtype ERβa are accompanied by highly conserved amino acid changes in the ligand binding domain. Endocrinology 145(6):2968–2977

Hawkins MB, Thornton JW, Crews D, Skipper JK, Dotte A, Thomas P (2000) Identification of a third distinct estrogen receptor and reclassification of estrogen receptors in teleosts. Proc Natl Acad Sci 97(20):10751–10756

Hecker M, Tyler CR, Hoffmann M, Maddix S, Karbe L (2002) Plasma biomarkers in fish provide evidence for endocrine modulation in the Elbe River, Germany. Environ Sci Technol 36(11):2311–2321

Hinck JE, Blazer VS, Denslow ND, Echols KR, Gross TS, May TW, Tillitt DE (2007) Chemical contaminants, health indicators, and reproductive biomarker responses in fish from the Colorado River and its tributaries. Sci Total Environ 378(3):376–402

Hiramatsu N, Matsubara T, Fujita T, Sullivan CV, Hara A (2006) Multiple piscine vitellogenins: biomarkers of fish exposure to estrogenic endocrine disruptors in aquatic environments. Mar Biol 149:35–47

Hiramatsu N, Luo W, Reading BJ, Sullivan CV, Mizuta H, Ryu YW, Hara A (2013) Multiple ovarian lipoprotein receptors in teleosts. Fish Physiol Biochem 39:29–32

Hu JY, Aizawa T, Ookubo S (2002) Products of aqueous chlorination of bisphenol A and their estrogenic activity. Environ Sci Technol 36(9):1980–1987

Ishibashi H, Tachibana K, Tsuchimoto M, Soyano K, Ishibashi Y, Nagae M, Arizono K (2001) In vivo testing system for determining the estrogenic activity of endocrine-disrupting chemicals (EDCs) in goldfish (*Carassius auratus*). J Health Sci 47(2):213–218

Jobling S, Casey D, Rodgers-Gray T, Oehlmann J, Schulte-Oehlmann U, Pawlowski S, Tyler CR (2003) Comparative responses of molluscs and fish to environmental estrogens and an estrogenic effluent. Aquat Toxicol 65(2):205–220

Johnson LL, Lomax DP, Myers MS, Olson OP, Sol SY, O'Neill SM, Collier TK (2008) Xenoestrogen exposure and effects in English sole (*Parophrysvetulus*) from Puget Sound, WA. Aquat Toxicol 88(1):29–38

Jones PD, De Coen WM, Tremblay L, Giesy JP (2000) Vitellogenin as a biomarker for environmental estrogens. Water Sci Technol 42(7–8):1–14

Kang IJ, Yokota H, Oshima Y, Tsuruda Y, Oe T, Imada N, Honjo T (2002) Effects of bisphenol A on the reproduction of Japanese medaka (*Oryziaslatipes*). Environ Toxicol Chem Int J 21(11):2394–2400

Kar N, Kar B, Kar S (2021) Stress and coping during COVID-19 pandemic: result of an online survey. Psychiatry Res 295:113598

Keum YH, Jee JH, Lee OH, Park SI, Kang JC (2005) In vivo effects of bisphenol A exposure on haematological parameters in Korean rockfish, *Sebastes schlegeli*. J Fish Pathol 18(3):293–300

Kime DE, Nash JP, Scott AP (1999) Vitellogenesis as a biomarker of reproductive disruption by xenobiotics. Aquaculture 177(1–4):345–352

Kirk CJ, Bottomley L, Minican N, Carpenter H, Shaw S, Kohli N, Harris RM (2003) Environmental endocrine disrupters dysregulate estrogen metabolism and Ca2+ homeostasis in fish and mammals via receptor-independent mechanisms. Comp Biochem Physiol A Mol Integr Physiol 135(1):1–8

Kishida M, Anderson TR, Specker JL (1992) Induction by β-estradiol of vitellogenin in striped bass (*Morone saxatilis*): characterization and quantification in plasma and mucus. Gen Comp Endocrinol 88(1):29–39

Kiyama R, Wada-Kiyama Y (2015) Estrogenic endocrine disruptors: molecular mechanisms of action. Environ Int 83:11–40

Kleinkauf A, Scott AP, Stewart C, Simpson MG, Leah RT (2004) Abnormally elevated VTG concentrations in flounder (*Platichthysflesus*) from the Mersey estuary (UK)—a continuing problem. Ecotoxicol Environ Saf 58(3):356–364

Knudsen FR, Pottinger TG (1998) Interaction of endocrine disrupting chemicals, singly and in combination, with estrogen-, androgen-, and corticosteroid-binding sites in rainbow trout (*Oncorhynchus mykiss*). Aquat Toxicol 44(3):159–170

Kramer VJ, Miles-Richardson S, Pierens SL, Giesy JP (1998) Reproductive impairment and induction of alkaline-labile phosphate, a biomarker of estrogen exposure, in fathead minnows (*Pimephalespromelas*) exposed to waterborne 17β-estradiol. Aquat Toxicol 40(4):335–360

Kuiper GG, Gustafsson JÅ (1997) The novel estrogen receptor-β subtype: potential role in the cell-and promoter-specific actions of estrogens and anti-estrogens. FEBS Lett 410(1):87–90

Lethimonier C, Flouriot G, Valotaire Y, Kah O, Ducouret B (2000) Transcriptional interference between glucocorticoid receptor and estradiol receptor mediates the inhibitory effect of cortisol on fish vitellogenesis. Biol Reprod 62(6):1763–1771

Li MH, Wang ZR (2005) Effect of nonylphenol on plasma vitellogenin of male adult guppies (*Poecilia reticulata*). Environ Toxicol Int J 20(1):53–59

Liang Y, Fang Z (2012) Molecular cloning and mRNA expression of the vitellogenin and nuclear receptor gene induced by 17β-estradiol in the mud carp, *Cirrhinusmolitorella*. Ecotoxicology 21:719–729

Lindholst C, Pedersen KL, Pedersen SN (2000) Estrogenic response of bisphenol A in rainbow trout (*Oncorhynchus mykiss*). Aquat Toxicol 48(2–3):87–94

Lindholst C, Wynne PM, Marriott P, Pedersen SN, Bjerregaard P (2003) Metabolism of bisphenol A in zebrafish (Danio rerio) and rainbow trout (*Oncorhynchus mykiss*) in relation to estrogenic response. Comp Biochem Physiol C Toxicol Pharmacol 135(2):169–177

Ma CH, Dong KW, Yu KL (2000) cDNA cloning and expression of a novel estrogen receptor β-subtype in goldfish (*Carassius auratus*). Biochim Biophys Acta 1490(1–2):145–152

Mandiki SN, Babiak I, Bopopi JM, Leprieur F, Kestemont P (2005) Effects of sex steroids and their inhibitors on endocrine parameters and gender growth differences in Eurasian perch (*Percafluviatilis*) juveniles. Steroids 70(2):85–94

Matozzo V, Gagné F, Marin MG, Ricciardi F, Blaise C (2008) Vitellogenin as a biomarker of exposure to estrogenic compounds in aquatic invertebrates: a review. Environ Int 34(4):531–545

McCormick SD, O'Dea MF, Moeckel AM, Lerner DT, Björnsson BT (2005) Endocrine disruption of parr-smolt transformation and seawater tolerance of Atlantic salmon by 4-nonylphenol and 17β-estradiol. Gen Comp Endocrinol 142(3):280–28

Mendel CM (1989) The free hormone hypothesis: a physiologically based mathematical model. Endocr Rev 10(3):232–274

Menuet A, Pellegrini E, Anglade I, Blaise O, Laudet V, Kah O, Pakdel F (2002) Molecular characterization of three estrogen receptor forms in zebrafish: binding characteristics, transactivation properties, and tissue distributions. Biol Reprod 66(6):1881–1892

Mikawa N, Utoh T, Horie N, Okamura A, Yamada Y, Akazawa A, Aoki T (2006) Cloning and characterization of vitellogenin cDNA from the common Japanese conger (*Conger myriaster*) and vitellogenin gene expression during ovarian development. Comp Biochem Physiol B: Biochem Mol Biol 143(4):404–414

Milligan SR, Khan O, Nash M (1998) Competitive binding of xenobiotic oestrogens to rat alpha-fetoprotein and to sex steroid binding proteins in human and rainbow trout (*Oncorhynchus mykiss*) plasma. Gen Comp Endocrinol 112(1):89–93

Mitchelmore CL, Rice CP (2006) Correlations of nonylphenol-ethoxylates and nonylphenol with biomarkers of reproductive function in carp (*Cyprinus carpio*) from the Cuyahoga River. Sci Total Environ 371(1–3):391–401

Mosselman S, Polman J, Dijkema R (1996) ERβ: identification and characterization of a novel human estrogen receptor. FEBS Lett 392(1):49–53

Naderi M, Mousavi SM, Safahieh A, Ghatrami ER, Zargham D (2014) Effects of 4-nonylphenol on balance of steroid and thyroid hormones in sexually immature male yellowfin seabream (*Acanthopagrus latus*). Environ Toxicol 29(4):459–465

Naderi M, Safahieh A, Madiseh SD, Zolgharnein H, Ghatrami ER (2015) Induction of vitellogenin synthesis in immature male yellowfin seabream (*Acanthopagrus latus*) exposed to 4-nonylphenol and 17β-estradiol. Toxicol Ind Health 31(3):209–220

Nagahama Y (1983) 6 the functional morphology of teleost gonads. In: Fish physiology, vol 9, pp 223–275

Nagler JJ, Ruby S, Idler DR, So YP (1987) Serum phosphoprotein phosphorus and calcium levels as reproductive indicators of vitellogenin in highly vitellogenic mature female and estradiol injected immature rainbow trout (*Oncorhynchus mykiss*). Can J Zool 65:2421–2425

Nagler JJ, Cavileer T, Sullivan J, Cyr DG, Rexroad C III (2007) The complete nuclear estrogen receptor family in the rainbow trout: discovery of the novel ERα2 and both ERβ isoforms. Gene 392(1–2):164–173

Nagler JJ, Cavileer TD, Verducci JS, Schultz IR, Hook SE, Hayton WL (2012) Estrogen receptor mRNA expression patterns in the liver and ovary of female rainbow trout over a complete reproductive cycle. Gen Comp Endocrinol 178(3):556–561

Nelson ER, Habibi HR (2013) Estrogen receptor function and regulation in fish and other vertebrates. Gen Comp Endocrinol 192:15–24

Neubert D (1997) Vulnerability of the endocrine system to xenobiotic influence. Regul Toxicol Pharmacol 26(1):9–29

Pakdel F, Féon S, Le Gac F, Le Menn F, Valotaire Y (1991) In vivo estrogen induction of hepatic estrogen receptor mRNA and correlation with vitellogenin mRNA in rainbow trout. Mol Cell Endocrinol 75(3):205–212

Pojana G, Bonfà A, Busetti F, Collarin A, Marcomini A (2004) Determination of natural and synthetic estrogenic compounds in coastal lagoon waters by HPLC-electrospray-mass spectrometry. Int J Environ Anal Chem 84(10):717–727

Ricciardi F, Matozzo V, Marin MG (2008) Effects of 4-nonylphenol exposure in mussels (*Mytilus galloprovincialis*) and crabs (*Carcinusaestuarii*) with particular emphasis on vitellogenin induction. Mar Pollut Bull 57(6–12):365–372

Robinson CD, Brown E, Craft JA, Davies IM, Moffat CF (2004) Effects of prolonged exposure to 4-tert-octylphenol on toxicity and indices of oestrogenic exposure in the sand goby (*Pomatoschistusminutus*, Pallas). Mar Environ Res 58(1):19–38

Rubin BS (2011) Bisphenol A: an endocrine disruptor with widespread exposure and multiple effects. J Steroid Biochem Mol Biol 127(1–2):27–34

Rykowska I, Wasiak W (2006) Properties, threats, and methods of analysis of bisphenol A and its derivatives. Acta Chromatogr 16:7

Sabo-Attwood T, Kroll KJ, Denslow ND (2004) Differential expression of largemouth bass (*Micropterus salmoides*) estrogen receptor isotypes alpha, beta, and gamma by estradiol. Mol Cell Endocrinol 218(1–2):107–118

Safe SH, Gaido K (1998) Phytoestrogens and anthropogenic estrogenic compounds. Environ Toxicol Chem Int J 17(1):119–126

Sánchez-Avila J, Tauler R, Lacorte S (2012) Organic micropollutants in coastal waters from NW Mediterranean Sea: sources distribution and potential risk. Environ Int 46:50–62

Schafer TE, Lapp CA, Hanes CM, Lewis JB, Wataha JC, Schuster GS (1999) Estrogenicity of bisphenol A and bisphenol A dimethacrylate in vitro. J Biomed Mater Res 45(3):192–197

Scott AP, Katsiadaki I, Witthames PR, Hylland K, Davies IM, McIntosh AD, Thain J (2006) Vitellogenin in the blood plasma of male cod (*Gadus morhua*): a sign of oestrogenic endocrine disruption in the open sea? Mar Environ Res 61(2):149–170

Segner H, Caroll K, Fenske M, Janssen CR, Maack G, Pascoe D, Wenzel A (2003) Identification of endocrine-disrupting effects in aquatic vertebrates and invertebrates: report from the European IDEA project. Ecotoxicol Environ Saf 54(3):302–314

Selman K, Wallace RA, Sarka A, Qi X (1993) Stages of oocyte development in the zebrafish, *Brachydanio rerio*. J Morphol 218(2):203–224

Sohoni PCRT, Tyler CR, Hurd K, Caunter J, Hetheridge M, Williams T, Sumpter JP (2001) Reproductive effects of long-term exposure to bisphenol A in the fathead minnow (*Pimephalespromelas*). Environ Sci Technol 35(14):2917–2925

Soto AM, Sonnenschein C, Chung KL, Fernandez MF, Olea N, Serrano FO (1995) The E-SCREEN assay as a tool to identify estrogens: an update on estrogenic environmental pollutants. Environ Health Perspect 103(suppl 7):113–122

Sumpter JP, Jobling S (1995) Vitellogenesis as a biomarker for estrogenic contamination of the aquatic environment. Environ Health Perspect 103(suppl 7):173–178

Tabata A, Miyamoto N, Ohnishi Y, Itoh M, Yamada T, Kamei T, Magara Y (2003) The effect of chlorination of estrogenic chemicals on the level of serum vitellogenin of Japanese medaka (*Oryziaslatipes*). Water Sci Technol 47(9):51–57

Teta C, Naik YS (2017) Vitellogenin induction and reduced fecundity in zebrafish exposed to effluents from the City of Bulawayo, Zimbabwe. Chemosphere 167:282–290

Tong Y, Shan T, Poh YK, Yan T, Wang H, Lam SH, Gong Z (2004) Molecular cloning of zebrafish and medaka vitellogenin genes and comparison of their expression in response to 17β-estradiol. Gene 328:25–36

Tyler CR, Sumpter JP, Handford RM (1990) The dynamics of vitellogenin sequestration into vitellogenic ovarian follicles of the rainbow trout, *Salmo gairdneri*. Fish Physiol Biochem 8:211–219

Vajda AM, Barber LB, Gray JL, Lopez EM, Bolden AM, Schoenfuss HL, Norris DO (2011) Demasculinization of male fish by wastewater treatment plant effluent. Aquat Toxicol 103(3–4):213–221

Van den Belt K, Berckmans P, Vangenechten C, Verheyen R, Witters H (2004) Comparative study on the in vitro/in vivo estrogenic potencies of 17β-estradiol, estrone, 17α-ethynylestradiol and nonylphenol. Aquat Toxicol 66(2):183–195

Verslycke T, Vandenbergh GF, Versonnen B, Arijs K, Janssen CR (2002) Induction of vitellogenesis in 17α-ethinylestradiol-exposed rainbow trout (*Oncorhynchus mykiss*): a method comparison. Comp Biochem Physiol C Toxicol Pharmacol 132(4):483–492

Versonnen BJ, Goemans G, Belpaire C, Janssen CR (2004) Vitellogenin content in European eel (*Anguilla Anguilla*) in Flanders, Belgium. Environ Pollut 128(3):363–371

Virk P, Al-Sakran AAM, Elobeid MA (2014) Effect of bisphenol A on the levels of vitellogenin and metallothionein in adult male carp, *Cyprinus carpio*. Trop J Pharm Res 13(7):1107–1112

Wahli W, Dawid IB, Ryffel GU, Weber R (1981) Vitellogenesis and the vitellogenin gene family. Science 212(4492):298–304

Wheeler JR, Gimeno S, Crane M, Lopez-Juez E, Morritt D (2005) Vitellogenin: a review of analytical methods to detect (anti) estrogenic activity in fish. Toxicol Mech Methods 15(4):293–306

World Health Organization (WHO) (2002) Global assessment of the state-of-the-science of endocrine disruptors. International Program on Chemical Safety

Yilmaz O, Patinote A, Nguyen T, Bobe J (2018) Multiple vitellogenins in zebrafish (Danio rerio): quantitative inventory of genes, transcripts and proteins, and relation to egg quality. Fish Physiol Biochem 44:1509–1525

Yokota H, Tsuruda Y, Maeda M, Oshima Y, Tadokoro H, Nakazono A, Kobayashi K (2000) Effect of bisphenol A on the early life stage in Japanese medaka (*Oryziaslatipes*). Environ Toxicol Chem Int J 19(7):1925–1930

Zaccaroni A, Gamberoni M, Mandrioli L, Sirri R, Mordenti O, Scaravelli D, Parmeggiani A (2009) Thyroid hormones as a potential early biomarker of exposure to 4-nonylphenol in adult male shubunkins (*Carassius auratus*). Sci Total Environ 407(10):3301–3306

Significance of Vitellogenin in Egg Yolk Production and Egg Quality

14

Viswanathan Vinotha and Vaseeharan Baskaralingam

Abstract

Vitellogenin is a protein precursor that plays a crucial role in egg yolk production and egg quality in female animals, particularly in egg-laying species. It is synthesized in the liver and secreted into the bloodstream, where it is taken up by developing oocytes (immature eggs) and processed into yolk proteins. The yolk, composed of vitellogenin-derived proteins and lipids, serves as a vital nutrient source for the developing embryo during early development. It provides a wide range of nutrients, including proteins, lipids, vitamins, minerals, and hormones, which are essential for embryonic growth, organ development, and energy metabolism. The synthesis and secretion of vitellogenin are under the control of reproductive hormones, primarily estrogen. Estrogen stimulates the production of vitellogenin by the liver and its uptake by the oocytes. Therefore, vitellogenin levels can serve as an indicator of reproductive status and hormonal balance in egg-laying animals. As a whole, vitellogenin plays a crucial role in egg yolk formation, nutrient supply to the developing embryo, egg size, fertility, and overall egg quality. Its synthesis, regulation, and deposition contribute to the successful reproduction and health of egg-laying species.

Keywords

Estrogen · Embryo · Fertility · Protein · Vitamins · Yolk

V. Vinotha
Biomaterials and Biotechnology in Animal Health Lab, Department of Animal Health and Management, Alagappa University, Karaikudi, Tamil Nadu, India

V. Baskaralingam (✉)
Department of Animal Health and Management, Alagappa University, Karaikudi, Tamil Nadu, India

V. Baskaralingam, R. Vanichviriyakit (eds.), *Vitellogenin in Fishes-Diversification, Biological Properties, and Future Perspectives*,
https://doi.org/10.1007/978-981-99-5340-0_14

14.1 Vitellogenin

"Vitellogenin"—a female specific lipoglycophosphoprotein, is usually expressed in the female at the time oogenesis process. Vitellogenin is the foremost precursor of egg yolk, which is crucial nutrient for the development of embryo. The accumulation of egg yolk in the oozyte is recognized as vitellogenesis (Pan et al. 1969; Hara et al. 2016; Li and Zhang 2017). At the time of vitellogenesis, in liver of vertebrates, the production of vitellogenin was induced by the estrogen released from the ovarian follicle. Hence, they were integrated into the oozyte through the blood stream and underwent to the cleavage process to produce lipovitellin and phosvitin (Hiramatsu et al. 2002; Patiño and Sullivan 2002; Farrell 2011; Hara et al. 2016; Li and Zhang 2017). Vitellogenin and its derivatives play a chief role in vitellogenesis process. Moreover, the quantity of vitellogenin involved in the determination of egg type and being a source of free amino acid to the embryo (Finn et al. 2009; Groh et al. 2011). It also acts as immune factors and antioxidant reagents toward egg and embryo (Ziv et al. 2008; Zhang et al. 2011, 2015).

14.2 Significance of Vitellogenin in Fish Egg Yolk Production

Vitellogenin is the chief precursor for the production of egg yolk, which is encoded by multigenes. The active part of *Vtg* genes differs from species to species. Vitellogenin enhances the size of the oozyte more than 90% during oogenesis process (King et al. 2003; Wu et al. 2013; Sun and Zhang 2015; Carducci et al. 2019). Vitellogenin has been synthesized in the liver and transmitted to the ovary through circulating system. The concentration of vitellogenin differs between blood stream and oozyte Because, only required quantity of vitellogenin gets into the oozyte by the support of ovarian receptors. After internalization, the vitellogenin undergoes to cleavage process and yields three types of derivatives including lipovitellin (Lv), phosvitin (Pv), and β-component (Williams et al. 2014; Sun and Zhang 2015). Lipovitellin (largest yolk protein derivative) and phosvitin (smaller yolk protein derivative) are participating in proteolytic process and has phosphorylated serine for the development of oozyte and stabilization of nascent vitellogenin (Romano et al. 2004; Yilmaz et al. 2015).

14.3 Factors Affecting the Functions of Vitellogenin

According to King et al. (2003), increase in the temperature (22 °C) could affect the level of plasma during vitellogenesis. This reduction leads to the reduced diameter in ovum and chorion damage. It may due to the direct influence of temperature or indirectly due to the stress caused by elevated temperature. Volkoff and London (2018) reported that malnutrition or insufficient food intake could affect the reproduction rate in salmon fish. Phosphate- and calcium-like minerals are required for the development of egg yolk protein. Whereas, deficiency of Vitamin C reduces the oogenesis process and number of egg production (Volkoff and London 2018). Similarly, ontogenicoozyte development and the previtellogenic phase had been affected by deficiency of vitamin A (Harlıoğlu and Farhadi 2017).

14.4 Factors Involved in Enhancement of Vitellogenin in Fish

Expression of VgA and VgB gene in *Thunnus thynnus* fish after healthy diet supplementation enhances the liver yolk production as well as accumulation of yolk content in egg (Pousis et al. 2011). Enhanced liver function, improved synthesis of vitellogenin, and egg production were observed in the experimental catfish after supplementation of turmeric mixed diet (Dewi et al. 2020; Kasiyati et al. 2016). Mixture of turmeric powder and thyroxine-based diet increase the amount of vitellogenin in fish reported by Rawung et al. (2020). Also, turmeric/curcumin-based diet supplementation increases the egg diameter in catfish (Lee and Yang 2002; Dewi et al. 2018). Similarly, Arfah et al. (2018) reported that PMSG hormone and turmeric-based diet supplemented catfish's gonads expressed well distinguished maturation. Pamungkas et al. (2020) reported that fatty acid-associated dietary supplementation enhances the reproduction rate in stripped catfish. Masoudi Asil et al. (2018) reported that dietary supplementation of essential fatty acid (i.e., arachidonic acid) increases the production of calcium, calcium, thyroid hormone, and cortisol levels in vitellogenesis in *Trichopodus trichopterus*. Dietary supplementation of *Spirulina* feed enriches the yolk globule production and vitellogenesis process in experimented *Danio rerio* (Calabrò et al. 2021). The level of vitellogenin and quality of egg have been enhanced by the increased level of feed protein (Hariani and Slamet 2019)

14.5 Conclusion

Overall, the vitellogenin is the significant precursor for the synthesis of egg yolk protein. The amount of vitellogenin is directly proportional to the quality of egg and the offspring. The vitellogenin synthesis can be increased by the compounds including fatty acid, vitamins, and plant derivatives like curcumin.

Acknowledgments First author (V. Vinotha-IF170397) sincerely acknowledges the financial support from the Department of Science and Technology, New Delhi, India, through INSPIRE scheme.

References

Arfah H, Sudrajat AO, Supriyadi MA, Zairin M (2018) Gonad maturation of female striped catfish *Pangasionodonhypophthalmus* (Sauvage, 1878) using a combination of pregnant mare serum gonadotropin+ antidopamine, vitamin E, and curcumin extract mixed feed outside its spawning season. Int J Fish Aquat Stud 6(5):52–57

Calabrò C, Bertuccio C, Gervasi T, Lauriano ER, Leonardi M, Cicero N, Lo CP (2021) Effects of spirulina diet on the oogenesis of zebrafish: morphological analysis and immunohistochemical determination of the vitellogenin. Nat Prod Res 35(22):4454–4459

Carducci F, Biscotti MA, Canapa A (2019) Vitellogenin gene family in vertebrates: evolution and functions. Eur Zool J 86(1):233–240

Dewi CD, Ekastuti DR, Sudrajat AO, Manalu W (2018) Improved vitellogenesis, gonad development and egg diameter in catfish (*Pangasianodonhypophthalmus*) supplemented with turmeric (*Curcuma longa*) powder. Aquac Res 49(2):651–658

Dewi CD, Manalu W, Ekastuti DR, Sudrajat AO (2020) The role of turmeric (Curcuma longa) powder in improving liver function to increase vitellogenin synthesis and deposition in the oocytes of catfish (*Pangasianodonhypophthalmus*). Jordan J Biol Sci 13:3

Farrell AP (2011) Encyclopedia of fish physiology: from genome to environment. Academic Press, London

Finn RN, Kolarevic J, Kongshaug H, Nilsen F (2009) Evolution and differential expression of a vertebrate vitellogenin gene cluster. BMC Evol Biol 9(1):1–2

Groh KJ, Nesatyy VJ, Segner H, Eggen RI, Suter MJ (2011) Global proteomics analysis of testis and ovary in adult zebrafish (*Danio rerio*). Fish Physiol Biochem 37:619–647

Hara A, Hiramatsu N, Fujita T (2016) Vitellogenesis and choriogenesis in fishes. Fish Sci 82:187–202

Hariani D, Slamet PW (2019) Combination of feed protein level and laserpuncture induction of broodstock catfish (*Clarias* sp.) to increase estrogen, vitellogenin, and egg quality. EurAsian J BioSci 13(2):769–779

Harlıoğlu MM, Farhadi A (2017) Factors affecting the reproductive efficiency in crayfish: implications for aquaculture. Aquac Res 48(5):1983–1997

Hiramatsu N, Matsubara T, Weber GM, Sullivan CV, Hara A (2002) Vitellogenesis in aquatic animals. Fish Sci 68(sup1):694–699

Kasiyati S, Ekastuti DR, Manalu W (2016) Roles of curcumin and monochromatic light in optimizing liver function to support egg yolk biosynthesis in Magelang ducks. Int J Poult Sci 15(10):414–424

King HR, Pankhurst NW, Watts M, Pankhurst PM (2003) Effect of elevated summer temperatures on gonadal steroid production, vitellogenesis and egg quality in female Atlantic salmon. J Fish Biol 63(1):153–167

Lee WK, Yang SW (2002) Relationship between ovarian development and serum levels of gonadal steroid hormones, and induction of oocyte maturation and ovulation in the cultured female Korean spotted sea bass *Lateolabrax maculatus* (Jeom-nong-eo). Aquaculture 207(1–2):169–183

Li H, Zhang S (2017) Functions of vitellogenin in eggs. Results Probl Cell Differ 63:389–401

Masoudi Asil S, Abedian Kenari AM, Van Der Kraak G (2018) Effect of different levels of dietary arachidonic acid on calcium, thyroid hormone, and cortisol levels in vitellogenesis and maturation stages of female blue gourami (*Trichopodustrichopterus*, Pallas, 1770). Fish Sci Technol 7(2):109–116

Pamungkas W, Jusadi D, Zairin M Jr, Setiawati M, Supriyono E, Imron I (2020) Effect of dietary essential fatty acids on level of oestradiol-17β and vitellogenin, reproductive performance and larval quality of striped catfish (*Pangasianodonhypophthalmus*) in out-of-spawning season. Aquac Res 51(9):3900–3909

Pan ML, Bell WJ, Telfer WH (1969) Vitellogenic blood protein synthesis by insect fat body. Science 165(3891):393–394

Patiño R, Sullivan CV (2002) Ovarian follicle growth, maturation, and ovulation in teleost fish. Fish Physiol Biochem 26:57–70

Pousis C, De Giorgi C, Mylonas CC, Bridges CR, Zupa R, Vassallo-Agius R, De La Gándara F, Dileo C, De Metrio G, Corriero A (2011) Comparative study of liver vitellogenin gene expression and oocyte yolk accumulation in wild and captive Atlantic bluefin tuna (*Thunnus thynnus* L.). Anim Reprod Sci 123(1–2):98–105

Rawung LD, Ekastuti DR, Junior MZ, Rahminiwati M, Sunarma A, Manalu W (2020) Reproductive performances and egg qualities in African catfish (*Clarias gariepinus*) broodstocks supplemented with curcumin and thyroxine hormone. Omni-Akuatika 16(1):32–47

Romano M, Rosanova P, Anteo C, Limatola E (2004) Vertebrate yolk proteins: a review. Mol Reprod Dev 69(1):109–116

Sun C, Zhang S (2015) Immune-relevant and antioxidant activities of vitellogenin and yolk proteins in fish. Nutrients 7(10):8818–8829

Volkoff H, London S (2018) Nutrition and reproduction in fish. Encycl Reprod 2:1–6

Williams VN, Reading BJ, Amano H, Hiramatsu N, Schilling J, Salger SA, Islam Williams T, Gross K, Sullivan CV (2014) Proportional accumulation of yolk proteins derived from multiple vitellogenins is precisely regulated during vitellogenesis in striped bass (*Morone saxatilis*). J Exp Zool A Ecol Genet Physiol 321(6):301–315

Wu LT, Hui JH, Chu KH (2013) Origin and evolution of yolk proteins: expansion and functional diversification of large lipid transfer protein superfamily. Biol Reprod 88(4):102

Yilmaz O, Prat F, Ibañez AJ, Amano H, Koksoy S, Sullivan CV (2015) Estrogen-induced yolk precursors in European sea bass, *Dicentrarchuslabrax*: status and perspectives on multiplicity and functioning of vitellogenins. Gen Comp Endocrinol 221:16–22

Zhang S, Wang S, Li H, Li L (2011) Vitellogenin, a multivalent sensor and an antimicrobial effector. Int J Biochem Cell Biol 43(3):303–305

Zhang S, Dong Y, Cui P (2015) Vitellogenin is an immunocompetent molecule for mother and offspring in fish. Fish Shellfish Immunol 46(2):710–715

Ziv T, Gattegno T, Chapovetsky V, Wolf H, Barnea E, Lubzens E, Admon A (2008) Comparative proteomics of the developing fish (zebrafish and gilthead seabream) oocytes. Comp Biochem Physiol D Genom Proteom 3(1):12–35

Concentration of Vitellogenin and Disruption of the Endocrine System in Fish

15

Anjugam Mahalingam and Santhanam Perumal

Abstract

Vitellogenin are a precursor to the protein found in egg yolks. It has long been believed that vitellogenin (Vg), the main precursor protein of the egg yolk, provides developing embryos and larvae with protein-and lipid-rich nutrition. In reaction to circulating endogenous estrogen, the liver of female oviparous vertebrates generally produces vitellogenin, a sizeable serum phospholipoglycoprotein. After being made in the liver, it enters the bloodstream and goes to the ovary, where it is absorbed and changed by ovulating eggs. A significant estrogen-inducible yolk precursor protein known as vitellogenin (Vg) has gained importance as a biomarker for determining the estrogenic potency of chemicals and the exposure of animals to estrogenic pollutants found in aquatic environments. These pollutants are referred to as endocrine-disruptive substances because they have the ability to impair the operation of the neuroendocrine system in vertebrates.

Keywords

Vitellogenin · Endocrine disruption · Endocrine disruptors

15.1 Introduction

Large serum proteins known as vitellogenins (Vgs) are produced by oviparous animals as precursors to the yolk. They have been extensively researched for their function in fish reproductive system as well as biomarkers for endocrine-disrupting

A. Mahalingam · S. Perumal (✉)
Department of Marine Science, School of Marine Sciences, Bharathidasan University, Thiruchirapalli, Tamil Nadu, India

V. Baskaralingam, R. Vanichviriyakit (eds.), *Vitellogenin in Fishes-Diversification, Biological Properties, and Future Perspectives*,
https://doi.org/10.1007/978-981-99-5340-0_15

xenobiotic (Tao et al. 1993; Hiramatsu et al. 2002). In teleosts, Vgs are complex glycolipophosphoproteins with molecular masses ranging from 300 to 640 kDa that are suspected to be heterodimers or to exist in several forms. In at least 17 species of teleosts, the various variants of Vg have been found (Tyler et al. 1996; Hiramatsu et al. 2006). In addition to supply the nutrients to the growing embryo, Vg also provides minerals in binding with Ca^{2+}, Mg^{2+}, and K^{+} to the developing fish. Male fishes and juveniles can also express the Vg gene, but they often do not have enough circulating estrogens to significantly increase the production of the protein. Males and young ones will, however, produce Vg if estrogen is given to them. Induction of Vg synthesis in males serves as an effective biomarker of endocrine-disrupting pollutants because this effect is also found with chemicals that resemble estrogen (Sumpter and Jobling 1995; Folmar et al. 2002). The change of typical hormone signaling by endogenous substances, frequently environmental pollutants, is referred to as endocrine disruption. Endocrine disrupters, also known as endocrine-disrupting chemicals, are substances that either replicate or counteract the effects of naturally occurring hormones. Any substance, including estrogens, androgens, and thyroid hormones, that modifies the function of hormones involved in growth, development, and reproduction is referred to as an endocrine disruptor (ED). Since the vertebrate neuroendocrine system controls important processes like development, growth, metabolism, and reproduction, it is now known that many chemical substances created over the past century have the potential to have an impact on it. Environmental toxicology has grown significantly over the past 10 years, and public interest has been piqued by research on the identification and consequences of such compounds (Hiramatsu et al. 2006). They include estrogenic EDCs (environmental estrogens), which function similarly to endogenous estradiol-17 to trigger an estrogenic response.

15.2 Vitellogenin

Vitellogenin was once thought to be a female-specific reproductive protein that cleaves into yolk proteins including phosvitin (Pv) and lipovitellin (Lv), which are stored in eggs and provide the building blocks for growing embryos. Vg has recently been demonstrated to be an immune-relevant molecule engaged in the host's defense against microorganisms like bacteria and viruses. Vg is often produced in an extra-ovarian organ, such as the liver of vertebrates, the hepatopancreas of crustaceans, or the fat body of insects (Lubzens et al. 2010; Girish et al. 2014). It is then carried to the ovary via the circulatory system, where it is internalized into developing oocytes by receptor-mediated endocytosis (Mak et al. 2005; Williams et al. 2014; Sun and Zhang 2015). When Vgs are internalized into the oocytes, the aspartic protease cathepsin D performs a proteolytic cleavage to produce yolk proteins like lipovitellin (Lv) subunits and phosvitin (Pv) and β-C. Cathepsin D is produced by the oocyte and stored in specific organelles that are located where the yolk precursors are integrated on the surface of the egg. The primary phosphoproteins in egg yolks are phosphovitin and lipovitellins a and b, and these proteins can be found in the yolk spheres as well. Additionally, the proteolytically cleaved products of Vg, Pv, and Lv, which are both present in developing embryos, function as defense proteins. Importantly, smaller peptides made from yolk protein also have antibacterial properties. Vgs were formerly thought to be a protein that was only found in females. However, synthesis has been demonstrated to occur in male and even sexually immature animals, albeit in smaller amounts, indicating that vitellogenin likely serve a more universal function that is gender-neutral (Nath and Sundararaj 1981). In male fish, the vitellogenin gene is typically silent but can be activated by estrogen exposure. Despite the fact that the vitellogenin gene in immature and male oviparous animals can be activated by estrogen exposure, vitellogenin is typically not detectable in their plasma because they lack circulating estrogen. Moreover, E2 and other synthetic compounds, such as ethinyl estradiol and nonylphenols, can cause Vg to be formed in male fish. The sensitivity of this reaction has led to the widespread adoption of Vg expression in male fish as a biomarker of exposure to environmental estrogen in several laboratory and field studies (Purdom et al. 1994). Indicators of exposure to exogenous estrogens or estrogen mimics in the aquatic environment are now frequently used to determine if male fish produce vitellogenin, which is a hormone produced by the body. Because there is no evidence on the connection between male plasma vitellogenin and reproductive end points in male and female fish, the usefulness of this indicator for predicting impacts on fish reproductive success is questionable (Mills et al. 2003). The conventional belief that Vtgs and yolk proteins were merely basic sources of nutrients for the growing embryos has been challenged by recent research that links both of these substances with the immune system and antioxidant activity in fish (Scharf et al. 2005). Curiously, the rates of distinct Vgs internalized by developing oocytes are not always equal to the rates of circulating Vgs in the blood, which may be attributable to the regulation of the system of multiple ovarian receptors involved in endocytosis of different Vg (Hiramatsu et al. 2015; Reading et al. 2011).

15.3 Concentration of Vitellogenin

Fish, Vtg is a protein that contains around 1% phosphorous, therefore after electrophoresis, Vtg in blood serum can be found by staining with phosphorus dye. Moreover, detecting alkaline-labile phosphorus in blood serum enables the indirect measurement of Vtg. Further, due to its large molecular weight, it is easily detectable by looking for high-molecular-weight peaks in gel-filtration chromatography (Akihiko and Sullivan 1993). Quantitative and qualitative evaluations based on immunological strategies using antibodies, however, are typically preferable. During oogenesis, the concentration of circulating Vg and their level of stimulation by estrogens seems to differ between the species and various forms of Vg within the species. In addition, fishes appear to respond differently to different types of Vg induced by estrogens depending on the environment (such as water, temperature and photoperiod), life history stage, estrogen concentration, and type.

15.4 Endocrine System of Fish

Endocrine system typically regulates physiological processes including digestion, metabolism, growth, development, reproduction, etc. as well as long-term actions of target organs. Endocrine glands are a specific type of gland that are found throughout the body of fish and are part of the endocrine system. Exocrine and endocrine glands are the two main types of glands found in animals. Ducts are used by exocrine glands to transport their secretions. There are no channels for the endocrine glands to transport their secretions. The endocrine glands are also known as ductless glands for this reason. Endocrinology is the study of endorcrine glands. A hormone is an identifiable end product (organic substance) of an endocrine gland that is produced into the circulation and transported to a specified area of the body where it has a specific physiological impact. The effect's mode of operation might be either excitatory or inhibitory. The way that hormones affect different organs varies. The hormones are sometimes known as autonomes or autocoids because they do not take part in biochemical processes. The maintenance of internal environmental parameters by hormones, such as blood glucose levels, ion and water balance, and temperature regulation, is referred to as homeostasis. These organs are known as target organs. The hormones thus serve as chemical messengers. Despite the fact that Vg is one of the most often used endpoints for endocrine disruption studies, there is no conclusive evidence linking any amount of exogenous Vg induction to a population-level effect. Males are not harmed by the induction of Vg, which is thought to be caused by anthropogenic chemicals rather than Vg itself. A link between elevated Vg concentrations and levels that are harmful to reproductive function would be ideal, allowing Vg to be utilized to assess the effects of pollutants. Vg and reproductive factors have been discovered to be correlated. Nevertheless, Vg levels often do not correspond well with the prevalence of intersex (Scott et al. 2006).

15.5 Endocrine Disruption in Fish

Over the past few decades, research on fish endocrine system disruption has grown in significance. An prominent ecotoxicological risk in the aquatic environment has been recognized as endocrine disruption, namely disturbance by estrogen-active substances (Casanova-Nakayama et al. 2011). Endocrine-disrupting substances are broadly referred to as environmental substances that interfere in some manner with normal endocrine function, whether they are created by nature or by humans. Endocrine disruption can be caused by a wide range of unfavorable environmental factors, such as parasites, low pH, unfavorable temperatures, and restricted food sources. Exoestrogens, which imitate the effects of endogenous estradiol to induce an estrogenic response, can be included in this varied collection of compounds. Several industrial, municipal, agricultural, and natural substances have been demonstrated to be estrogenic or are at least suspected of being so. These substances include the biodegradation of by-products such as polychlorinated biphenyls, dichlorodiphenyltrichloroethane (DDT), chlordecone, and methoxychlor, as well as natural and manufactured estrogens (17-estradiol, ethinyl estradiol), and alkylphenol polyethoxylates (Arcand-Hoy and Benson 1998). In fish, this can cause females to lay fewer eggs or have smaller gonads, or it might feminize genetically male fish. Although it is well known that male fish exposed to estrogenic chemicals exhibit enhanced vitellogenin production, the biological significance of elevated vitellogenin levels has, for the most part, only been hypothesized. Endocrine-disrupting chemical concentrations and their potential have impacts on whole animals that has also been the subject of extensive discussion and controversy. It has been demonstrated that populations of wild freshwater fish around the world have endocrine system disruption. Endocrine-disrupting substances may change the hormone pathways that control sexual function. Consequences can range from minor adjustments to fish's physiology and sexual behavior to lasting alterations in sexual differentiation, gonad development impairment, and/or impaired fertility (Jobling and Tyler 2003). Global reports of effects in wild fish populations include stunted growth, disruptions in reproduction, and altered sexual development (Matthiessen 2000; Vos et al. 2000).

15.6 Endocrine Disruptors

Endocrine disruptors (EDs) are synthetic or naturally occurring chemical molecules in the environment that have the potential to have a negative impact on the homeostasis of the endocrine axis, resulting in neurological, developmental, immunological, and reproductive dysfunction at the organismal level (Kar et al. 2021). Endocrine disruption, in particular disruption by estrogen-active compounds, has been identified as an important ecotoxicological hazard in the aquatic environment (Casanova-Nakayama et al. 2011). Pesticides, hormone-mimicking substances, heavy metals, polychlorinated biphenyls (PCBs), phthalates, organic solvents, flame retardants, surfactants, medicines, and other substances are examples of EDs (Fig. 15.1). In addition, certain EDs are produced artificially, such as polycyclic aromatic hydrocarbons, while others are produced naturally, such as phyto- and myco-estrogens, which are produced by fungus and plants (Liu et al. 2010). Endocrine disruptors have been demonstrated to change mating behaviors in addition to their physiological impacts. When fish spawn, they engage in intricate and crucial courtship activities. It is believed that these are some of the most delicate endpoints, but also the most challenging to research.

In natural species, EDs are known to cause irreversible reproductive consequences (Tubbs and McDonough 2018). In order to respond appropriately to environmental cues such as physical, chemical, or biological ones and to preserve internal homeostasis, organisms have evolved sensitivity to both endogenous and external chemical signals. Estrogen-active chemicals, or molecules that bind as agonists to estrogen receptors, have received the most attention in studies on the effects of EDCs on fish. EDs can exert their effects in a variety of ways, including by binding to hormone receptors, changing endogenous hormone levels, or influencing

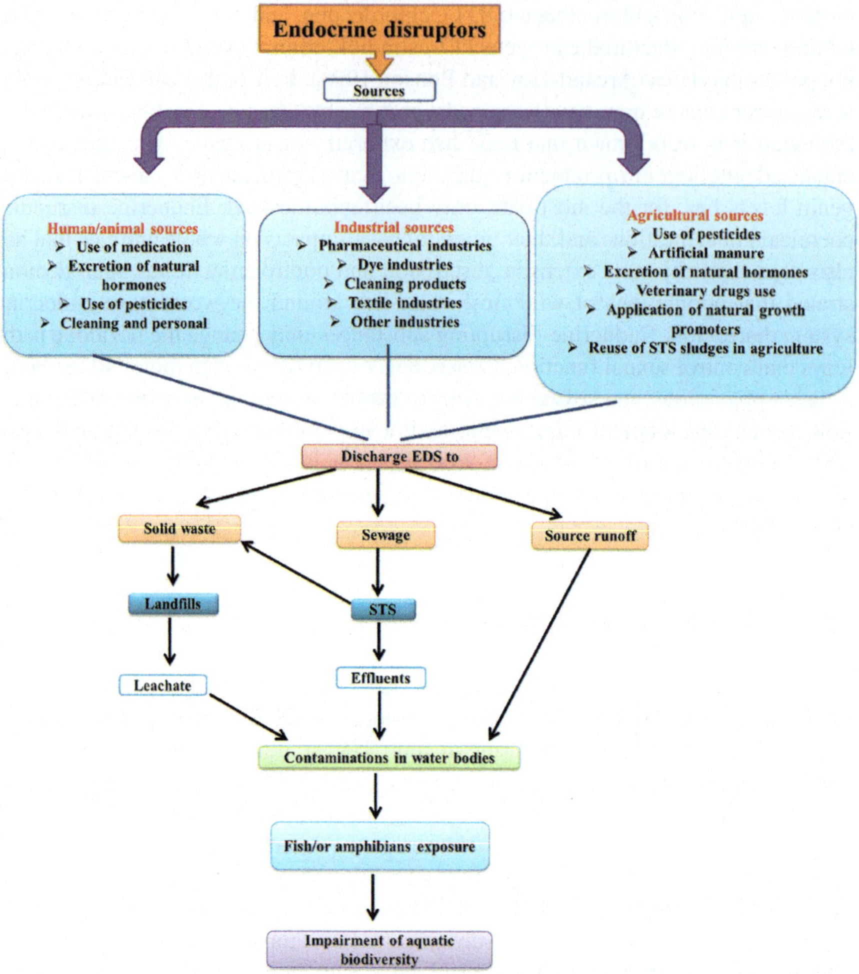

Fig. 15.1 Types of endocrine disruptors in water bodies which affect the aquatic biodiversity

gene networks. This affects normal parameters and functioning by stimulating or inhibiting downstream cellular and molecular processes. The control of sexual differentiation and reproduction is one of estrogens' major physiological functions. Thus, the great bulk of research on the effects of estrogen-active EDCs on fish concentrated on reproductive traits including fecundity, fertility, or offspring survival. These traits included sexual differentiation, sexual behavior, gonad development and vitellogenesis, and gonad size (Ankley et al. 2009; Segner 2011). Nevertheless, estrogens are not only play a role in reproductive physiology but they also have pleiotropic actions that target a range of physiological processes outside the reproductive system. For instance, estrogens are involved in the growth hormone/insulin-like growth factor system (Filby et al. 2006; Shved et al. 2007, 2008), the stress response (Pottinger et al. 1996), osmoregulation (Madsen et al. 2004), and the differentiation of neurosensory systems (Froehlicher et al. 2009). Therefore, it is important to consider how environmental estrogens may affect processes other than reproduction. In addition to causing reproductive abnormalities, EDs are also known to disrupt other endocrine systems across a variety of axes, such as the thyroid, hypothalamo-hypophyseal-gonadal (HHG) axis and other cellular systems by exerting antagonistic or agonistic effects upon binding to hormone receptors. Additionally, if exposed at a given dose and period, some of these adverse effects (seen in wildlife fauna/experimental organisms) may also predominate in humans, leading to endocrine dysregulation. Many studies have been conducted on the effects of ED exposure in aquatic animals, particularly fish. Although hundreds of EDs have been found so far, only a small number have been evaluated.

15.7 Conclusion

As a result of causing disruptions in fish's hormone balance, certain chemicals can undoubtedly induce physiological abnormalities on their own. In some instances, these combinations might have harmful impacts on health. It is necessary to conduct carefully focused investigations, which have not yet been done, to gain a better understanding of how contaminants interact to affect the endocrine system in fish.

Acknowledgments MA gratefully acknowledges the Science and Engineering Research Board (SERB), New Delhi, India, for the financial assistance Rendered (Ref: PDF/2021/001905).

References

Akihiko H, Sullivan CV (1993) Isolation and some characterization of vitellogenin and its related egg yolk proteins from coho salmon (Oncorhynchus kisutch). Zool Sci 10:245–256

Ankley GT, Bencic DC, Breen MS, Collette TW, Conolly RB, Denslow ND, Edwards SW, Ekman DR, Garcia-Reyero N, Jensen KM, Lazorchak JM (2009) Endocrine disrupting chemicals in fish: developing exposure indicators and predictive models of effects based on mechanism of action. Aquat Toxicol 92:168–178

Arcand-Hoy LD, Benson WH (1998) Fish reproduction: an ecologically relevant indicator of endocrine disruption. Environ Toxicol Chem Int J 17:49–57

Casanova-Nakayama A, Wenger M, Burki R, Eppler E, Krasnov A, Segner H (2011) Endocrine disrupting compounds: can they target the immune system of fish? Mar Pollut Bull 63:412–416

Filby AL, Thorpe KL, Tyler CR (2006) Multiple molecular effect pathways of an environmental oestrogen in fish. J Mol Endocrinol 37:121–134

Folmar LC et al (2002) A comparison of the estrogenic potencies of estradiol, ethynylestradiol, diethylstilbestrol, nonylphenol and methoxychlor *in vivo* and *in vitro*. Aquat Toxicol 60:101–110

Froehlicher M, Liedtke A, Groh K, López-Schier H, Neuhauss SC, Segner H, Eggen RI (2009) Estrogen receptor subtype β2 is involved in neuromast development in zebrafish (*Danio rerio*) larvae. Dev Biol 330:32–43

Girish BP, Swetha CH, Reddy PS (2014) Hepatopancreas but not ovary is the site of vitellogenin synthesis in female fresh water crab, Oziothelphusa senex senex. Biochem Biophys Res Commun 447:323–327

Hiramatsu N, Matsubara T, Hara A, Donato DM, Hiramatsu K, Denslow ND, Sullivan CV (2002) Identification, purification and classification of multiple forms of vitellogenin from white perch (Morone americana). Fish Physiol Biochem 26:355–370

Hiramatsu N, Matsubara T, Fujita T, Sullivan CV, Hara A (2006) Multiple piscine vitellogenins: biomarkers of fish exposure to estrogenic endocrine disruptors in aquatic environments. Mar Biol 149:35–47

Hiramatsu N, Todo T, Sullivan CV, Schilling J, Reading BJ, Matsubara T, Ryu YW, Mizuta H, Luo W, Nishimiya O, Wu M (2015) Ovarian yolk formation in fishes: molecular mechanisms underlying formation of lipid droplets and vitellogenin-derived yolk proteins. Gen Comp Endocrinol 221:9–15

Jobling S, Tyler CR (2003) Endocrine disruption, parasites and pollutants in wild freshwater fish. Parasitology 126:S103–S107

Kar S, Sangem P, Anusha N, Senthilkumaran B (2021) Endocrine disruptors in teleosts: evaluating environmental risks and biomarkers. Aquac Fish 6:1–26

Liu ZH, Kanjo Y, Mizutani S (2010) A review of phytoestrogens: their occurrence and fate in the environment. Water Res 44:567–577

Lubzens E, Young G, Bobe J, Cerdà J (2010) Oogenesis in teleosts: how fish eggs are formed. Gen Comp Endocrinol 165:367–389

Madsen SS, Skovbølling S, Nielsen C, Korsgaard B (2004) 17-β estradiol and 4-nonylphenol delay smolt development and downstream migration in Atlantic salmon, *Salmo salar*. Aquat Toxicol 68:109–120

Mak AS, Choi CL, Tiu SH, Hui JH, He JG, Tobe SS, Chan SM (2005) Vitellogenesis in the red crab Charybdis feriatus: Hepatopancreas-specific expression and farnesoic acid stimulation of vitellogenin gene expression. Mol Reprod Dev 70:288–300

Matthiessen P (2000) Is endocrine disruption a significant ecological issue? Ecotoxicology 9:21–24

Mills LJ, Gutjahr-Gobell RE, Horowitz DB, Denslow ND, Chow MC, Zaroogian GE (2003) Relationship between reproductive success and male plasma vitellogenin concentrations in cunner, *Tautogolabrusadspersus*. Environ Health Perspect 111:93–100

Nath P, Sundararaj BI (1981) Isolation and identification of female-specific serum lipophosphoprotein (vitellogenin) in the catfish, *Heteropneustesfossilis* (Bloch). Gen Comp Endocrinol 43:184–190

Pottinger TG, Carrick TR, Hughes SE, Balm PH (1996) Testosterone, 11-ketotestosterone, and estradiol-17β modify baseline and stress-induced interrenal and corticotropic activity in trout. Gen Comp Endocrinol 104:284–295

Purdom CE, Hardiman PA, Bye VV, Eno NC, Tyler CR, Sumpter JP (1994) Estrogenic effects of effluents from sewage treatment works. Chem Ecol 8:275–285

Reading BJ, Hiramatsu N, Sullivan CV (2011) Disparate binding of three types of vitellogenin to multiple forms of vitellogenin receptor in white perch. Biol Reprod 84:392–399

Scharf ME, Wu-Scharf D, Zhou X, Pittendrigh BR, Bennett GW (2005) Gene expression profiles among immature and adult reproductive castes of the termite *Reticulitermes flavipes*. Insect Mol Biol 14:31–44

Scott AP, Katsiadaki I, Kirby MF, Thain J (2006) Relationship between sex steroid and vitellogenin concentrations in flounder (Platichthysflesus) sampled from an estuary contaminated with estrogenic endocrine-disrupting compounds. Environ Health Perspect 114:27–31

Segner H (2011) Reproductive and developmental toxicity in fishes. In: Reproductive and developmental toxicology. Academic Press, London, pp 1145–1166

Shved N, Berishvili G, D'Cotta H, Baroiller JF, Segner H, Eppler E, Reinecke M (2007) Ethinylestradiol differentially interferes with IGF-I in liver and extrahepatic sites during development of male and female bony fish. J Endocrinol 195:513–524

Shved N, Berishvili G, Baroiller JF, Segner H, Reinecke M (2008) Environmentally relevant concentrations of 17α-ethinylestradiol (EE2) interfere with the growth hormone (GH)/insulin-like growth factor (IGF)-I system in developing bony fish. Toxicol Sci 106:93–102

Sumpter JP, Jobling S (1995) Vitellogenesis as a biomarker for estrogenic contamination or the aquatic environment. Environ Health Perspect 103:173–178

Sun C, Zhang S (2015) Immune-relevant and antioxidant activities of vitellogenin and yolk proteins in fish. Nutrients 7:8818–8829

Tao Y, Hara A, Hodson RG, Woods LC, Sullivan CV (1993) Purification, characterization and immunoassay of striped bass (*Morone saxatilis*) vitellogenin. Fish Physiol Biochem 12:31–46

Tubbs CW, McDonough CE (2018) Reproductive impacts of endocrine-disrupting chemicals on wildlife species: implications for conservation of endangered species. Annu Rev Anim Biosci 6:287–301

Tyler CR, Van Der Eerden B, Jobling S, Panter G, Sumpter JP (1996) Measurement of vitellogenin, a biomarker for exposure to oestrogenic chemicals, in a wide variety of cyprinid fish. J Comp Physiol B 166(7):418–426

Vos JG, Dybing E, Greim HA, Ladefoged O, Lambré C, Tarazona JV, Brandt I, Vethaak AD (2000) Health effects of endocrine-disrupting chemicals on wildlife, with special reference to the European situation. Crit Rev Toxicol 30:71–133

Williams VN, Reading BJ, Hiramatsu N, Amano H, Glassbrook N, Hara A, Sullivan CV (2014) Multiple vitellogenins and product yolk proteins in striped bass, *Morone saxatilis*: molecular characterization and processing during oocyte growth and maturation. Fish Physiol Biochem 40:395–415

Biological Activities of Vitellogenin and Its Mechanism

16

Ramachandran Ishwarya, Govindan Tamilmani, and Rengarajan Jayakumar

Abstract

Over the past 10 years, our understanding of the roles played by vitellogenin (Vtg) in reproduction has established. Vtg was once believed to be a female-specific reproductive protein that breaks down into yolk proteins including phosvitin (Pv) and lipovitellin (Lv), which are kept in eggs and give early embryos nourishment. Recent research has shown that Vtg functions as an immuno-component factor that can protect the host against attack by bacteria and viruses. Moreover, in developing embryos, Pv and Lv, which are also proteolytically cleaved products of maternal Vtg, have an antibacterial role, and Vtg has antioxidant characteristics that can protect cells from free radical damage. These findings show that Vtg not only contributes to the production of yolk proteins but also plays non-nutritional roles by acting as immune-relevant molecules and antioxidant agents. Vtgs are discovered to exhibit antioxidant activity, to shield the host from oxidative stress, in addition to these immune-relevant tasks. Understanding the physiological activities that the molecules play on non-nutritional functions also establishes a strong foundation for the molecules' potential usage in the future to advance human health.

Keywords

Vitellogenin · Antioxidant · Antimicrobial · Biological activity · Mechanism

R. Ishwarya · G. Tamilmani
Mandapam Regional Centre, ICAR—Central Marine Fisheries Research Institute, Mandapam, Tamil Nadu, India

R. Jayakumar (✉)
ICAR—Central Institute of Brackishwater Aquaculture (CIBA), Chennai, Tamil Nadu, India

V. Baskaralingam, R. Vanichviriyakit (eds.), *Vitellogenin in Fishes-Diversification, Biological Properties, and Future Perspectives*, https://doi.org/10.1007/978-981-99-5340-0_16

Abbreviation

DNA	Deoxy ribonucleic acid
DUF	Domain of unknown factor
E2	Estradiol-17β
EDCs	Endocrine-disrupting chemicals
EEDCs	Oestrogenic-like endocrine-disrupting chemicals
ELISA	Enzyme-linked immunosorbant assay
ERE	Oestrogen response element
ERs	Oestrogen receptors
FSH	Follicle-stimulating hormone
GnRH	Gonadotropin-releasing hormone
IPNV	Infectious pancreas necrosis virus
Ldlr	Low-density lipoprotein receptor
LPS	Lipopolysaccharide
LTA	Lipoteichoic acid
Lv	Lipovitellin
PAMPs	Pathogen-associated molecular patterns
PGN	Peptidoglycan
Pv	Phosvitin
RNA	Ribonucleic acid
TTX	Tetraodotoxin
VLDL	Very-low-density lipoprotein
Vtg	Vitellogenin
vWD	Willebrand factor type domain
YPs	Yolk proteins

16.1 Introduction

There are about 26,000 known fish species, the vast majority of which are in the infraclass, and Teleosts exhibits a broad array of morphological, physiological and behavioural characters. One of the most varied families of vertebrates is the teleosts. They use a variety of reproductive techniques to get over the obstacles posed by the various ecological niches they occupy. Oviparity is the most prevalent reproductive strategy in teleosts. Oviparous creatures produce eggs that have little to no embryonic development when laid by the mother. There is an increasing need to comprehend fish population dynamics and the underlying mechanisms that control their capacity to rebound because many fisheries throughout the world are currently overfished. The process of reproduction and subsequent recruitment of juveniles to the fishery is of considerable significance to fisheries biologists. Fish reproduction comprises a wide variety of methods, which include gender differentiation, fertilisation techniques and spawning cycles. There are both biological and behavioural adaptation, viz., pair selection, cotious behaviour and parental care. The majority of invertebrates are oviparous, which means that their eggs are fertilised externally. A viable embryo is produced after fertilisation from an egg, or haploid reproductive cell, which is the product of oocyte development and differentiation (Lubzens et al. 2010). Early in the twentieth century, immunological techniques were used to pinpoint a specific antigen in the blood of gravid female fish during investigations on sex discrimination. The principal precursor of egg yolk protein, which is currently known as vitellogenin and is generated in the female liver before being secreted into the circulation and integrated into the egg, is currently identified as this particular antigen. Numerous vitellogenin variations have recently been discovered thanks to protein and gene analysis. Additionally, choriogenin was discovered as a novel precursor of egg envelope proteins in the 1980s. Like vitellogenin, choriogenin is secreted into the blood in response to oestrogen stimulation. In addition to having significant functions in the process of oogenesis, these two proteins are used as effective biomarkers for assessing the impact of oestrogen-like endocrine-disrupting chemicals (environmental hormones) in aquatic settings. The females of egg-laying vertebrates, including the majority of fish species, go through stages in their life cycles where their oocytes mature in preparation for ovulation and spawning. Growing follicles synthesise and expel steroid hormones into the systemic circulation, which regulates a multitude of various metabolic processes, under the multifaceted impact of hormonal centres like the hypothalamus and pituitary gland. The high yolk mass found in teleost eggs provides a nutrient rich, protein and lipid rich substrate for larval growth and embryonic development. All nutritional needs for oviparous animals developing young are met entirely by stored egg yolk. The ovulated egg must have enough nutrients to support progeny growth from the time of fertilisation to the start of exogenous feeding. The process of vitellogenesis involves feed for developing oocytes with the essential yolk nutrients so they can mature into eggs. Proteins, carbohydrates, lipids, vitamins, minerals and ions are among the maternally produced molecules that make up these nutrients; they are all carried from the liver to the ovary in the form of circulating vitellogenins, which are precursors to the yolk. The main component of the yolk of vertebrate eggs is vitellogenin. As essential nutrients for developing embryos, all fishes manufacture vitellogenins. The majority of the components required to create and support a new life are delivered to the ooplasm of oocytes by vitellogenin when they develop by orders of magnitude during oogenesis. Over the past 20 years, there has been a significant advancement in our understanding of the diversity, evolution, structure and functions of fish vitellogenins and their receptors, particularly from a molecular perspective and with an eye towards using the knowledge to enhance finfish aquaculture, fisheries management and biomedical research.

16.2 Vitellogenin

A glycolipophosphoprotein termed vitellogenin (Vtg) is conserved in almost all oviparous species, including fish, amphibians and the majority of invertebrates. It has been 33 years since Pan et al. 1969 initially used the name 'vitellogenin' to refer to the precursor to insect yolk proteins (YPs). The manufacture, blood circulation, uptake into oocytes, proteolytic cleavage into YPs and usage of YPs by developing embryos were the main topics of early investigations of vertebrate Vtg. In 1935, Laskowski mixed the serum of data suggesting similarities between yolk and blood contents in the laying hens with water and got a precipitate. This was the first time vitellogenin was isolated in crude form from any animal (Laskowski 1935). Because of the dominance in the laying hen, he gave this substance the name serumvitellin. Laskowski's serumvitellin could be separated into at least two components in an ultracentrifuge, as McIndoe (1960) later shown. It is normally found in females but is also present in small levels in men. In these animals, the female liver is normally where Vtg genes are expressed and produced. The protein is subsequently delivered to the ovary via the circulation. The Vtg protein is split up into subdomains in the ovary and then absorbed into the egg yolk. A possible candidate for one of the maternal nutrients provided to the intraovarian embryo in fishes. The transport molecule for the numerous kinds of chemicals that the developing oocyte accumulates is vitellogenin, and it has a considerable protein chain as its backbone (molecular weight: 250,000–600,000), but it also contains large amounts of lipid, glucose, phosphate and mineral salts. The transport molecule vitellogenin is broken down and accumulates as egg-specific yolk components, such as phosvitin and lipovitellin, after being selectively taken up into the oocyte. The ovary receives Vtgs from the liver via blood, where oocytes absorb them and transform them into their derivative YPs. After vitellogenesis is complete, the ovary is loaded with fully yolked eggs, which subsequently go through maturation and ovulation. In some species, the Vtg-derived yolk, which can account for as much as 80–90% of the dry mass of an ovulated egg, can significantly contribute to oocyte growth.

The following succinctly describes the general properties of Vtg in oviparous vertebrates:

- When male or young fish are given oestrogen, plasma Vtg is produced.
- Vtg is a high-molecular-mass complex protein made up of sugar, lipid and phosphorus and binding other elements like calcium, iron and zinc.
- Vtg is a precursor to egg yolk proteins that react with antibodies made against egg extracts.

Vtg has gained interest as a biomarker for assessing the impact of endocrine system-disrupting substances found in diverse waterways since the mid-1990s. Male rainbow trout stocked close to a wastewater treatment facility in the UK were the first to exhibit environmental induction of Vtg (Sumpter and Jobling 1995). Later, numerous research primarily from Europe and North America that connected Vtg expression to oestrogenic-like endocrine-disrupting chemicals (EEDCs) were published (Sumpter 1997; Giesy and Snyder 1998). It is now known that the vertebrate neuroendocrine system, which regulates vital processes like development, growth, metabolism and reproduction, is vulnerable to the effects of a number of chemicals developed over the past century. Environmental toxicology now includes a significant amount of study on the identification and consequences of such compounds, and this work has drawn considerable public interest. Endocrine disrupters, also known as endocrine-disrupting chemicals, are typically defined as substances that either mimic or oppose the actions of naturally occurring hormones (EDCs). They include oestrogenic EDCs (environmental oestrogens), which act to initiate an estrogenic response akin to endogenous estradiol-17 (E2).

16.3 Biological Activities of Vitellogenin

The Vtg gene family underwent processes during the evolution of vertebrates that resulted in a diverse range of genes in different species. The primary role of vitellogenin proteins is to provide as an early developmental stage source of yolk nutrients. However, the existence of numerous vitellogenin genes raises fresh concerns regarding the potential uses of distinct Vtgs and the yolk protein derived from them. Additionally, a growing research interest has identified a number of Vtg non-nutritional functions (Fig. 16.1).

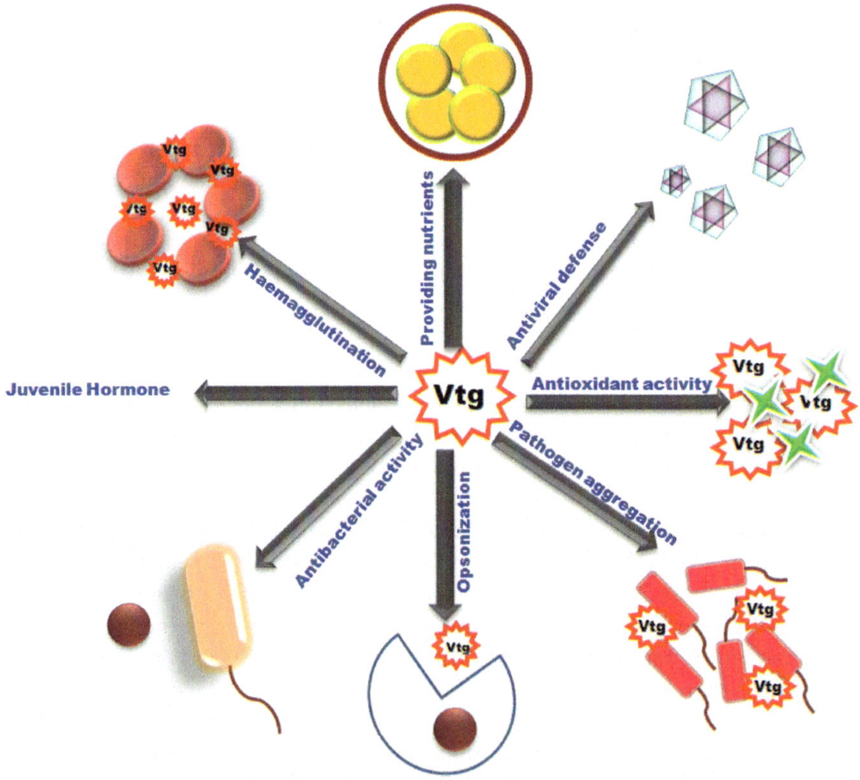

Fig. 16.1 Biological activities of vitellogenin in fishes

16.4 Immune-Relevant Activities of Vtg

The majority of fish and aquatic invertebrates release and fertilise their eggs externally, exposing the developing embryos to a hostile aquatic environment full of possible pathogens that can cause a variety of diseases and even cause death. Their embryos also have a restricted or nonexistent capacity to generate endogenous immune-relevant chemicals in the early stages of development, and their immune-relevant cells and tissues are not fully developed (Ellis 1988; Magnado'ttir et al. 2004). It is familiar that fish and aquatic invertebrates generate eggs that are fully developed fish embryos in an aquatic environment, including all the necessary nutrition and defence mechanisms. It has been demonstrated that Vtgs contribute to embryo protection. Vg has a newly discovered role connected to the host immune system. Zhang et al. 2015 were the first to discover that Vg had hemagglutinating and antibacterial properties in the amphioxus *Branchiostoma japonicum.* The first concrete proof that Vtg has an immune-relevant function was found by Zhang et al. in 2015, who found that Vtg purified from the ovaries of the protochordate amphioxus (*Branchiostoma japonicum*) exhibited haemagglutinating activity against chick, toad and grass carp erythrocytes as well as antibacterial activity against the Gram-negative bacterium *E. coli.* The infectious pancreas necrosis virus (IPNV) was recently found to be susceptible to neutralisation by the serum of Atlantic salmon (*Salmon salar* L.) (Liu et al. 2009) and by mosquito Vg interfering with Anopheles gambiae's anti-Plasmodium response (Zhang et al. 2005). Additionally, it was discovered that VWD and DUF1943 and DUF1944 helped VWD act as a pattern recognition receptor function as an opsonin (Sun et al. 2013). All of these findings suggest that Vg is an immunocomponent molecule that plays a critical role in fish defence against microbes, including bacteria and viruses. It does this by acting as a pattern recognition receptor that can distinguish between Gram-negative and Gram-positive bacteria, an effector molecule that can kill bacteria and neutralise viruses, and an opsonin that can improve phagocytosis. Intriguingly, it has been demonstrated that honey bee Vg can identify endogenous damage-associated molecular patterns (DAMPs), including phosphatidylserine, indicating that Vg may function as an anti-inflammatory agent like many other plasma proteins. Later, it was shown that the rosy barb (*Puntius conchonius*) Vtg and males create Vtgs in response to bacterial exposure and have haemagglutinating roles in blood coagulation. Additionally, it was discovered that *Hexagrammosotakii* Vtg could connect to macrophage cell surfaces but not red blood cell surfaces (Lu et al. 2013; Li et al. 2008). Since they imply that the opsonisation of Vtg was not species specific, Liu et al. findings from 2011 showing that the *H. otakii* Vtg could opsonise the fungus *Pichia pastoris* for phagocytosis by macrophages isolated from sea bass *Lateolabrax japonicas* are significant. All of these findings suggest that Vtg is an opsonin that bridges the gap between the host macrophages and the invader pathogens, promoting increased phagocytosis (Liu et al. 2011). The various domains of Vtg contribute to its wide range of immune-relevant functions. Following this interaction, Vtg may operate as effector-disrupting/destabilising cell walls or as a bridge molecule promoting phagocytosis through opsonisation (Li et al. 2008; Zhang et al. 2011). It is worthwhile to investigate if fish Vg can function as a DAMP receptor. The mala zebrafish (*Danio rerio*) was injected with LPS or LTA, and Tong et al. recently demonstrated that this resulted in a quick and considerable upregulation of Vg at both the transcriptional and translational levels. Additionally, the serum Vg generated was capable of binding to both *E. coli* and *Staphylococcus aureus* (Gram-positive bacterium) and to dose dependently suppress their growth. These findings imply that Vg may act in vivo as an acute phase protein, resulting in the destruction of invasive pathogens. Notably, numerous investigations reveal that fish do not just synthesise Vg in the liver.The heart and brain of the Chinese rare minnow *Gobiocypris rarus* (Ma et al. 2009), the ovary, gill and testis of the white cloud mountain minnow *Tanichthys albonubes* (Wang et al. 2010), and the gill, heart, white adipose tissue and skin of the zebra fish, for instance, have all been documented to express Vtg (Wang et al. 2005; Islinger et al. 2003; Yin et al. 2009). The widespread extra hepatic expression of vg may imply that vg has roles at local areas following the commencement of illness or injury in addition to its functions in the circulating blood, lymph and body fluid (Lu et al. 2011).

16.5 Antimicrobial-Relevant Activities

In all oviparous organisms, vitellogenin (Vg), a phospholipoglycoprotein, serves as the precursor of the main yolk proteins.Vg is typically produced by the hepatocytes of female oviparous vertebrates, including fish, under the control of oestrogen, modified post-translationally in the liver (phosphorylation, glycosylation and lipidation), released into the bloodstream and transported to the ovary. There, it is taken up by the developing oocytes and proteolytically cleaved to form the yolk proteins, which later serve as energy reserves for developing embryos and larvae (Hiramatsu et al. 2005; Shyu et al. 1986; Dhadialla and Raikhel 1990). The roles of Vg, however, seem to go beyond these tasks. In the mature eusocial honeybee, Vg has been shown to be associated with a variety of biological processes, such as social organisation, temporal division of labour and foraging specialisation, modulation of hormone dynamics and changes in gustory response (Amdam et al. 2006; Guidugli et al. 2005; Nelson et al. 2007). Vg's contribution to host immune defence is another unique function it performs. Both hemagglutinating and bactericidal properties of Vg have been demonstrated in the protochordate amphioxus (*Branchiostomabelcheri*) and the bony fish rosy barb (*Puntius conchonius*) (Zhang et al. 2005; Shi et al. 2006). Additionally, the ability of the male rose barb to create Vg in response to an *Escherichia coli* challenge suggests that Vg may be connected to an infection-resistant response. Recent research has shown that piscine Vg functions as an opsonin that can improve macrophage phagocytosis and is a multivalent pattern recognition molecule that can recognise non-self components such glucan, lipopolysaccharide, peptidoglycan and lipoteichoic acid (Li et al. 2008; Liu et al. 2009). Increasing evidence indicates that Vtg are connected to the host's antibacterial function against microorganisms like bacteria and viruses (Garcia et al. 2010; Zhang et al. 2011). It has been demonstrated that the Vtg of zebra fish (*Danio rerio*), carp

(*Cyprinus carpio*) and greeling (*H. otakii*) exhibit multivalent pattern recognition and bind to a variety of pathogen associated structures, such as bacterial lipopolysaccharide, peptidoglycan and lipoteichoic acid, in a bacteriostatic manner It is indeed bacteriocidal for purified green Vtg to interact with lipopolysaccharide (LPS) and lipoteichoic acid (LTA) in the cell wall. Vtg has been demonstrated to work as an acute phase protein in zebrafish, enabling the eradication of invasive microorganisms like *Escherichia coli* and *Staphylococcus aureus* (Tong et al. 2010). In addition, Lu et al. (2012, 2013) demonstrated that zebrafish skin was challenged with the Gram-negative bacterium *Citrobacter freundii*, which resulted in upregulated expression of Vtg. Shi et al. (2006) demonstrated that intraperitoneal injection of *E. coli* could increase the level of serum Vtg in male rosy barbs *P. conchonius*. After that, it was discovered that Vtg purified from the rosy barb *P. conchonius* was able to inhibit the growth of the Gram-positive bacteria *Staphylococcus aureus, Bacillus subtilis* and *Streptococcus pyogenes* as well as the Gram-negative bacteria *E. coli, E. aerogenes* and *Pseudomonas putida* (Shi et al. 2006). In reality, Tong et al. 2010 demonstrated that Vtg is an acute phase reactant with antibacterial activity against *E. coli* and *S. aureus* that is formed in male zebrafish as a result of stimulation by LPS and LTA (Fischer et al. 2013). Li et al. provided evidence that the marine fish *H. otakii's* Vtg could bind to both Gram-positive and Gram-negative bacteria, including *S. aureus* and *Pichia pastoris* (Li et al. 2008). The ability to aggregate pathogens and identify invasive germs is provided by the binding of Vtgs to bacteria (Liu et al. 2009). The molecules conserved in a class of microbes known as pathogen-associated molecular patterns (PAMPs), such as LPS of Gram-negative bacteria, LTA of Gram-positive bacteria, peptidoglycan (PGN) of Gram-negative and positive bacteria and glucan of fungi, were found to have specific affinities for Vtgs in additional research using an ELISA assay. The ability of the Vtg of Atlantic salmon to neutralise the contagious pancreatic necrosis virus was demonstrated by Garcia et al. in 2010, demonstrating that Vtg is also involved in host antiviral immunity. These findings have shown that Vtg possesses antiviral activity in addition to antibacterial activity, which calls for further investigation. All of these facts imply that Vtg actively defend the host against infection. It has been demonstrated that the Vtg of Atlantic salmon (*Salmo salar*) can defend against some viruses. There is strong evidence that Vtgs play a role in the host's antimicrobial defence against widespread bacteria and viruses. Vtgs have been demonstrated to function as a multivalent pattern recognition receptor that can bind to virions, LPS, LTA, PGN and glucan as well as a bactericidal molecule that can damage bacterial cell walls and an opsonin that can improve the phagocytosis of bacteria by macrophages (Garcia et al. 2010; Li et al. 2008; Sun and Zhang 2015; Zhang et al. 2011, 2015).

16.6 Antioxidant-Relevant Activities

Vtg has been discovered to possess antioxidant activity in addition to the immunological functions (Sun and Zhang 2015), which is essential for defence against oxidative damage (Li and Zhang 2017). A chemical process called oxidation can generate free radicals, which can set off a cascade of events that seriously harm DNA, proteins and lipids. Antioxidant defence is therefore regarded to be crucial for all stages of an organism's life. This is also true during the growth and development of the embryo because the strong metabolism of the embryo results in a significant amount of oxidising chemicals being produced. A fascinating subject in the fields of ecological evolution and animal production is how swiftly developing embryos defend themselves against damage from free radicals (Müller et al. 2012; Selim et al. 2012). It has been demonstrated that oviparous animals' eggs contain significant levels of maternally derived antioxidants. Mothers provide a variety of antioxidants, including vitamin A, vitamin E and beta-carotene, in their eggs (Barim-Oz and Sahin 2016; Dale et al. 2017). Particular antioxidants found in egg yolk are crucial for embryonic growth. The ability of Vtg from the eel *Anguilla japonica* to withstand copper-induced oxidation and to shield the very low density lipoprotein (VLDL) from copper-induced oxidation was initially demonstrated by Ando and Yanagida 1999. This was the first account of Vtg's role as an antioxidant agent and its ability to reduce the effects of free radicals on fish oocytes. It is undeniable that in vertebrates also Vtg has the antioxidant properties.

16.7 Reproductive-Relevant Activity

Additionally, after initial processing, Vtgs go through a second proteolysis that varies depending on whether fish produce pelagic or demersal eggs or if their embryos develop quickly or slowly (Finn and Kristoffersen 2007). The heavy chain of VtgAa lipovitellin is severely broken down during oocyte maturation in acanthomorph fish that lay pelagic eggs, creating a pool of free amino acids that creates an osmotic gradient that can pull water. The increased oocyte hydration that results has an impact on the buoyancy of the egg. This is related to the effect of water salinity on the proportionality of VtgAa, VtgAb and VtgC (Reading and Sullivan 2011). In contrast, LvH generated from VtgAb undergoes less proteolysis during the development and maturation of the oocyte and is utilised in the late larval stages, along with VtgC. Salmonids lack evidence of this second proteolysis, likely because they lay their eggs in freshwater. The third proteolysis takes place during development, but there is little information given in the literature. The action of a vitellogenin subdomain as a binding protein capable of transferring tetraodotoxin (TTX) from liver to ovary in *Takifugu pardalis* has recently been reported as a new function. This poison, which builds up in the eggs, serves as both a deterrent against predators and an attractive scent for males. The findings of a non-gender-related expression of Vtg further challenge the conventional understanding of Vtg as merely a source of nutrition for the growing embryos, directing research towards the discovery of non-nutritional roles of Vtg. The first experimental evidence of selective deletion of several Vtg forms in zebrafish was recently published by (Yilmaz et al. 2019). Their research has uncovered novel regulatory effects on fecundity and fertility in addition to the involvement of Vtg in embryonic and larval development. They demonstrated that fecundity was doubled in Vtg1 knock out females and was 50% lower in

Vtg3 knock out females using multiple CRISPR/Cas9 genome editing. Additionally, mortality rises in Vtg1–1 and Vtg-3 knockout embryos as well as in eggs and embryos. These recent discoveries first determined that vitellogenins are crucial, acting at various phases of reproduction and embryonic development. Overall, exposure to oestrogens and endocrine-disrupting chemicals (EDCs), which are typically found in contaminated environments, can cause the synthesis of Vtg. Numerous chemical substances with oestrogen-like properties are exclusively linked to human activity and are most prevalent in aquatic environments (Hara et al. 2016). Since environmental oestrogens can be harmful, vitellogenin has become an important biomarker for evaluating the effects of EDC on teleosts. Research conducted over the past 20 years has documented how different fish species respond to exposure to endocrine disruptors. Additionally, the simultaneous development of new Vtg based bioassays useful for quickly detecting environmental pollution was made possible by the utilisation of Vtg and yolk proteins in the detection of EDC contamination (Wang et al. 2017). In some fish, Vtgs have also been shown to control their own synthesis. The injection of heterologous Vtg purified from Indian mrigal carp (*Cirrhinusmrigala*) has been shown to promote vitellogenin production in female walking cat fish (*Clarias batrachus*). High levels of Vtg inside the oocyte in rainbow trout can change the way Vtg is made in the liver by preventing the ovary from producing E2. It is yet unknown what exact varieties of Vtgs perform these numerous potential tasks.

16.8 Mechanism of Vitellogenin

Gonadotropins regulate the cyclical or seasonal nature of vitellogenesis. Many internal and external factors affect the brain's (hypothalamus) ability to produce gonadotropin-releasing hormone, including inherent biorhythms, nutritional status, and seasonal changes in day length and water temperature (GnRH). In response to GnRH, pituitary gonadotrophs emit follicle-stimulating hormone (FSH), which causes theca and granulose cells of the ovarian follicle to secrete estradiol-17b (E2). As a result, the liver is given the go-ahead to produce Vtgs and release them into the bloodstream. In some species, the follicular synthesis of E2 is also triggered by the pituitary's release of luteinising hormone in response to GnRH. Growth hormone, thyroid hormone, cortisol, androgens, as well as other oestrogenic steroids like estrone may also help to induce hepatic vitellogenesis in some species. The vitellogenesis of hagfish (*Eptatretus stoutii*) is often less responsive to E2 and may be influenced by other factors, such as the time after feeding. Circulating E2 binds to one of the three oestrogen receptors (ERs) in the liver cells and is likely to do so in the Mozambique tilapia, *Oreochromis mossambicus* (Davis et al. 2007). The ER can bind protein factors necessary for specific transcriptional activation once the E2 is bound due to conformational changes. The ER is translocated to the nucleus, where it dimerises, binds an anestrogen response element (ERE), a particular genomic DNA sequence motif located close to a basal Vtg gene promoter, and starts transcription. Vtg-encoding RNAs are translated by ribosomes on the rough endoplasmic reticulum, and the translated polypeptides are then transported to the endomembrane system. Significant posttranslational modifications are made to the Vtg. They need to have their Lv domains lipidated, Pv domains phosphorylated, 15–16 residue signal peptides broken and carbohydrate moieties linked to their glycosylation sites. They also need to be appropriately folded and dimerised. The trans-Golgi network is where mature Vtg are exported before being bundled into secretory vesicles and discharged into the bloodstream. Ribosomes on the rough endoplasmic reticulum translate Vtg-coding RNAs, and the translated polypeptides are then transferred to the endomembrane system.

Mature Vtgs are exported into secretory vesicles and released into the circulation from the trans-Golgi network. VtgAa and VtgC are secreted at a slower rate than VtgAb, which may explain why there are comparatively high amounts of VtgAb in the serum. Similar findings have been made with the blood plasma of female striped bass (*Morone saxatilis*) and white perch (*Morone americana*), where VtgC is minor and VtgAa is moderate (Reading et al. 2011). VtgAa reaches peak levels from mid-tolate vitellogenesis in these Moronid species, whereas peak plasma VtgAb levels are obtained during early- to mid-vitellogenesis. Throughout the whole process of vitellogenesis, the VtgC is expressed at low levels. The follicular theca's capillaries, basement membrane and intercellular gaps are where circulating Vtgs travel between granulosa cells to reach the granulosa epithelium's extracellular matrix, where it comes into contact with the oolemma. The Vtgs gather in clathrin-coated pits that invade and are endocytosed after becoming coupled to Vtg. Due to a slight acidity of the vesicle in the peripheral ooplasm, the clathrin coat is peeled off and Vtg dissociates from the Vtg. Vesicles are then grouped into an early endosome population, which fuses with organelles resembling lysosomes to create multivesicular bodies. Colocalised with the imported Vtg are inactive cathepsin Dzymogens, and V-class vacuolar ATPases further acidify the lumen of the many vesicular bodies. There have also been reports of the oocyte amassing several Vtg in disparate ways. The ratio of VtgAa: VtgAb: VtgC in oocytes is 4:13.3:1 in grey mullet (*Mugil cephalus*), an acanthomorph species that lays spelagic eggs, whereas it is 9:15:1 in barfin flounder. The composition of Vtg C in the total Vtg-derived yolk is B5% in both of these species, which is identical to what has been observed in white perch, a species that lays sticky eggs. The final yolk composition of eggs can differ significantly among fish species, as evidenced by the fact that the Vtg C component of the yolk can reach as high as 25% in mosquito fish and striped bass, two species that spawn neutrally buoyant eggs (Williams et al. 2017). The early life history characteristics of fish, such as egg buoyancy or nutritional requirements for larvae, may be related to this variance. White perch raised in brackish water (7–15 ppt) have a proportional VtgAa:VtgAb:VtgC ratio of 5:36:1, whereas those raised in near fresh water have an 8:16:1 ratio. This suggests that the Vtg-yolk system may be sensitive to water salinity and may be physiologically regulated for the production of eggs with the proper buoyancy based on the water's specific gravity (Reading et al. 2009). White perch and striped bass both take 3–5 days and 7–9 days post-fertilisation, respectively, before they begin to feed. This suggests that striped bass have a longer developmental window, which may be related to their higher VtgC-derived yolk storage. Variations in Vtg yolk accumulation also imply the existence of a system for differentiating Vtg oocyte uptake, which may be receptor mediated (Fig. 16.2).

It is becoming more and more clear that each fish species demonstrates a unique composition of proteins produced from Vtg that are deposited in the egg yolk. This custom yolk appears to be produced in a species-specific manner using a combination of intricate regulatory systems, such as differential expression of numerous Vtg genes, translation and disparate deposition of the resulting multiple Vtg proteins via subtype-specific and universal Vtg receptors, followed by selective processing of the Vtgs and their yolk protein products during oocyte growth and maturation. It is most likely that the taxonomic richness and complexity of the reproductive techniques used by fishes to enhance their reproductive success in a variety of habitats are supported by the diversity and complexity of these mechanisms governing yolk formation. Among recent findings, the discovery of multiple Vtg receptors as well as knowledge of the phylogenic distribution of multiple Vtg subtypes among teleosts set the stage for elucidation of molecular mechanisms by which Vtg-derived yolk products are properly accumulated and processed in a species-specific manner. Further research should be directed at identifying remaining unknown Vtg receptors and at elucidating the molecular mechanisms regulating Vtg receptor gene expression. Likewise, regulation of the translation, translocation and activation of Vtg receptor proteins begs further study, including investigation of the endocrine control

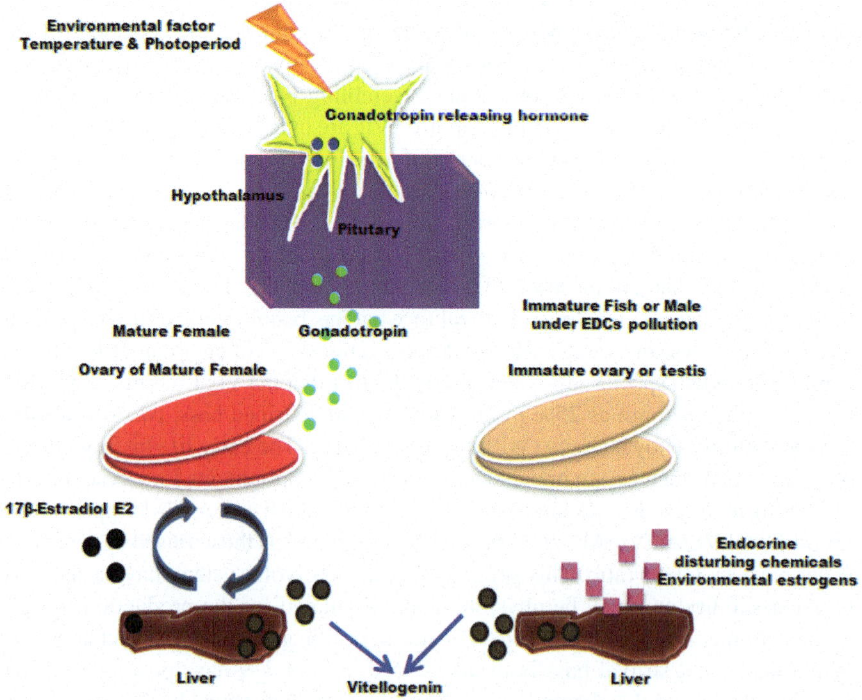

Fig. 16.2 Schematic diagram for mechanism of vitellogenin in fishes

of these processes. In order to uncover the actions and interactions of additional receptor-associated and adapter proteins that may be required for the oocyte to develop competence for Vtg endocytosis, the process of receptor-mediated endocytosis of Vtg also requires inspection at a finer level of molecular detail. Even though our understanding of the molecular pathways behind oocyte lipidation is still limited, it is now expanding quickly. Results from in vivo and in vitro studies on salmonids, as well as certain other marine and freshwater species, show that maternal VLDL is a major source of ooplasm lipids and the lipase-dependent, non-endocytotic pathway is the predominant mechanism for the uptake of VLDL-associated lipids. Although the creation of ooplasm lipid droplets does not appear to include the lipoprotein receptor-mediated pathway, there is still a chance that Ldlr is connected to this process in some species, including anguillid eels. The expression profiles of numerous lipid transporter genes point to a potential connection between the development of ooplasm lipid droplets and studies on oocyte lipid transporters, which have recently started. The innovative findings discussed here pave the way for elucidating the molecular specifics of neutral lipid accumulation in fish oocytes, a field of study that has largely remained unexplored up until now. A useful general model of teleost oocyte lipidation must be applicable to species with or without ooplasm lipid droplets and will likely include considerable diversity in molecular mechanisms for acquisition of maternal lipoprotein-associated lipids, as already evidenced by comparisons of trout and eels. Comparative studies of fish species with few or no ooplasm lipid droplets, perhaps involving experiments tracking the fate of fluorescent-labelled lipoproteins, will be especially useful as we seek to define unifying principles broadly applicable to teleost fishes. As shown by comparisons of trout and eels, a viable generic model of teleost oocyte lipidation must be applicable to species with or without ooplasm lipid droplets and will likely incorporate a wide range of molecular processes for acquisition of maternal lipoprotein-associated lipids. In order to establish general guidelines that apply to all teleost fishes, comparative studies of fish species with few or no ooplasm lipid droplets will be especially helpful. These studies may involve experiments that track the fate of fluorescent-labelled lipoproteins.

16.9 Conclusion

Over the past 10 years, there has been a radical change in our knowledge of how Vtgs act in animal reproduction. The current status of our understanding of Vtgs in fish reproduction was summed up in this chapter. Recently, though, it was discovered that Vtg also functions as an immune-relevant molecule that aids in the host's defence against microorganisms including viruses and bacteria. Additionally, Vtgs exhibit an antibacterial action in developing embryos and antioxidant agents that can shield cells from free radical damage. In addition to improving and deepening our understanding of the physiological roles played by the molecules, these non-nutritional functions also offer a solid foundation for the molecules' potential use in promoting human health.

Acknowledgement The author gratefully acknowledges the Science and Engineering Research Board, India, New Delhi, India, for the financial assistance Rendered [Ref: PDF/2020/001027].

References

Amdam GV, Norberg K, Page RE, Erber J, Scheiner R (2006) Down regulation of vitellogenin gene activity increases the gustatory responsiveness of honey beeworkers (Apis mellifera). Behav Brain Res 169:201–205

Ando S, Yanagida K (1999) Susceptibility to oxidation of copper-induced plasma lipoproteins from Japanese eel: protective effect of vitellogenin on the oxidation of very low density lipoprotein. Comp Biochem Physiol C Pharmacol Toxicol Endocrinol 123:1–7

Barim-Oz O, Sahin H (2016) The influence of dietary antioxidant on ovarian eggs and levels of vitamin E, C, a, astaxanthin, β-carotene and oxidative stress in tissues of *Astacusleptodactylus* (Eschscholtz) during reproduction. Cell Mol Biol 62:1–10

Dale K, Rasinger J, Thorstensen K, Penglase S (2017) Ellingsen S vitamin E reduces endosulfan induced toxic effects on morphology and behavior in early development of zebrafish (*Danio rerio*). Food Chem Toxicol 101:84–93

Davis LK, Hiramatsu N, Hiramatsu K (2007) Induction of three vitellogenins by 17 beta-estradiol with concurrent inhibition of the growth hormone-insulin like growth factor 1 axis in a euryhaline teleost, the tilapia (*Oreochromis mossambicus*). Biol Reprod 77:614–625

Dhadialla TS, Raikhel AS (1990) Biosynthesis of mosquito vitellogenin. J Biol Chem 265:9924–9933

Ellis AE (1988) Fish vaccination. Academic Press, New York, NY

Finn RN, Kristoffersen BA (2007) Vertebrate vitellogenin gene duplication in relation to the 3R hypothesis: correlation to the pelagic egg and the oceanic radiation of teleosts. PLoS One 2:169

Fischer M, Regitz C, Kull R, Boll M, Wenzel U (2013) Vitellogenins increase stress resistance of *Caenorhabditis elegans* after *Photorhabdusluminescens* infection depending on the steroid-signaling pathway. Microbes Infect 15:569–578

Garcia J, Munro ES, Monte MM, Fourrier MC, Whitelaw J, Smail DA, Ellis AE (2010) Atlantic salmon (*Salmo salar* L.) serum vitellogenin neutralises infectivity of infectious pancreatic necrosis virus (IPNV). Fish Shellfish Immunol 29:293

Giesy JP, Snyder EM (1998) Xenobiotic modulation of endocrine function in fish. In: Kendall RJ, Dickerson RL, Giesy JP, Suk WA (eds) Principles and processes for evaluating endocrine disruption in wildlife. SETAC Press, Pensacola, FL, pp 155–237

Guidugli KR, Nascimento AM, Amdam GV, Barchuk AR, Omholt S, Simões ZL, Hartfelder K (2005) Vitellogenin regulates hormonal dynamics in the worker caste of a eusocial insect. FEBS Lett 579:4961–4965

Hara A, Hiramatsu N, Fujita T (2016) Vitellogenesis and choriogenesis in fish. Fish Sci 82:187–202

Hiramatsu N, Cheek AO, Sullivan CV, Matsubara T, Hara A (2005) Vitellogenesis and endocrine disruption. In: Mommsen TP, Moon TW (eds) Biochemistry and molecular biology of fishes, vol 6. Elsevier Science, London, pp 431–471

Islinger M, Willimski D, Völkl A, Braunbeck T (2003) Effects of 17α-ethinylestradiolon the expression of three estrogen-responsive genes and cellular ultrastructure of liver and testes in male Zebrafish. Aquat Toxicol 62:85–103

Laskowski M (1935) Der Stellungsmethode des Serum Vitellins. Biochem Z 278:345–348

Li H, Zhang S (2017) Functions of vitellogenin in eggs. Oocytes 63:389–401

Li Z, Zhang S, Liu Q (2008) Vitellogenin functions as a multivalent pattern recognition receptor with an opsonic activity. PLoS One 3:1940

Liu QH, Zhang SC, Li ZJ, Gao CR (2009) Characterization of a pattern recognition molecule vitellogenin from carp (*Cyprinus carpio*). Immunobiology 214:257–267

Liu M, Pan J, Ji H, Zhao B, Zhang S (2011) Vitellogenin mediates phagocytosis through interaction with FcR. Mol Immunol 49:211–218

Lu A, Hu X, Xue J, Zhu J, Wang Y, Zhou G (2011) Gene expression profiling in the skin of zebrafish infected with *Citrobacter freundii*. Fish Shellfish Immunol 32:273–283

Lu A, Hu X, Xue J, Zhu J, Wang Y, Zhou G (2012) Gene expression profiling in the skin of zebrafish infected with *Citrobacter freundii*. Fish Shellfish Immunol 32:273–283

Lu A, Hu X, Wang Y, Shen X, Zhu A, Shen L, Ming Q, Feng Z (2013) Comparative analysis of the acute response of zebra fish *Danio rerio* skin to two different bacterial infections. J Aquat Anim Health 25:243–251

Lubzens E, Young G, Bobe J, Cerdà J (2010) Oogenesis in teleosts: how eggs are formed. Gen Comp Endocrinol 165:367–389

Ma L, Li D, Wang J, He J, Yin Z (2009) Effects of adrenergic agonists on the extra hepatic expression of vitellogenin Ao1 in heart and brain of the Chinese rare minnow (*Gobiocyprisrarus*). Aquat Toxicol 18:19–23

Magnado'ttir B, Lange S, Steinarsson A, Gudmundsdo'ttir S (2004) The ontogenic development of innate immune parameters of cod (*Gadus morhua* L). Comp Biochem Physiol B Biochem Mol Biol 139:217–224

McIndoe WM (1960) Changes in the protein content of yolk during chick embryogenesis. Development 8:47–53

Müller W, Vergauwen J, Eens M, Blount JD (2012) Environmental effects shape the maternal transfer of carotenoids and vitamin E to the yolk. Front Zool 9:1–11

Nelson CM, Ihle KE, Fondrk MK, Page RE, Amdam GV (2007) The gene vitellogenin has multiple coordinating effects on social organization. PLoS Biol 5:0673–0677

Pan ML, Bell WJ, Telfer WH (1969) Vitellogenic blood protein synthesis by insect fat body. Science 165:393–394

Reading BJ, Hiramatsu N, Sawaguchi S (2009) Conserved and variant molecular and functional features of multiple egg yolk precursor proteins (vitellogenins) in white perch (*Morone americana*) and other teleosts. Mar Biotechnol 11:169–187

Reading BJ, Hiramatsu N, Sullivan CV (2011) Disparate binding of three types of vitellogenin to multiple forms of vitellogenin receptor in white perch. Biol Reprod 84:392–399

Reading BJ, Sullivan CV (2011) The reproductive organs and processes|Vitellogenesis in fishes. Encyclopedia of fish physiology:635–646

Selim SA, Gaafar KM, Elballal SS (2012) Influence of in-ovo administration with vitamin E and ascorbic acid on the performance of Muscovy ducks. Emirates J Food Agric 24:264–271

Shi X, Zhang S, Pang Q (2006) Vitellogenin is a novel player in defense reactions. Fish Shellfish Immunol 20:769

Shyu AB, Raff RA, Blumenthal T (1986) Expression of the vitellogenin gene in female and male sea urchin. Proc Natl Acad Sci U S A 83:3865–3869

Sumpter JP (1997) Environmental control of fish reproduction: a different perspective. Fish Physiol Biochem 17:25–31

Sumpter JP, Jobling S (1995) Vitellogenesis as a biomarker for estrogenic contamination of the aquatic environment. Environ Health Perspect 103:173–178

Sun C, Hu L, Liu S, Gao Z, Zhang S (2013) Functional analysis of domain of unknown function (DUF) 1943, DUF1944 and von Willebrand factor type D domain (VWD) in vitellogenin 2 in zebrafish. Dev Com Immunol 41:469–476

Sun C, Zhang S (2015) Immune-relevant and antioxidant activities of vitellogenin and yolk proteins in fish. Nutrients 7:8818–8829

Tong Z, Li LR, Zhang S (2010) Vitellogenin is an acute phase protein with bacterial-binding and inhibiting activities. Immunobiology 215:898–902

Wang H, Tan JTT, Emelyanov A, Krozh V, Gong Z (2005) Hepatic and extrahepatic expression of vitellogenin genes in the zebrafish, *Danio rerio*. Gene 356:91–100

Wang R, Gao Y, Zhang L, Zhang Y, Fang Z, He J, Zhang W, Ma G (2010) Cloning, expression, and induction by 17-β estradiol (E2) of a vitellogenin gene in the white cloud mountain minnow *Tanichthysalbonubes*. Fish Physiol Biochem 36:157–164

Wang J, Ma S, Zhang Z, Zheng M, Dong Y, Ru S (2017) Vitellogenin induction in caudal fin of guppy (*Poecilia reticulata*) as a less invasive and sensitive biomarker for environmental estrogens. Sci Rep 7:7647

Williams VN, Reading BJ, Amano H (2017) Proportional accumulation of yolk proteins derived from multiple vitellogeninsis precisely regulated during vitellogenesis in striped bass(*Morone saxatilis*). J Exp Zool A Ecol Genet Physiol 321:301–315

Yilmaz O, Patinote A, Nguyen T, Com E, Pineau C, Bobe J (2019) Genome editing reveals reproductive and developmental dependencies on specific types of vitellogenin in zebrafish (*Danio rerio*). Mol Reprod Dev 86:1168–1188

Yin N, Jin X, He J, Yin Z (2009) Effects of adrenergic agents on the expression of zebrafish (Danio rerio) vitellogenin Ao1. Toxicol Appl Pharmacol 238:20–26

Zhang SC, Sun YN, Pang QX, Shi XD (2005) Hemagglutinating and antibacterial activities of vitellogenin. Fish Shellfish Immunol 19:93–95

Zhang S, Wang S, Li H, Li L (2011) Vitellogenin, a multivalent sensor and an antimicrobial effector. Int J Biochem Cell Biol 43:303–305

Zhang S, Dong Y, Cui P (2015) Vitellogenin is an immunocompetent molecule for mother and offspring in fish. Fish Shellfish Immunol 46:710

Multivalent Properties of Vitellogenin in Marine and Freshwater Fishes

17

Chinnah Amutha, Dharshana Dhinesh, and Ardhra Gopan

Abstract

The synthesis and secretion of vitellogenins (Vtgs) is induced by the presence of estrogen hormone (E2), with the liver being the site of their production. These proteins are subsequently transported through the bloodstream to the ovary, where they are taken up by developing oocytes via receptor-mediated endocytosis. Upon entry into the oocyte, the Vtgs are enzymatically cleaved to form yolk proteins, which play a crucial role in providing nourishment for fish embryos during the process of embryogenesis. The yolk proteins are stored in the ooplasm or cytoplasm of the egg or ovum, serving as a significant source of nutrients for the developing embryo. This chapter deals with the properties of vitellogenin in marine and freshwater fish.

Keywords

Ovary · Embryo · Oocyte · Multivalent · Endocytosis · Nourishment

17.1 Introduction

Recent investigations have placed significant emphasis on the use of fish as indicator organisms for xenoestrogenic compounds in the aquatic environment. Vitellogenin (Vtg), a phosphoglycolipoprotein synthesized in the liver of many oviparous organisms, is produced in a variety of fish species, including brown trout (Noberg and Haux 1988), rainbow trout (Donohoe and Curtis 1996), sole (Rainuzzo et al. 1989), sea bass (Mananos et al. 1997), salmon (Soetal. 1985), carp (Tyler and

C. Amutha (✉) · D. Dhinesh · A. Gopan
Department of Animal Behaviour and Physiology, School of Biological Sciences, Madurai Kamaraj University, Madurai, Tamil Nadu, India

V. Baskaralingam, R. Vanichviriyakit (eds.), *Vitellogenin in Fishes-Diversification, Biological Properties, and Future Perspectives*,
https://doi.org/10.1007/978-981-99-5340-0_17

Sumpter 1990), and tilapia (Buerano et al. 1995). Its potential use as an early indicator of exposure to xenoestrogens has been proposed in multiple studies, such as those by Monosson et al. (1996), Tyler et al. (1996), and Soimasuo et al. (1998). Indeed, according to Tyler et al. (1996) and Kime (1999), the effect of estrogens is readily detectable in fish, since they are the only known stimulant of the yolk protein Vtg (Encyclopedia of Reproduction (Second Edition) 2018).

Therefore, production of this protein in male or immature fish provides a useful indicator for detecting the presence of environmental pollutants with estrogenic activity. Today, there is no doubt that contamination of aquatic ecosystems has occurred in many countries as a result of the wide use of chemical compounds in agricultural and industrial activities.

In many African countries, because the economy largely depends on agriculture (Rajendran and Subramanian 1999), pesticides and herbicieds are likely to represent an important source of xenoestrogens in contaminated rivers and lagoons. Herbicides are mostly applied in agricultural places. They are also used for controlling uncontrolled weeds (Abate et al. 2000). In addition, a quantity of the herbicides present in rivers could come from the banned herbicides stocked in many countries. According to Wodageneh and Van der Wulp (1996), these herbicides are stored under conditions that rarely meet internationally accepted standards. Due to decay of their packaging, the pesticides can be delivered to the soil, thereby aggravating the contamination problem. Among these herbicides are atrazine, 2,4-D. Many xenoestrogens have been shown to be present in aquatic ecosystems, where they pose a potential health hazard to wildlife (Colborn et al. 1993; Sumpter and Jobling 1995).

The objective of this paper is to review the present understanding of the effects of herbicides on vitellogenesis in fish. After briefly describing the vitellogenic cycle in fish, this review discusses several aspects of the impairment of this cycle in fish exposed to herbicides and other organic compounds, in laboratory experiments.

17.2 Significance of Vitellogenin Studies

The function of vitellogenin (Vtg) molecules in fish reproduction has been extensively researched, particularly regarding their association with seasonal reproductive activity. During vitellogenesis, there is a rapid increase in plasma levels of Vtg, which continues throughout the growth phase and correlates with oocyte growth in female serum. Consequently, Vtg is an effective indicator for detecting the onset of puberty and monitoring the progress of maturation in female fishes, both in laboratory studies and in the fields of aquaculture and fisheries research. The measurement of plasma vitellogenin levels is a useful method for determining the reproductive status of female fish, as embryos and larvae rely heavily on vitellogenin and its products before transitioning to external feeding. A better understanding of the biochemistry of fish vitellogenin can enhance our knowledge of fish reproduction and the nutritional requirements of larvae. An analysis of the amino acid composition of fish vitellogenin can aid in the development of starter diets for fish larvae. The discharge of diverse chemicals into the environment has been shown to mimic or oppose hormone actions, and these endocrine-disrupting chemicals (EDCs) can interfere with physiological systems, resulting in developmental, growth, and reproductive alterations in organisms living in the affected environment. Effects of EDCs include delayed sexual maturation, reduced gonad size, suppressed steroid hormone release, and impaired pituitary hormone release. In some species, these physiological effects are associated with reduced spawning ability, reduced number and size of fry, intersexuality, and feminization of male fish. The concentration of estrogenic pollutants is vast and makes it challenging to create assay systems for detecting and measuring each pollutant individually. Therefore, indirect indicators, such as biomarker studies, are more useful and have been more commonly employed in environmental monitoring programs. Fish have been a frequently employed model to investigate estrogenic endocrine disruption because the production of vitellogenin (Vtg), which serves as a precursor to egg yolk proteins, is induced by estrogen or estrogen mimics in both male and female fish. The production of serum vitellogenin (Vtg) in fish is negligible or low in males and immature females, but mature female fish exhibit a seasonal pattern in Vtg levels, with a maximum value of several milligrams per milliliter. This indicates that Vtg is a highly specific indicator of estrogen exposure in fish.

Vitellogenesis is a sequential process incorporating the following events:

1. induction of vitellogenin production and its subsequent release into the bloodstream,
2. the transportation of vitellogenin to the blood stream to the target tissue,
3. the absorption of vitellogenin by developing oocytes and,
4. the transformation of vitellogenin into storable forms (Ho 1987).

17.3 From: Developments in Aquaculture and Fisheries Science, 1996

The development of ovarian follicles can be categorized into two stages: primary growth, also known as previtellogenesis, and secondary growth, also known as vitellogenesis. Throughout these stages, the oocyte accumulates significant amounts of protein and lipid nutrients that are essential for the development of the embryo and larvae. During the previtellogenic stage, neutral lipids are stored in the ooplasm in the form of lipid droplets. During vitellogenic growth, oocytes also accumulate phospholipid-rich yolk protein (YP) precursors termed vitellogenins (Vtgs). Vtgs are synthesized by the liver and transported through the bloodstream to the ovary, where they are taken up by oocytes and processed into their derivative YPs. Following the completion of vitellogenesis, the ovary contains fully yolked oocytes that undergo subsequent maturation and ovulation. The contribution to oocyte growth of Vtg-derived yolk can be substantial, comprising up to 80–90% of the dry mass of an ovulated egg in some species.

17.4 Characterization of Fish Vitellogenin

Pan et al. (1969) initially employed the term "vitellogenin" to denote a female-specific protein found in the hemolymph of the Cecropia moth. In chum salmon (Oncorhynchus keta) and rainbow trout (O. mykiss), Hara (1976) detected a female-specific serum protein (FSSP) that exhibited iron-binding properties. Following the purification of trout FSSP, Vtg was identified for the first time in a teleost fish. Numerous biochemical and immunological techniques have been utilized to isolate, identify, and characterize Vtg in a variety of fish species (review: Wallace 1985; Mommsen and Walsh 1988; Specker and Sullivan 1994).

In general, all Vtg proteins have the following characteristics: they are

1. female-specific serum or plasma proteins,
2. induced by estrogen,
3. glyco-lipo phosphor-proteins with molecular masses ranging from 300 to 600 kDa,
4. function as carrier proteins and contain both a lipid and an ionic component, including calcium, zinc, cadmium, iron, and others,
5. precursors to the major yolk proteins and exhibit immunological cross-reactivity to them (Hiramatsu et al. 2006).

Vitellogenin in teleost species is composed of around 80% protein and 20% lipid, and although they share general characteristics, there are notable differences among species in terms of molecular weight, electrophoretic patterns, and post-translational modifications. Initially, vitellogenin was thought to be a unique lipoprotein with similar amino acid chains and fatty acid compositions across different teleost species.

In teleost species, vitellogenin has a proximate composition of approximately 80% protein and 20% lipid, with phospholipids being the dominant lipid class. Phospholipids constitute 60–90% of the total lipids in vitellogenin (Norberg and Haux 1985; Norberg 1995). Recent research has confirmed that teleost vitellogenins are high molecular weight phospholipoglycoproteins (Hiramatsu et al. 2006). Various studies have demonstrated the phospholipoglycoprotein nature of teleost vitellogenins (Emmersen and Petersen 1976; Campbell and Idler 1980; Nath and Sundraraj 1981; Roy et al. 2004). Goldfish vitellogenin has been reported to contain carbohydrate and bind calcium and serve as a carrier for multivalent metals such as Fe++, Ca++, and Mg++ by utilizing the metal complex-forming ability of phosphoserines (Grogan and Taborsky 1987; Nagler and Idler 1990). The amino acid composition of Vtg was found to be similar among most teleosts, with nonpolar amino acid residues comprising more than one-third of the total amino acids present. Alanine was the most abundant amino acid in many cases (Maltais and Roy 2009). The high proportion of nonpolar amino acids in Vtgs may be related to the lipoprotein composition (Tao et al. 1993). Cysteine residues are present in small amounts (0.1–2% of total amino acids) in Vtgs of many fish species (Hara et al. 1980; Tao et al. 1993; Yao and Crim 1996). The lipid content of fish vitellogenins (Vtgs) is higher compared to other vertebrates, with various teleosts having approximately 20% lipid content (Norberg and Haux 1985; MacKay and Lazier 1993). Vtgs from different fish species, such as goldfish, rainbow trout, sea trout, Atlantic halibut, and Atlantic cod, have been reported to contain lipid components ranging from 18 to 21. The predominant lipid components in Vtgs are phospholipids, accounting for approximately 60–70% of the total lipid content, while the remaining portion primarily comprises triacylglycerols and cholesterol (Norberg 1995; Ohkubo et al. 2004). Vtg is a significant transporter of lipids into fish eggs, according to Leger et al. (1981) and Mommsen and Walsh (1988). In halibut eggs, the total lipid content accounts for about 12% of the dry weight, with polar lipids, especially phosphatidylcholine, representing around 70%, and neutral lipids, mainly triacylglycerols and cholesterol, making up the remaining 30% (Falk-Petersen et al. 1986).

Norberg (1995) demonstrated that the ratio of phospholipids to total lipids was similar in halibut egg yolk and Vtg. Lipids have a key role in nutrition, since they constitute the most efficient energy storage form. Moreover, lipids are important cell components and both phospholipids and cholesterol are critical constituents of biological membranes. An adequate lipid supply is thus crucial for early embryonic and larval development to proceed normally. The phosphorus content of fish Vtgs is often lower than in other vertebrate Vtgs (Norberg and Haux 1985). Varying phosphorous content in fish Vtgs have been reported, e.g., 0.63% for rainbow trout, *Salmo gairdneri* and 0.58% for sea trout, *Salmo trutta* (Norberg and Haux 1985), 0.6% for Atlantic halibut, *Hippoglossus hippoglossus* (Norberg 1995), 3.56% for oceanic pout, *Macrozoarces americanus* and 2.15% for Atlantic cod, Gadus morhua (Yao and Crim 1996), 0.68% for grouper, *Epinephelus malabaricus* (Utarabhand and Bunlipatanon 1996), and 0.92% for murrel, *Channa punctatus* (Sehgal and Goswami 2005). The biological function of the inorganic phosphorus which is bound to Vtg may serve to provide the high-energy bonds that are required for embryonic development; however, the precise biological function of this compound has not yet been determined (Rosenstein and Taborsky 1970; Murakami et al. 1991). Alternatively, high-energy phosphate bonds may have an important function during final maturation and ovulation of pelagic eggs.

During the final maturation of pelagic eggs in several marine fish species, specific yolk proteins are broken down into small peptides and amino acids while the eggs are simultaneously undergoing massive hydration. As a result, the volume of the oocytes increases by four to seven times its original size (Greeley et al. 1986; Craik and Harvey 1987; Norberg 1987). While it is not entirely clear which yolk proteins are responsible for the breakdown observed during the final maturation and massive hydration of pelagic eggs in various marine fish species, it appears that phosvitins play a crucial role in this process (Craik 1982; Wallace and Begovac 1985).

The energy provided by the release of phosphorus from phosvitin may be essential for the normal uptake of water (Craik and Harvey 1987). Different fish species exhibit considerable variation in the carbohydrate component of their Vtgs, with levels ranging from 8.1 to 20.8 μg/mg protein in grouper, *Epinephelus malabaricus* (Utarabhand and Bunlipatanon 1996), 17.3–37.4 μg/mg protein in stickleback (Zanuy et al. 1987), to as high as 132 μg/mg protein in medaka, Oryziaslatipe

(Hamazaki et al. 1987). Relatively few studies have measured the isoelectric point (pI) of fish vitellogenins.

When teleost Vtgs are analyzed using isoelectric focusing, they are often separated into multiple components with pI values ranging from 3.8 to 6.7 (Maltais and Roy 2009). These different bands may represent Vtg isomers with varying charges, which is known as charge heterogeneity. Sehgal and Goswami (2005) observed that Vtg in murrel (*Channa punctatus*) separated into three charged isomers with pI values of 5.9, 4.6, and 3.8. They suggested that this variation in charge may be due to differences in the degree of phosphorylation. Various factors contribute to the presence of multiple isomers of native Vtg during isoelectric focusing (IEF). These factors include variations in polypeptide sequences, which are encoded by different genes, as well as differences in the spatial orientation of protein side groups, interactions between amino acid residues, posttranslational modifications such as phosphorylation, glycosylation, and lipidation, and the presence of noncovalently bonded lipids and metal ions.

Vitellogenin has been studied in various fish species, and its molecular weight has been determined. Initially, many fish species were reported to have only one form of vitellogenin, such as goldfish (*Carassius auratus*) with a molecular weight of 380 kDa, (de Vlaming et al. 1980), trout (Salmo gairdneri) with 470 kDa, catfish (*Heteropneustes fossilis*) with 550 kDa, sea trout (Salmo trutta) with 440 kDa (Norberg and Haux 1985; Campbell and Idler 1980), viviparous blenny (*Zoarces viviparous*) with 500 kDa (Korsgaard and Pedersen 1998), smooth flounder (*Pleuronectes putnami*) with 520 kDa (Roy et al. 2004), and African catfish (*Clarias gariepinus*) with 520 kDa (Manohar et al. 2005). It has been observed that teleost vitellogenins are present in plasma as dimeric proteins, and their molecular weight ranges from approximately 300 to 600 kDa (Hiramatsu et al. 2006).

17.5 Classification of Vitellogenin

The evolutionary history of fish Vtgs is characterized by a complex pattern of structural and functional variations. The nomenclature for Vtgs that is currently accepted was developed by R.N. Finn and colleagues and is based on the existence of multiple Vtg types that arose as a result of whole genome duplications (WGDs) during the evolutionary history of vertebrates. The scenario suggests that the ancestral chordate Vtg (Vtg ABCD), found in lancet, *Amphioxus floridae* and silver lamprey, Ichthyomyzonunicupsis, evolved into two types of Vtg: VtgAB (Chondrostean vitellogenin) present in Chondrostean fishes, amphibians, and avian species, and VtgCD. Furthermore, VtgAB gave rise to two forms, VtgA, which is found universally in teleosts, and VtgB, which is no longer present in most derived fishes.

The genome sequence of the spotted gar (*Lepisosteus oculatus*) contains a copy of the VtgB locus, suggesting that some extant fish species that are less evolved may still express it. However, the spotted gar is currently the only known species that exhibits this characteristic. Vtg CD has given rise to two different types of vitellogenin genes in teleosts, namely VtgC (also known as Vtg3 or phosvitinlessVtg) and

VtgD. VtgC is found in various phylogenetic lineages of teleosts, while VtgD is now extinct. In addition, there have been other duplications of Vtg genes in different fish lineages, not necessarily involving whole-genome duplication (WGD). For example, some fish lineages, such as Protacanthopterygii, Ostariophysi, and Elopomorpha, have multiple A-type Vtgs, such as VtgAsa and VtgAsb in salmonids, VtgAo1 and VtgAo2 in Ostariophysian fish, and VtgAe1, VtgAe2, and VtgAe3 in Elopomorph fish.

17.6 Endocrine Control of Vitellogenesis

Vitellogenesis is a seasonal or cyclic process depending on gonadotropins. The production of gonadotropin-releasing hormone (GnRH) in the hypothalamus is influenced by various intrinsic and extrinsic factors such as innate biorhythms, nutritional status, and changes in seasonal day length and water temperature. GnRH stimulates the secretion of follicle-stimulating hormone (FSH) from pituitary gonadotrophs, which in turn prompts theca and granulosa cells of the ovarian follicle to release estradiol-17b (E2). E2 instructs the liver to synthesize vitellogenin's (Vtgs) and release them into the bloodstream. In some species, luteinizing hormone produced by the pituitary in response to GnRH also elicits follicular production of E2. Growth hormone, thyroid hormone, cortisol, androgens, and other estrogenic steroids, such as estrone, may also contribute to induction of hepatic vitellogenesis in some species. In hagfish (Eptatretusstoutii), vitellogenesis is generally less responsive to E2 and might be regulated by other factors also.

The primary event during oogenesis is the synthesis by the liver and uptake by the oocytes of the yolk precursor protein vitellogenin (Wiegand 1982; De Vlaming et al. 1984; Johnson et al. 1991). Oogenesis is triggered by environmental cues and controlled by a series of regulating hormones (Epler 1974; Billard et al. 1978) as illustrated by Fig. 17.1. Under the effect of temperature and/or photoperiod has the capacity to induce the secretion of peptide hormones known as gonadotropins by the pituitary gland, which are responsible for regulating the reproductive function in vertebrates (De Vlaming 1972; Campbell and Idler 1976; Campbell and Blobel 1976), and they also promote meiotic maturation and ovulation (Harmin and Crim 1992).

The pituitary gland's secretory functions are regulated by brain neurohormones, including gonadotropin-releasing hormone (GnRH), which is released by the hypothalamus (GnRH), which is released by the hypothalamus (Browder et al. 1991). Upon initiation of oogenesis, the release of gonadotropins occurs, and these hormones are transported via the bloodstream to the ovaries, where they promote the growth of oocytes and the subsequent process of ovulation. Additionally, they stimulate the synthesis of estrogens by the follicle cells (primarilyestradiol) (Van der kraak et al. 1990; Singh and Singh 1991; Nagler and Idler 1992). The estradiol is then released into the serum where it is bound by steroid-binding proteins or albumins (Lazier et al. 1985; Pottinger 1986; Lazier and MacKay).

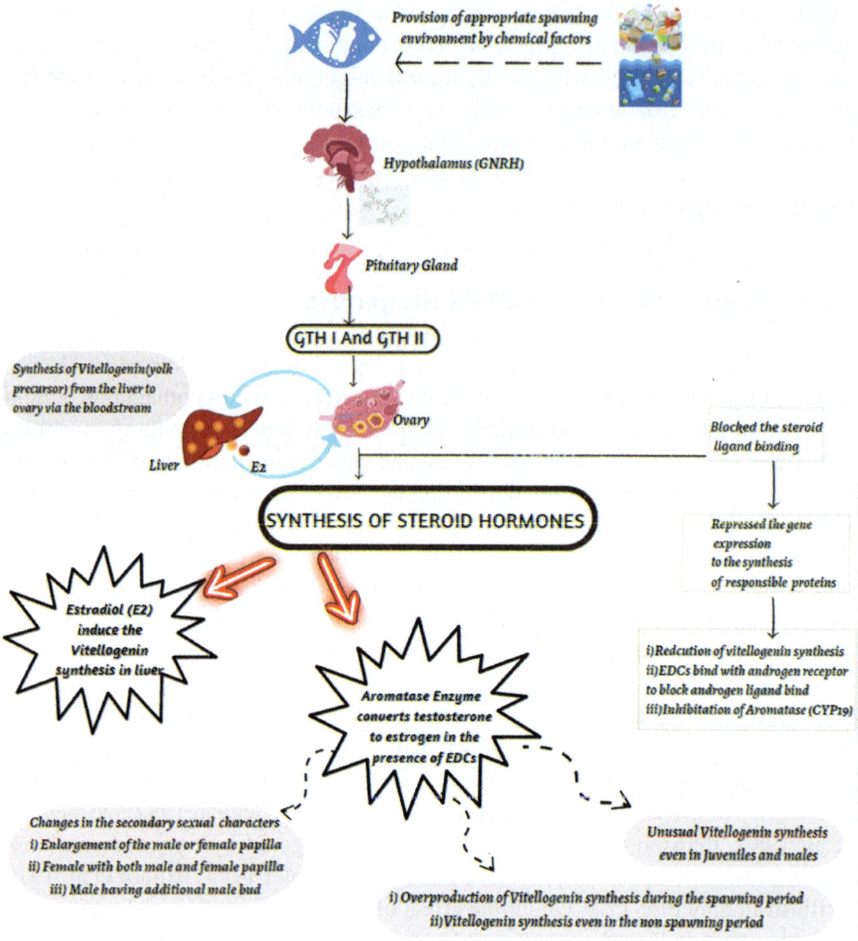

Fig. 17.1 Strategies for endocrine disruption in fish reproduction

Estradiol's function includes the stimulation of vitellogenin synthesis by the liver (Emmersen and Petersen 1976; Emmersen et al. 1979; LeMennetal. 1980; Wallace 1985; MacKay and Lazier 1993). Vitellogenin is a serum protein that is specific to females and serves as the precursor to egg yolk in most oviparous vertebrates. It contains a range of essential nutrients, such as lipids, carbohydrates, phosphorus, calcium, and iron (Craik 1978; Wiley et al. 1979; Nagler and Idler 1990). Estradiol diffuses into liver cells and is subsequently sequestered within target cells through its high-affinity binding to a specific receptor protein (ER, estrogen receptor) (Lazier and Haggarty 1979; Turner et al. 1981; Sloop et al. 1983; Lazier et al. 1985; Pottinger 1986; McPherson et al. 1988; Smith and Thomas 1990). This results in the activation of the transcription of the vitellogenin loci (Wiskocil et al. 1980; Tata et al. 1987; Pakdel et al. 1991). Binding affinity of ER for estradiol increases with higher doses of estradiol (Lazier et al. 1985; Anderson et al. 1996; Donohoe and Curtis 1996).

Environmental estrogenic pollutants, particularly exogenous estrogenic substances, can have significant adverse effects on both fish populations and human health. Oviparous fish possess a unique biomarker called vitellogenin (Vtg) that can be used to detect exposure to these toxins. Recent research has shown that Vtg is a reliable biomarker of exposure to environmental estrogen in a range of fish species. An important dose–response association was discovered when studied between Vtg levels and estradiol concentration in studies that looked at Vtg levels in Male and female Tilapia (*Oreochromis mossambicus*) exposed to various BPA, 2–2 dichloropropane, and Glyphosate concentrations, of strong estrogenic chemicals that Vtg is a sensitive biomarker of exposure to this substance. The presence of VTG mRNA in male fish can be utilized as a biomarker of exposure to estrogenic substances in the environment, which can harm fish populations and human health. Estrogenic pollutants can be identified via VTG, and their presence in aquatic environments can be monitored to help protect fish populations and public health.

17.7 Function of Vitellogenin (Vtg) in Ovarian Follicles and Embryonic Development

Vitellogenin is a yolk precursor protein produced in the liver of female fishes and utilized to support developing oocytes and also in the heart, spleen, kidney, skin, muscle, gill, eye, and brain (Zhong et al. 2014). Female-specific proteins expressed in the blood or bodily fluids during the process of oogenesis in females were designated as "vitellogenin" (derived from the words "vitelline" and "genin"), which refers to their role as a source of egg yolk. Vtgs are extra ovarian dimeric proteins (in the liver or bloodstream) that comprise phosphate, lipide, carbohydrate, and proteins components for embryogenesis and then transported through the bloodstream to the ovary, where they are taken up by oocytes and transmitted between ovarian follicular cells via receptor-mediated endocytosis. In fish, the primarily exogenous synthesis of vitellogenin is initiated by gonadotropins and regulated by estrogens. The synthesis, transport, and ingestion of yolk proteins into the oocyte are known as vitellogenesis, whereas zona genesis is the production of eggshell zona radiata proteins. The synthesis, transport, and ingestion of yolk proteins into the oocyte are known as vitellogenesis, whereas zona genesis is the production of eggshell zona radiata proteins (Arukwe and Goksøyr 2003). VTGs are phosphorylated, glycosylated, and have lipid groups added to them. They are transported through the bloodstream to developing oocytes, where they attach to receptors via clathrin-mediated endocytosis (Patiño and Sullivan 2002).

VTG has a molecular weight of 325 kilodaltons with a 20% lipid content and divides into two main peptides with 190 and 160 kD. Vtg comes in two varieties: complete and incomplete. Yolk protein is composed of five distinct components heavy-chain lipovitellin, light-chain lipovitellin, phosvitin component, and carboxy-terminal component (Gupta et al. 2021; Canapa et al. 2012). During vitellogenesis,

Vtg is found in endosomes of ovarian follicles and Cathepsin D cleaves it into major yolk components, lipovitellin, phosvitin, and β-c, which are stored in the cell and acidified by a proton pump. There is a highly variable cleavage site between LvL and '-c terminal, and other cleavage locations exist in the Vtg peptides that are involved in the secondary dissociation phase (Finn 2007).

17.8 Strategies for Endocrine Disruption in Fish Reproduction

Based on their fundamental structure and physiological function, these Vg subtypes were classified and labeled as VgA, VgB, and VgC (Hiramatsu et al. 2002, 2005, 2006). In zebrafish, chromosome 22 is connected to 14 genes that are involved in vitellogenesis, while chromosome 11 is home to the phosvitinless gene (vtg3) Ziv et al. (2008). After this initial processing, Vtg undergoes a second proteolysis to produce pelagic or demersal eggs or have rapid or slow embryonic development. Finally, a third proteolysis occurs during embryogenesis (Finn 2007). Vitellogenin 1 (VtgAo1), a Vtg protein, has three primary classes (with five corresponding genes, Vtgs 1, 4, 5, 6, and 7). Eight significant genes comprising vtg1 and vtg3, nots, syne1, fst1, nosip, grik1, and esr1 exhibit distinctive variations in their expression patterns in response to estrogen (Arukwe and Goksøyr 2003).

17.9 The Role of Estrogen and Its Receptors in Vitellogenesis Regulation

The vitellogenin gene family is composed of various gene numbers in oviparous and ovoviviparous animals that encode polypeptides precursors of yolk proteins and derivatives. Gene expression profiling of the zebrafish liver has revealed 1046 transcripts that are differentially expressed during the vitellogenesis phase. The ovarian steroid hormone 17-estradiol (E2) is the main hormone of vitellogenin (Vtg) expression (Polzonetti-Magni et al. 2004). It is highly conserved across different species of fish and is regulated by female hormones such as estradiol. Its expression is regulated by regulatory elements, and its amount of production is directly related to estradiol levels in the bloodstream. Fish, mammals, and birds all express the two main ERs, ER alpha, and ER beta. E2 concentration is essential for controlling the female reproductive endocrine axis. Ovarian follicles create E2 during vitellogenesis in response to FSH signaling (Lubzens et al. 2010). Estrogen receptor (ER) subtypes ER and ER are expressed in the livers of most teleost species, with levels peaking during oocyte formation and increasing during ovary recruitment. 17-estradiol (E2) is the primary hormone of vitellogenin (Vtg) expression, and estrogen receptors (ERs) must connect with the promoter area at particular palindromicestrogen response element (ERE) sequences (Gruber et al. 2004). E2 is the most effective agent for stimulating Vtg expression, and seasonal fluctuations in Vtg levels in plasma correspond to E2 levels in fish. E2 administration dramatically raises Vtg levels in fish, amphibians, reptiles, and birds (Li et al. 2006). Estrogen receptors (ERs) attach to the promoter region of estrogen-responsive genes at certain palindromic estrogen response element (ERE) sequences, causing coactivators or corepressors to be recruited to the promoter. Estrogen receptors mediate the signaling pathways in oocyte maturation and E2-induced hepatic vitellogenin synthesis (Hu et al. 2018). Two major ERs, ER alpha, and ER beta, have been identified (Hawkins et al. 2000; Menuet et al. 2002). The late follicular phase of the cycle's sustained estradiol increases cause extended GnRH/LH surges, which cause a switch in the production of E2 to 17, 20-dihydroxy-4-pregnen-3-one as ovarian maturation advances. This drop in E2 synthesis and the corresponding rise in maturational progestogens leads to an LH-dependent gain of oocyte maturational competence and subsequent ovulation (Nagahama and Yamashita 2008) (Fig. 17.2).

The knockdown of ER, ER1, and ER2 in goldfish and zebrafish indicated that expression levels of Vtg genes were considerably downregulated, indicating that E2 and its receptors play a vital role in vitellogenesis regulation (Sabo-Attwood et al. 2004; Nelson et al. 2004). E2 regulates nuclear estrogen receptor expression, cyp19a1, and E2, and feedback regulation along the brain–pituitary–gonadotropic axis (Rather et al. 2020; Nelson and Habibi 2013).

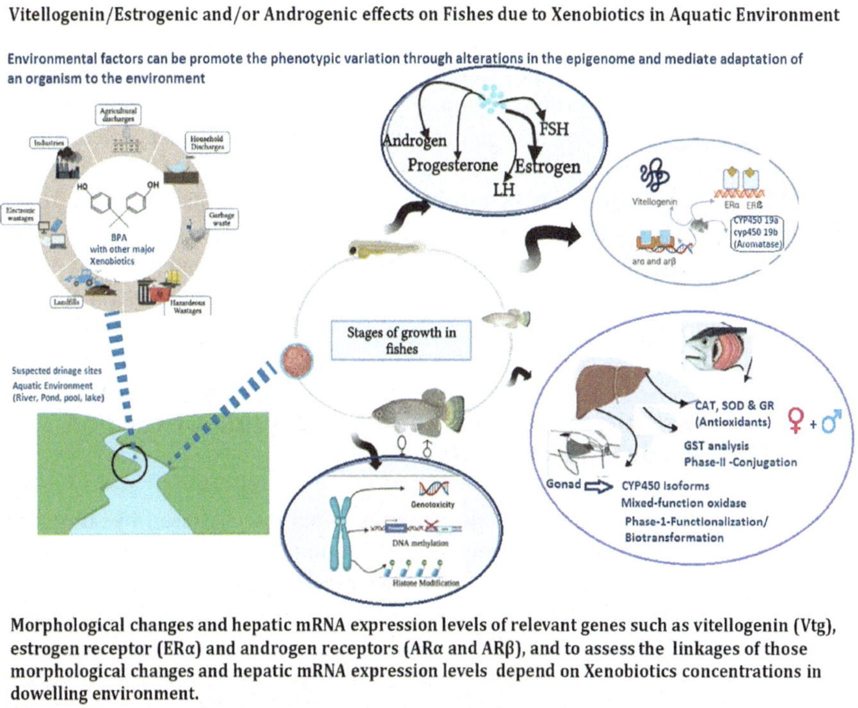

Fig. 17.2 Role of vitellogenin and its receptor in fish reproduction

17.10 Role of Vitellogenin and Its Receptor in Fish Reproduction

Vitellogenin (VTG) production is seasonal or cyclic, depending on gonadotropins, dietary conditions, seasonal variations, and water temperature fluctuations. Plasma VTG levels are influenced by factors such as age, reproductive status, and season. Steroid hormones, non-steroid hormones and non-hormonal factors are all able to influence plasma Vtg levels. VTG mRNA (Vitellogenin mRNA) is a transcript of the vitellogenin gene, which is involved in the synthesis of yolk protein in female fish livers. The mRNA is subsequently carried to the ovaries' developing oocytes, where it is translated into the vitellogenin protein, which provides necessary nutrients for embryonic development. GnRH stimulates the hypothalamus to create FSH, which increases pituitary FSH production. Estradiol-17 (E2) binds to estrogen receptors on hepatocytes, inducing gene expression and VTG transcription in the liver. The process of vitellogenesis is governed by hypothalamic–pituitary–gonadal (HPG) axis through the increased levels of estradiol and the gene, which is responsible for vitellogenin synthesis is not normally expressed in males (Ankley et al. 2005).

The tilapia was used to confirm the binding affinity between VtgAb and the Vtg receptor, which is considered an equivalent of the very-low-density lipoprotein (VLDL) receptor found in mammals (Li et al. 2003). Mizuta et al. (2014) used immune biochemical methods to demonstrate that the gene encoding the Lr8-type Vtg receptor translated into an approximately 100 kDa receptor protein. Immunohistochemical techniques indicate that the receptor protein is predominantly localized at the outermost region of the oocyte membranes within the ovary (Mizuta et al. 2013). Reading et al. experiments with white perch supported the idea that VtgAa and VtgAb might bind to distinct receptors and suggested that the new lipoprotein receptor Lrp13 may be involved in Vtg binding (Reading et al. 2014).

Vtg is also an antioxidant reagent used to reduce free radical reactions in fish oocytes, and honey bee vtg is used to identify cell damage through membrane binding and shielding living cells from reactive oxygen species (Havukainen et al. 2013; Ando and Yanagida 1999).

17.11 Vitellogenin as a Biomarker for Endocrine Disruption in Aquatic Environments

Endocrinological disruption is a major environmental issue, caused by substances from natural and human-made resources being dumped in aquatic environments. After exposure to high concentrations of some endocrine disrupters, there have been observed effects on the endocrine system in individual species and populations (Schwaiger and Negele 1998). Xenobiotics can affect the endocrine system by affecting transcription and signal transduction and can act through receptor-mediated or non-receptor-mediated mechanisms (Van der kraak et al. 1992). Exposure to estrogenic substances can cause an increase in the expression of VTG mRNA in male fishes, leading to the generation of vitellogenin and the feminization of male fish. The presence of VTG mRNA in male fish can be utilized to predict the presence of estrogenic pollutants in the environment, which can harm fish populations and human health.

Evidence suggests that Vtg levels are affected by seasonal, environmental, and physiological/biorhythm factors in feral and tank-reared fish (Rice and Xiang 2000; Larsson et al. 2002). Environmental chemicals have an impact on the reproductive health of many wildlife and people, and biomarkers can be used as early indicators of exposure and potential longer-term consequences (Hutchinson et al. 2006). EDCs with estrogenic and anti-androgenic activity can have harmful effects on fish, including the induction of vitellogenin in young males or juveniles, the delay or absence of secondary sexual characteristics and behavior in males, intersex condition, skewed sex ratio, and eventually population extinction (Lawrence and Hemingway 2008). Fish are the most used model due to their unique characteristics, and aquatic species are more exposed to EDCs than terrestrial animals. The vitellogenic response exhibited by fishes in reaction to exogenous estrogens represents a valuable biomarker for the assessment of exposure. The induction of vitellogenin (Vtg) serves as a valuable indicator of exposure to estrogenic compounds, but the connection between Vtg protein induction and negative impacts at higher levels of biological organization has not been conclusively established (Mills and Chichester 2005). Most xenobiotic estrogens and their metabolites are stable and lipophilic, which causes many of them to bioaccumulate and biomagnify (Caliani et al. 2021). Fish VTGs are found in the plasma of adult female fish and can be activated by exogenous estrogenic substances, making them a reliable bioindicator for tracking estrogenic contamination of aquatic habitats (Hutchinson et al. 2000). In female egg-laying vertebrates, exogenous estrogens can promote the development of vitellogenin, a precursor to the egg yolk protein that can be utilized as a biomarker of exposure to estrogenic substances in the environment (Nicolas 1999). In general, the induction of vitellogenin (Vtg) in fish is triggered at lower levels of exposure to endocrine-disrupting compounds (EDCs) than those required for the manifestation of reproductive abnormalities. Conversely, the EDC exposure level at which Vtg induction occurs is equal to or greater than the level that leads to the development of reproductive abnormalities in immature fish (Hashimoto 2005). Mixtures of ERa agonists have additive effects on Vtg synthesis (Thorpe et al. 2003, Brian et al.).

Vtg, a protein that is unique to females and is a precursor to egg yolk, which is typically made by female hepatocytes in response to endogenous estrogens. Men also have Vtg genes, which can be activated by xenoestrogens and antiandrogens (Thomas-Jones et al. 2003; Bhatia et al. 2014). A male individual exhibits a transient induction of Vtg that is above the baseline level; however, the effects on reproduction (such as abnormalities in gonad morphology, reproductive behavior, fertilization, development, and sexual differentiation) are highly individualized and depend on a variety of factors, including age. Because males have no mechanism for excreting VTG (e.g., through deposition in developing oocytes), very high levels of VTG have been associated with renal pathology and death (Folmar et al. 2001). Females' plasma VTG concentrations decreased due to chemicals tested, resulting in fewer oocytes being deposited in developing oocytes and less egg production (Ankley et al. 2009). Male fish residing in bodies of water contaminated with estrogen, caged fish exposed to wastewater treatment plant effluents, and rainbow trout have been found to exhibit increased levels of plasma Vtg (Allen et al. 1999; Harris et al. 2006). Elevated levels of Vtg in males have the potential to cause adverse effects, such as kidney malfunction that can result in mortality. Furthermore, high concentrations of plasma Vtg are linked with reduced fitness and increased mortality in Fathead minnows (Schmid et al. 2002). The concentration of plasma Vtg in males decreases after exposure to clean water, but a reduction to baseline concentrations may take many months due to slow Vtg clearance in males (Rodgers-Gray et al. 2001). The induction of Vtg is concentration-dependent until reliant on the concentration of exogenous estrogens until it reaches its maximal levels are reached. Maximal Vtg induction can be achieved through through either intermittent or continuous exposure to exogenous estrogens (Panter et al. 2000).

Vtg mRNA is a useful marker of the disrupting effect of estrogenic ECDs in male animals. It is induced rapidly and is useful for short-term screenings of ecosystems. Frenzilli et al. (2008) and Ankley et al. (2009) use biomarkers in sentinel species to evaluate the health of aquatic ecosystems. Fish VTGs are found in the plasma of adult female fish and can be activated by exogenous estrogenic substances, making them a reliable bioindicator for tracking estrogenic contamination of aquatic habitats (Hutchisnon 2000). Vitellogenin can also be elicited in the bloodstream of male or immature fish through exposure to exogenous estrogen, and its occurrence in the blood of males or juveniles has been widely acknowledged as a biomarker of their exposure to endocrine-disrupting chemicals (EEDCs) (Hiramatsu et al. 2002, 2004). The utilization of Vtg and yolk proteins in the detection of EDC contamination has permitted the development of new Vtg-based bioassays. Studies have demonstrated that many fish species exhibit Vtg responses to exposure to endocrine disruptors. In the vitellogenin protein (VTG) and messenger RNA (mRNA) precursor (Vtg) (Jones et al. 2000). Expression-based exposure biomarkers are routinely used to detect estrogenic exposure using both deployed and laboratory-exposed fish. The identification of Vtg produced in the blood of male or juvenile fish, which may be identified by electrophoresis and comparable techniques, is one of the most reliable techniques for assessing EEDC activity in water. Yet, several Vtg subtypes frequently have molecular weights that are equal or very similar, making it challenging to differentiate or distinguish each subtype individually at the protein level. To determine the main Vtg subtypes and quantify their subtype-specific mRNA, quantitative real-time PCR can be utilized. The ability to identify Vtg protein in the blood for a longer time after estrogen exposure than Vtg mRNA in the liver suggests that each measurement reflects a unique exposure history. Vtg levels in male blood that are higher than normal indicate that the intended has experienced or is currently experiencing exposure to EEDCs within a relatively recentpast. Ideally, the identification of both commodities would facilitate the chronological evaluation of exposure history to EEDC.

According to recent studies regarding the physiological roles of various types of vitellogenin (Vtg) during oocyte maturation and embryonic development after fertilization, it has been proposed that the quality of eggs is closely linked to the relative abundance or accumulation ratio of each Vtg in the ovulated eggs (Williams et al. 2014). Field research has shown that exposure to (EEDCs) has the potential to anticipate the onset of reproductive abnormalities in juvenile or immature fish. Moreover, the identification of several polymorphic forms of Vitellogenin (Vtg) in fish may provide fresh insights into the detection and quantification of EEDC contamination via Vtg measurements.

17.12 Our Studies on Vitellogenin Expression in Fish Exposed to Estrogenic Compounds

Vtg production is normally restricted to mature females and present very little amount in males or sexually immature females. However, the exposure of fishes of estrogenic compounds like BPA can trigger vtg expression in males, since they are having the vtg gene. Male fish also possess the hepatocyte estrogen receptor (ER) and results in vitellogenin gene expression when exposed to E2 or other estrogen mimetics. Its metabolism in males is slow. Thus, its presence in oviparous fish, particularly in males, makes it an ideal biomarker for studies on the effect of 162 estrogenic EDCs on fish.

Environmental estrogens are chemicals that can mimic the effects of natural estrogens in animals and humans and can have adverse effects on reproduction, development, and physiology. Vitellogenin induction is a protein produced by female animals to make egg yolk. It can be caused by different types of environmental estrogens, such as natural hormones, synthetic hormones, phytoestrogens, and xenoestrogens. It is used to assess estrogenic contamination in aquatic environments and evaluate the potential risks of environmental estrogen on wildlife and human health.

BPA is often referred to as "environmental estrogen" due to its structural resemblance to estrogen (Yoneda et al. 2001). Our lab conducted studies that have been conducted on the effect of BPA on vitellogenin gene expression in *Oreochromis mossambicus*. Results showed that male fishes were exposed to BPA but were treated with Laccase nanoparticle conjugates. In female fishes, BPA-exposed female fishes showed maximum level of gene expression, but in conjugate-treated fishes, there was a considerable reduction in gene expression. Additionally, quantitative RT PCR revealed that solely BPA-treated individuals' vitellogenin-related gene expression was high. This suggests that BPA triggers early maturation in the fingerling stage of *Oreochromis mossumbicus,* leading to increased vitellogenesis and aromatase production (Manna and Amutha 2016; Manna et al. (2018).

The vitellogenin-based study was done to fishes collected from plastic industry effluent. The female fingerlings of *Orrechrombis mossambicus* manifested early maturation due to endocrine disruption. The fingerlings exposed paper mill industry effluent compound was significantly higher in vitellogenin and aromatase value. qRT PCR of also revealed that the vitellogenin gene expression in fingerlings was upregulated. The male fingerlings responded with lower expression of vitellogenin when compared with high levels in female fishes. Vitellogenin protein expression was high in pulp mill effluent treated female and male fishes of *Oreochromis mosambicus* by ELISA and FPLC-based detection of Vtg expression (Sakthivel and Amutha 2022).

In the study area, it has been observed that male tilapia fish have displayed elevated levels of vitellogenin (Vtg), a protein typically associated with female egg production. Such an occurrence may suggest the existence of estrogenic substances within their environment. Elevated levels of vitellogenin (Vtg) in this particular species of fish have been associated with pathological conditions affecting both the renal and gonadal systems, such as the presence of enlarged and abnormal glomeruli, the development of unilateral intersex gonads, and the occurrence of bilateral ovarian follicular structures within the testes. Elevated vitellogenin levels were observed in the presence of aromatase activity, which facilitates the conversion of testosterone to estrogen. High glyphosate concentration increased VTG levels after 40 days. Endocrine-disrupting chemicals (EDCs), such as glyphosate, may have adverse effects on the estradiol receptors of hepatocytes and disrupt the synthesis of vitellogenin (VTG). The results of this study indicate that the observed elevation in vitellogenin (Vtg) levels among male fish could have a notable impact on the reproductive dynamics of the population in the river, potentially causing changes to the overall composition of the fish population (Dharshana and Amutha 2021).

References

Abate T, van Huis A, Ampofo JKO (2000) Pest management strategies in traditional agriculture: an African perspective. Annu Rev Entomol 45(1):631–659

Allen Y, Matthiessen P, Scott AP, Haworth S, Feist S, Thain JE (1999) The extent of oestrogenic contamination in the UK estuarine and marine environments—further surveys of flounder. Sci Total Environ 233(1–3):5–20

Anderson MJ, Olsen H, Matsumura F, Hinton DE (1996) In vivo modulation of 17β-Estradiol-induced Vitellogenin synthesis and Estrogen receptor in rainbow trout (Oncorhynchus mykiss) liver cells by β-Naphthoflavone. Toxicol Appl Pharmacol 137(2):210–218

Ando S, Yanagida K (1999) Susceptibility to oxidation of copper-induced plasma lipoproteins from Japanese eel: protective effect of vitellogenin on the oxidation of very low density lipoprotein. Comp Biochem Physiol C Pharmacol Toxicol Endocrinol 123(1):1–7

Ankley GT, Jensen KM, Durhan EJ, Makynen EA, Butterworth BC, Kahl MD, Wilson VS (2005) Effects of two fungicides with multiple modes of action on reproductive endocrine function in the fathead minnow (Pimephalespromelas). Toxicol Sci 86(2):300–308

Ankley GT, Bencic DC, Breen MS, Collette TW, Conolly RB, Denslow ND, Watanabe KH (2009) Endocrine disrupting chemicals in fish: developing exposure indicators and predictive models of effects based on mechanism of action. Aquat Toxicol 92(3):168–178

Arukwe A, Goksøyr A (2003) Eggshell and egg yolk proteins in fish: hepatic proteins for the next generation: oogenetic, population, and evolutionary implications of endocrine disruption. Comp Hepatol 2(1):1–21

Bhatia H, Kumar A, Ogino Y, Du J, Gregg A, Chapman J, Iguchi T (2014) Effects of the commercial antiandrogen flutamide on the biomarkers of reproduction in male murray rainbowfish (Melanotaeniafluviatilis). Environ Toxicol Chem 33(5):1098–1107

Billard R, Breton B, Fostier A, Jalabert B, Weil C (1978) Endocrine control of the teleost reproductive cycle and its relation to external factors: salmonid and cyprinid models

Browder LW, Erickson CA, Jeffery WR (1991) Oogenesis. In: College S (ed) Developmental biology, 3rd. Publishing, USA, pp 55–115

Buerano CC, Inaba K, Natividad FF, Morisawa M (1995) Vitellogenins of Oreochromis niloticus: identification, isolation, and biochemical and immunochemical characterization. J Exp Zool 273(1):59–69

Caliani I, Campani T, Conti B, Cosci F, Bedini S, D'Agostino A, Casini S (2021) First application of an integrated biological response index to assess the ecotoxicological status of honeybees from rural and urban areas. Environ Sci Pollut Res 28:47418–47428

Campbell CM, Idler DR (1980) Characterization of an estradiol-induced protein from rainbow trout serum as vitellogenin by the composition and radioimmunological cross reactivity to ovarian yolk fractions. Biol Reprod 22(3):605–617

Campbell CM, Idler DR (1976) Hormonal control of vitellogenesis in hypophysectomized winter flounder (Pseudopleuronectes americanus Walbaum). Gen Comp Endocrinol 28(2):143–150

Campbell PN, Blobel G (1976) The role of organelles in the chemical modification of the primary translation products of secretory proteins. FEBS Lett 72(2):215–226

Canapa A, Olmo E, Forconi M, Pallavicini A, Makapedua MD, Biscotti MA, Barucca M (2012) Composition and phylogenetic analysis of vitellogenin coding sequences in the I ndonesian coelacanth L atimeriamenadoensis. J Exp Zool B Mol Dev Evol 318(5):404–416

Craik JC (1982) Levels of phosphoprotein in the eggs and ovaries of some fish species. Comparative biochemistry and physiology. B, comparative. Biochemistry 72(4):507–510

Craik JCA (1978) An annual cycle of vitellogenesis in the elasmobranch Scyliorhinus canicula. J Mar Biol Assoc U K 58(3):719–726

Craik JCA, Harvey SM (1987) The causes of buoyancy in eggs of marine teleosts. J Mar Biol Assoc U K 67(1):169–182

Dharshana K, Amutha C (2021) Tissue damaging effects on tilapia (*Oreochromis mossambicus*) due to prolonged exposure to glyphosate herbicide. J Exp Zool India 24:387–394

Donohoe RM, Curtis LR (1996) Estrogenic activity of chlordecone, o, p′-DDT and o, p′-DDE in juvenile rainbow trout: induction of vitellogenesis and interaction with hepatic estrogen binding sites. Aquat Toxicol 36(1–2):31–52

Emmersen BK. Petersen IM (1976) Natural occurrence and experimental induction by estradiol-17fl, of a lipophosphoprotein in flounder (Platichthys fiesus L). Comp Biochem Physiol 54B, 443–446

Emmersen J, Karsgoard B, Petersen I (1979) Dose response kinetics of serum vitellogenin by estradiol 17□ in made flounders Platichthys flessus L. Comp Biochemical Physical 63:B1–B6

Epler P. (1974) Daily changes in blood serum levels of 17 b-estradiol and 11- ketotestosterone in the mature carp cyprinus carpio l. Chronobiologia 13(23):23

Falk-Petersen IB, Frivoll V, Gulliksen B, Haug T (1986) Occurrence and size/age relations of polar cod, Boreogadus saida (Lepechin). Spitsbergen coastal waters Sarsia 71(3–4):235–245

Finn RN (2007) Vertebrate yolk complexes and the functional implications of phosvitins and other subdomains in vitellogenins. Biol Reprod 76(6):926–935

Folmar LC, Gardner GR, Schreibman MP, Magliulo-Cepriano L, Mills LJ, Zaroogian G, Denslow ND (2001) Vitellogenin-induced pathology in male summer flounder (Paralichthysdentatus). Aquat Toxicol 51(4):431–441

Frenzilli G, Falleni A, Scarcelli V, Del Barga I, Pellegrini S, Savarino G, Nigro M (2008) Cellular responses in the cyprinid Leuciscus cephalus from a contaminated freshwater ecosystem. Aquat Toxicol 89(3):188–196

Greeley MS Jr, Calder DR, Taylor MH, Hols H, Wallace RA (1986) Oocyte maturation in the mummichog (Fundulus heteroclitus): effects of steroids on germinal vesicle breakdown of intact follicles in vitro. Gen Comp Endocrinol 62(2):281–289

Grogan J, Taborsky G (1987) Iron binding by phosvitins: variable mechanism of iron release by phosvitins of diverse species characterized by different degrees of phosphorylation. J Inorg Biochem 29(1):33–47

Gruber CJ, Gruber DM, Gruber IM, Wieser F, Huber JC (2004) Anatomy of the estrogen response element. Trends Endocrinol Metabol 15(2):73–78

Gupta G, Kumar M, Rani S, Mohanta B (2021) Vitellogenesis and their endocrine control in fishes. In: Recent updates in molecular endocrinology and reproductive physiology of fish, pp 23–34. https://doi.org/10.1007/978-981-15-8369-8_2

Hamazaki TS, Iuchi I, Yamagami K (1987) Purification and identification of vitellogenin and its immunohistochemical detection in growing oocytes of the teleost, Oryzias latipes. J Exp Zool 242(3):333–341

Hara A (1976) Iron-binding activity of female-specific serum proteins of rainbow trout (Salmo gairdneri) and chum salmon (Oncorhyncus Keta). Biochimica et Biophysica Acta (BBA)-protein. Structure 427(2):549–557

Hara K, Tydeman P, Kirschner M (1980) A cytoplasmic clock with the same period as the division cycle in Xenopus eggs. Proc Natl Acad Sci 77(1):462–466

Harris J, Schwinn N, Mahoney JA, Lin HH, Shaw M, Howard CJ, Gordon S (2006) A vitellogenic-like carboxypeptidase expressed by human macrophages is localized in endoplasmic reticulum and membrane ruffles. Int J Exp Pathol 87(1):29–39

Harmin SA, Crim LW (1992) Gonadotropic hormone-releasing hormone analog (GnRH-A) induced ovulation and spawning in female winter flounder, Pseudopleuronectes americanus (Walbaum). Aquaculture 104(3–4):375–390

Hashimoto S, Horiuchi A, Yoshimoto T, Nakao M, Omura H, Kato Y, Giesy JP (2005) Horizontal and vertical distribution of estrogenic activities in sediments and waters from Tokyo Bay, Japan. Arch Environ Contam Toxicol 48:209–216

Havukainen H, Münch D, Baumann A, Zhong S, Halskau Ø, Krogsgaard M, Amdam GV (2013) Vitellogenin recognizes cell damage through membrane binding and shields living cells from reactive oxygen species. J Biol Chem 288(39):28369–28381

Hawkins MB, Thornton JW, Crews D, Skipper JK, Dotte A, Thomas P (2000) Identification of a third distinct estrogen receptor and reclassification of estrogen receptors in teleosts. Proc Natl Acad Sci 97(20):10751–10756

Hiramatsu N, Chapman RW, Lindzey JK, Haynes MR, Sullivan CV (2004) Molecular characterization and expression of vitellogenin receptor from white perch (Morone americana). Biol Reprod 70(6):1720–1730

Hiramatsu N, Matsubara T, Weber GM, Sullivan CV, Hara A (2002) Vitellogenesis in aquatic animals. Fish Sci 68(sup1):694–699

Hiramatsu N, Cheek AO, Sullivan CV, Matsubara T, Hara A (2005) Biochemistry and molecular biology of fishes. In: Mommsen TP, Moon TV (eds) Vitellogenesis and endocrine disruption, vol 6. Elsevier, Amsterdam, pp 432–433

Hiramatsu N, Matsubara T, Fujita T, Sullivan CV, Hara A (2006) Multiple piscine vitellogenins: biomarkers of fish exposure to estrogenic endocrine disruptors in aquatic environments. Mar Biol 149:35–47

Ho SM (1987) Endocrinology of vitellogenesis. In: Hormones and reproduction in fishes, amphibians, and reptiles, pp 145–169

Hu P, Meng Z, Jia Y (2018) Molecular characterization and quantification of estrogen receptors in turbot (Scophthalmus maximus). Gen Comp Endocrinol 257:38–49

Hutchisnon TH, Brown R, Brugger KE, Campbell PM, Holt M, Länge R, van Egmond R (2000) Ecological risk assessment of endocrine disruptors. Environ Health Perspect 108(11):1007–1014

Hutchinson TH, Ankley GT, Segner H, Tyler CR (2006) Screening and testing for endocrine disruption in fish—biomarkers as "signposts," not "traffic lights," in risk assessment. Environ Health Perspect 114(Suppl 1):106–114

Jones PD, De Coen WM, Tremblay L, Giesy JP (2000) Vitellogenin as a biomarker for environmental estrogens. Water Sci Technol 42(7–8):1–14

Kime DE, Nash JP, Scott AP (1999) Vitellogenesis as a biomarker of reproductive disruption by xenobiotics. Aquaculture 177(1–4):345–352

Korsgaard B, Pedersen KL (1998) Vitellogenin in Zoarces viviparus: purification, quantification by ELISA and induction by estradiol-17β and 4-nonylphenol. Comp Biochem Physiol C: Pharmacol Toxicol Endocrinol 120(1):159–166

Larsson DJ, Mayer I, Hyllner SJ, Förlin L (2002) Seasonal variations of vitelline envelope proteins, vitellogenin, and sex steroids in male and female eelpout (Zoarcesviviparus). Gen Comp Endocrinol 125(2):184–196

Lawrence AJ, Hemingway KL (eds) (2008) Effects of pollution on fish: molecular effects and population responses. John Wiley & Sons, Chichester

Lazier CB, Haggarty AJ (1979) A high-affinity oestrogen-binding protein in cockerel liver cytosol. Biochem J 180(2):347–353

Lazier CB, Lonergan K, Mommsen TP (1985) Hepatic estrogen receptors and plasma estrogen-binding activity in the Atlantic salmon. Gen Comp Endocrinol 57(2):234–245

Leger C, Fremont L, Marion D, Nassour I, Desfarges MF (1981) Essential fatty acids in trout serum lipoproteins, vitellogenin and egg lipids. Lipids 16:593–600

Li A, Sadasivam M, Ding JL (2003) Receptor-ligand interaction between vitellogenin receptor (VtgR) and vitellogenin (Vtg), implications on low density lipoprotein receptor and apolipoprotein B/E: the first three ligand-binding repeats of VtgR interact with the amino-terminal region of Vtg. J Biol Chem 278(5):2799–2806

Li K, Chen L, Zhou Z, Li E, Zhao X, Guo H (2006) The site of vitellogenin synthesis in Chinese mitten-handed crab Eriocheir sinensis. Comp Biochem Physiol B Biochem Mol Biol 143(4):453–458

Lubzens E, Young G, Bobe J, Cerdà J (2010) Oogenesis in teleosts: how fish eggs are formed. Gen Comp Endocrinol 165(3):367–389

Mackay ME, Lazier CB (1993) Estrogen responsiveness of vitellogenin gene expression in rainbow trout (Oncorhynchus mykiss) kept at different temperatures. Gen Comp Endocrinol 89(2):255–266

Maltais D, Roy RL (2009) Purification and partial characterization of vitellogenin from shorthead redhorse (Moxostoma macrolepidotum) and copper redhorse (Moxostoma hubbsi) and detection in plasma and mucus with a heterologous antibody. Fish Physiol Biochem 35:241–254

Mommsen TP, Walsh PJ (1988) 5 Vitellogenesis and oocyte assembly. In: Fish physiology, vol 11. Academic Press, pp 347–406

Manna A, Amutha C (2018) Early maturation and liver necrosis in the fingerling stage of Oreochromis mossambicus due to BPA can cause an ecological imbalance. RSC Adv 8(23):12894–12899

Manna A, Geetha S, Tamilzhalagan S, Amutha C (2016) The in vivo estrogenic modulatory effect of bisphenol a (BPA) on Oreochromis mossambicus and prevention of early maturation of ovary by conjugates of intracellular laccase and silica nanoparticles. RSC Adv 6(103):101560–101570

Manohar D, Rao GD, Sreenivasulu G, Senthilkumaran B, Gupta AD (2005) Purification of vitellogenin from the air breathing catfish, Clarias gariepinus. Fish Physiol Biochem 31:235–239

McPherson JM, Sawamura SJ, Condell RA, Rhee W, Wallace DG (1988) The effects of heparin on the physicochemical properties of reconstituted collagen. Coll Relat Res 8(1):65–82

Monosson E, Hogson RG, Fleming WJ, Sullivan CV (1996) Blood plasma levels of sex steroid hormones and vitellogenin in striped bass (Morone saxatilis) exposed to 3, 3′, 4, 4′-tetrachlorobiphenyl (TCB). Bull Environ Contam Toxicol 56:782–787

Mañanós EL, Zanuy S, Carrillo M (1997) Photoperiodic manipulations of the reproductive cycle of sea bass (Dicentrarchus labrax) and their effects on gonadal development, and plasma 17ß-estradiol and vitellogenin levels. Fish Physiol Biochem 16:211–222

Menuet A, Pellegrini E, Anglade I, Blaise O, Laudet V, Kah O, Pakdel F (2002) Molecular characterization of three estrogen receptor forms in zebrafish: binding characteristics, transactivation properties, and tissue distributions. Biol Reprod 66(6):1881–1892

Mills LJ, Chichester C (2005) Review of evidence: are endocrine-disrupting chemicals in the aquatic environment impacting fish populations? Sci Total Environ 343(1–3):1–34

Mizuta H, Luo W, Ito Y, Mushirobira Y, Todo T, Hara A, Hiramatsu N (2013) Ovarian expression and localization of a vitellogenin receptor with eight ligand binding repeats in the cutthroat trout (Oncorhynchus clarki). Comp Biochem Physiol B Biochem Mol Biol 166(1):81–90

Murakami M, Iuchi I, Yamagami K (1991) Partial characterization and subunit analysis of major phosphoproteins of egg yolk in the fish, Oryzias latipes. Comparative Biochemistry and Physiology Part B: Comparative Biochemistry 100(3):587–593

Nagahama Y, Yamashita M (2008) Regulation of oocyte maturation in fish. Develop Growth Differ 50:S195–S219

Nagler JJ, Idler DR (1990) Ovarian uptake of vitellogenin and another very high density lipoprotein in winter flounder (Pseudopleuronectes americanus) and their relationship with yolk proteins. Biochem Cell Biol 68(1):330–335

Nagler JJ, Idler DR (1992) In vitro ovarian estradiol-17β and testosterone responses to pituitary extract and corresponding serum levels during the prespawning to vitellogenic phases of the reproductive cycle in winter flounder (Pseudopleuronectes americanus). Comp Biochem Physiol A Physiol 101(1):69–75

Nath P, Sundararaj BI (1981) Isolation and identification of female-specific serum lipophosphoprotein (vitellogenin) in the catfish, Heteropneustes fossilis (Bloch). Gen Comp Endocrinol 43(2):184–190

Nelson ER, Wiehler WB, Cole WC, Habibi HR (2004) Homologous regulation of estrogen receptor subtypes in goldfish (Carassius auratus). Molecular reproduction and development 74(9):1105–12

Nelson ER, Habibi HR (2013) Estrogen receptor function and regulation in fish and other vertebrates. Gen Comp Endocrinol 192:15–24

Nicolas JM (1999) Vitellogenesis in fish and the effects of polycyclic aromatic hydrocarbon contaminants. Aquat Toxicol 45(2–3):77–90

Norberg B (1995) Atlantic halibut (Hippoglossus hippoglossus) vitellogenin: induction, isolation and partial characterization. Fish Physiol Biochem 14:1–13

Norberg, B. (1987, July). Vitellogenin and egg proteins in three marine fish species. In proceedings of the 3rd international symposium on reproductive physiology of fish, St. John's (pp. 212-213)

Norberg B, Haux C (1988) An homologous radioimmunoassay for brown trout (Salmo trutta) vitellogenin. Fish Physiol Biochem 5:59–68

Norberg B, Haux C (1985) Induction, isolation and a characterization of the lipid content of plasma vitellogenin from two Salmo species: rainbow trout (Salmo gairdneri) and sea trout (Salmo trutta). Comparative biochemistry and physiology. B, comparative. Biochemistry 81(4):869–876

Ohkubo N, Andoh T, Mochida K, Adachi S, Hara A, Matsubara T (2004) Deduced primary structure of two forms of vitellogenin in Japanese common goby (Acanthogobius flavimanus). Gen Comp Endocrinol 137(1):19–28

Pakdel F, Féon S, Le Gac F, Le Menn F, Valotaire Y (1991) In vivo estrogen induction of hepatic estrogen receptor mRNA and correlation with vitellogenin mRNA in rainbow trout. Mol Cell Endocrinol 75(3):205–212

Pan ML, Bell WJ, Telfer WH (1969) Vitellogenic blood protein synthesis by insect fat body. Science 165(3891):393–394

Panter GH, Thompson RS, Sumpter JP (2000) Intermittent exposure of fish to estradiol. Environ Sci Technol 34(13):2756–2760

Patiño R, Sullivan CV (2002) Ovarian follicle growth, maturation, and ovulation in teleost fish. Fish Physiol Biochem 26:57–70

Polzonetti-Magni AM, Mosconi G, Soverchia L, Kikuyama S, Carnevali O (2004) Multihormonal control of vitellogenesis in lower vertebrates. Int Rev Cytol 239(Supplement C):1–46

Pottinger TG (1986) Estrogen-binding sites in the liver of sexually mature male and female brown trout, Salmo trutta L. Gen Comp Endocrinol 61(1):120–126

Rainuzzo JR, Olsen Y, Rosenlund G (1989) The effect of enrichment diets on the fatty acid composition of the rotifer Brachionus plicatilis. Aquaculture 79(1–4):157–161

Rajendran RB, Subramanian AN (1999) Chlorinated pesticide residues in surface sediments from the river Kaveri, South India. J Environ Sci Health B 34(2):269–288

Rather MA, Dutta S, Guttula PK, Dhandare BC, Yusufzai SI, Zafar MI (2020) Structural analysis, molecular docking and molecular dynamics simulations of G-protein-coupled receptor (Kisspeptin) in fish. J Biomol Struct Dyn 38(8):2422–2439

Roy RL, Morin Y, Courtenay SC, Robichaud P (2004) Purification of vitellogenin from smooth flounder (Pleuronectes putnami) and measurement in plasma by homologous ELISA. Comp Biochem Physiol B: Biochem Mol Biol 139(2):235–244

Reading BJ, Hiramatsu N, Schilling J, Molloy KT, Glassbrook N, Mizuta H, Sullivan CV (2014) Lrp13 is a novel vertebrate lipoprotein receptor that binds vitellogenins in teleost fishes [S]. J Lipid Res 55(11):2287–2295

Rice CD, Xiang Y (2000) Immune function, hepatic CYP1A, and reproductive biomarker responses in the gulf killifish, Fundulus grandis, during dietary exposures to endocrine disrupters. Mar Environ Res 50(1–5):163–168

Rodgers-Gray TP, Jobling S, Kelly C, Morris M, Brighty G, Waldock M et al (2001) Exposure of juvenile roach (Rutilus rutilus) to treated sewage effluent induces dose-dependent and persistent disruption in duct development. Environ Sci Technol 35:462–470

Rosenstein RW, Taborsky G (1970) Mechanism of the oxidative dephosphorylation of the phosphoprotein phosvitin. Biochemistry 9(3):649–657

Sabo-Attwood T, Kroll KJ, Denslow ND (2004) Differential expression of largemouth bass (Micropterus salmoides) estrogen receptor isotypes alpha, beta, and gamma by estradiol. Mol Cell Endocrinol 218(1–2):107–118

Sakthivel K and Amutha C (2022) Heavy metal accumulation and their toxic effect in commercially available fishes. J. Exp. Zool. India 25, 2157-2161

Schmid T, Gonzalez-Valero J, Rufli H, Dietrich DR (2002) Determination of vitellogenin kinetics in male fathead minnows (Pimephales promelas). Toxicol Lett 131(1–2):65–74

Schwaiger J, Negele RD (1998) Plasma vitellogenin–a blood parameter to evaluate exposure of fish to xenoestrogens. Acta Vet Brno 67(4):257–264

Sehgal N, Goswami SV (2005) Vitellogenin exists as charge isomers in the Indian freshwater murrel, Channa punctatus (Bloch). Gen Comp Endocrinol 141(1):12–21

Sinha N, Lal B, Singh TP (1991) Pesticides induced changes in circulating thyroid hormones in the freshwater catfish Clarias batrachus. Comparative biochemistry and physiology. C, comparative. Pharmacol Toxicol 100(1–2):107–110

Smith JS, Thomas P (1990) Binding characteristics of the hepatic estrogen receptor of the spotted seatrout. Cynoscion nebulosus General and comparative endocrinology 77(1):29–42

Sloop TC, Clark JC, Rumbaugh RC, Lucier GW (1983) Imprinting of hepatic estrogen-binding proteins by neonatal androgens. Endocrinology, 112(5), 1639–1646

Soimasuo MR, Karels AE, Leppänen H, Santti R, Oikari AOJ (1998) Biomarker responses in whitefish (Coregonus lavaretus L. sl) experimentally exposed in a large lake receiving effluents from pulp and paper industry. Arch Environ Contam Toxicol 34:69–80

Specker JL, Sullivan CV (1994) Vitellogenesis in fishes: status and perspectives. In: Davey KG, Peter RE, Tobe SS (eds) Perspectives in comparative endocrinology. National Research Council, Ottawa, pp 304–315

Sumpter JP, Jobling S (1995) Vitellogenesis as a biomarker for estrogenic contamination of the aquatic environment. Environ Health Perspect 103(suppl 7):173–178

Tao Y, Hara A, Hodson RG, Woods LC, Sullivan CV (1993) Purification, characterization and immunoassay of striped bass (Morone saxatilis) vitellogenin. Fish Physiol Biochem 12:31–46

Tata JR, Ng WC, Perlman AJ, Wolffe AP. (1987). Activation and regulation of the vitellogenin gene family. In Gene regulation by steroid hormones III (pp. 205–233). New York, NY: Springer New York

Thomas-Jones E, Thorpe K, Harrison N, Thomas G, Morris C, Hutchinson T, Tyler C (2003) Dynamics of estrogen biomarker responses in rainbow trout exposed to 17β-estradiol and 17α-ethinylestradiol. Environmental Toxicology and Chemistry: An International Journal 22(12):3001–3008

Thorpe KL, Cummings RI, Hutchinson TH, Scholze M, Brighty G, Sumpter JP, Tyler CR (2003) Relative potencies and combination effects of steroidal estrogens in fish. Environ Sci Technol 37(6):1142–1149

Tyler CR, Sumpter JP, Witthames PR (1990) The dynamics of oocyte growth during vitellogenesis in the rainbow trout (Oncorhynchus mykiss). Biol Reprod 43(2):202–209

Tyler CR, Van der Eerden B, Jobling S, Panter G, Sumpter JP (1996) Measurement of vitellogenin, a biomarker for exposure to oestrogenic chemicals, in a wide variety of cyprinid fish. J Comp Physiol B 166:418–426

Turner RT, Dickhoff WW, Gorbman A (1981) Estrogen binding to hepatic nuclei of Pacific hagfish, Eptatretus stouti. Gen Comp Endocrinol 45(1):26–29

Utarabhand P, Bunlipatanon P (1996) Plasma vitellogenin of grouper (Epinephelus malabaricus): isolation and properties. Comp Biochem Physiol C: Pharmacol Toxicol Endocrinol 115(2):101–110

deVlaming V, Fitzgerald R, Delahunty G, Cech JJ Jr, Selman K, Barkley M (1984) Dynamics of oocyte development and related changes in serum estradiol-17β, yolk precursor, and lipid levels in the teleostean fish, Leptocottus armatus. Comp Biochem Physiol A Physiol 77(4):599–610

de Vlaming VL (1972) Reproductive cycling in the estuarine gobiid fish. Gillichthys mirabilis Copeia:278–291

de Vlaming VL, Wiley HS, Delahunty G, Wallace RA (1980) Goldfish (Carassius auratus) vitellogenin: induction, isolation, properties and relationship to yolk proteins. Comparative Biochemistry and Physiology Part B: Comparative Biochemistry 67(4):613–623

Wallace RA (1985) Vitellogenesis and oocyte growth in nonmammalian vertebrates. Oogenesis:127–177

Wallace RA, Begovac PC (1985) Phosvitins in Fundulus oocytes and eggs. Preliminary chromatographic and electrophoretic analyses together with biological considerations. J Biol Chem 260(20):11268–11274

Wiegand MD (1982) Vitellogenesis in fishes. Reproductive physiology of fish:136–146

Wiley HS, Opresko L, Wallace RA (1979) New methods for the purification of vertebrate vitellogenin. Anal Biochem 97(1):145–152

Williams VN, Reading BJ, Hiramatsu N, Amano H, Glassbrook N, Hara A, Sullivan CV (2014) Multiple vitellogenins and product yolk proteins in striped bass, Morone saxatilis: molecular characterization and processing during oocyte growth and maturation. Fish Physiol Biochem 40:395–415

Wiskocil R, Bensky P, Dower W, Goldberger RF, Gordon JI, Deeley RG (1980) Coordinate regulation of two estrogen-dependent genes in avian liver. Proc Natl Acad Sci 77(8):4474–4478

Wodageneh A, Van der Wulp H (1996) Obsolete pesticides in developing countries. Pesticides News (United Kingdom)

Yao Z, Crim LW (1996) A biochemical characterization of vitellogenins isolated from the marine fish ocean pout (Macrozoarces americanus L.), lumpfish (Cyclopterus lumpus) and Atlantic cod (Gadus morhua). Comp Biochem Physiol B: Biochem Mol Biol 113(2):247–253

Yoneda M, Tokimura M, Fujita H, Takeshita N, Takeshita K, Matsuyama M, Matsuura S (2001) Reproductive cycle, fecundity, and seasonal distribution of the anglerfish Lophius litulon in the East China and yellow seas. Fish Bull 99(2):356–356

Zanuy S, Carrillo M, Prat F, Copeland P, Covens M (1987) Aislamiento y caracterizacion de la vitelogenina en lubina, Dicentrarchus labrax (Linneo, 1766). Teleostei, Serranidae

Zhong L, Yuan L, Rao Y, Li Z, Zhang X, Liao T, Dai H (2014) Distribution of vitellogenin in zebrafish (Danio rerio) tissues for biomarker analysis. Aquat Toxicol 149:1–7

Ziv T, Gattegno T, Chapovetsky V, Wolf H, Barnea E, Lubzens E, Admon A (2008) Comparative proteomics of the developing fish (zebrafish and gilthead seabream) oocytes. Comp Biochem Physiol D Genom Proteom 3(1):12–35

Effect of Vitellogenin in the Growth of Ornamental Fishes

18

V. Ramasubramanian, M. S. Shabana, and C. Ragunath

Abstract

Fish egg quality has been a focus of research in aquaculture and fisheries for decades since it is essential for successful captive reproduction and recruitment and constitutes a significant life history feature. Since most fish are oviparous and their developing progeny are fully dependent on stored egg yolk for nutritional sustenance, adequate yolk formation is a crucial factor impacting egg quality. These nutrients, which are received from the mother, travel from the liver to the ovary in the form of lipoprotein particles like vitellogenins and are made up of proteins, carbohydrates, lipids, vitamins, minerals, and ions. The food, husbandry, and other intrinsic and extrinsic factors of the brood stock may have an impact on the yolk composition. Additionally, several maternal variables that may affect egg quality, such as gene transcripts that control early embryonic development, are also stored in eggs. Poor quality eggs and failure to flourish within hours of fertilization may be caused by dysfunctional control of gene or protein expression. The gene transcripts may serve as significant markers. In addition to these intrinsic variables, stress can affect fish behavior, fecundity, and ovulation rate and cause ovarian atresia or reproductive failure. This chapter gives a brief overview of the reproductive biology of ornamental fishes before going into detail about the main issues that exist in culture and the hormonal interventions created in recent years to deal with these dysfunctions. Future developments in the field of spawning induction technologies are also taken into account.

Keywords

V. Ramasubramanian (✉) · M. S. Shabana · C. Ragunath
Unit of Aquatic Biotechnology, Department of Zoology, Bharathiar University, Coimbatore, Tamil Nadu, India

V. Baskaralingam, R. Vanichviriyakit (eds.), *Vitellogenin in Fishes-Diversification, Biological Properties, and Future Perspectives*,
https://doi.org/10.1007/978-981-99-5340-0_18

Vitellogenins · Fecundity · Spawning · Antiestrogens · Embryonic development

18.1 Introduction

Fish growth is influenced by intrinsic (sex hormones, growth hormone, insulin-like growth factor-1, and leptin) and extrinsic (photoperiod, temperature, and food availability) factors, as in other species. Although fish can grow throughout their lives, due to energy partitioning, gaining weight slows significantly during gonadal development and gamete production (Enberg et al. 2008). Reproductive fitness is an essential factor in both wild and domestic animal survival. The quality of an egg is challenging and can vary depending on the situation and on a wide range of internal and external animal elements. The egg quality is the capacity of developmentally competent eggs to produce viable embryos after sperm fertilization (Chapman et al. 2014; Bobe 2015; Sullivan et al. 2015). Although straightforward in context, this concept is far more challenging to grasp in practice. This chapter will give an overview of vitellogenesis in ornamental fish.

Products made from fish roe benefit the food business because they have health benefits beyond simple nutrition. A fish's ability to produce high-quality egg masses assures that the species will continue to exist through successful reproduction. This is mainly because, after fertilization and ovulation, the egg must function autonomously to meet the nutritional demands of the developing embryo. As a result, female fish aggressively amass nutrients inside developing eggs to create the egg yolk, serving as the embryo's future food source.

Oocytes in the lipid stage are more significant (110–220 m in diameter), have poor ooplasm basophilia, and have a rising number of tiny lipid droplets. The ooplasm contains dictyosomes, mitochondria, and ribosomes. The nucleus includes finely granular chromatin, and the nuclear membrane has many holes. As the zone radiata thickens, Pas 1 material can be seen under light microscopy. Uptake of vitellogenin and yolk protein generated from vitellogenin deposited in the ooplasm as yolk granules are two characteristics of vitellogenic oocytes, which can be up to 500 m in diameter (platelets). Anti-vitellogenin antibodies can be used in immunohistochemical staining to detect minute levels of yolk proteins in oocytes with a minimum diameter of 220 m (Susca et al. 2001).

Yolk globules gradually fill the ooplasm in oocytes with advanced vitellogenesis as they grow in size and number. Oocyte microvilli expand into the intercellular spaces of granulosa cells after passing via the zona radiata pore canals. Transmission electron microscopy data has shown a strong oocyte uptake of exogenous materials at this stage of vitellogenesis. The ooplasm is abundant with clathrin-coated vesicles containing electron-dense material and is thought to originate from membrane invaginations (receptor-mediated endocytosis). Endocytotic vesicles shed their clathrin coating and combine to create yolk platelets, which get bigger and bigger as they grow. The yolk platelets contain several large lipid droplets.

Oocytes begin post-vitellogenesis, also known as oocyte maturation, once vitellogenesis is finished. During this phase, meiosis is resumed, including the nucleus migrating toward the animal pole and the nuclear envelope breaking down (BD). The core of the oocyte is occupied by a single, sizable lipid globule that is formed when lipid droplets coalesce (coalescence).

Convoluted post-ovulatory follicles are made up of follicular cells after ovulation and are distinguished by hypertrophy granulosa cells that define an uneven lumen. Post-ovulatory follicles are transient, rapidly degenerating structures that eventually blend into the connective stroma and atretic follicles. Atretic follicles can be seen in the ovaries in the ordinary course of vitellogenin and post-vitellogenic development. Follicles in the atretic stage of atresia are easily distinguished due to their atypical form and zona radiata disintegration. An atretic oocyte's nuclear membrane disintegrates, the yolk granules lose their structural integrity, and granulosa cells that infiltrate the egg gradually phagocytize the granules. Atretic follicles, also known as atretic follicles, are composed only of disordered granulosa and thecal cells with pyknotic nuclei after the oocyte has completely degraded.

18.2 Vitellogenesis

The liver produces vitellogenins, phospholipid-rich precursors to yolk proteins collected by the oocyte during vitellogenesis. The reproductive hypothalamic–pituitary–gonadal neuroendocrine axis controls vitellogenesis as a seasonal or cyclical process (Fig. 18.1). The fish's intrinsic biorhythm, bioenergetic state, and seasonal variations in photoperiod and water temperature are all endogenous and exogenous cues that enhance the production of gonadotropin-releasing hormone (GnRH) by the hypothalamus of the brain. Pituitary gonadotrophs produce follicle-stimulating hormone (FSH) in response to GnRH, which causes the ovarian follicle to emit estradiol-$_{17}$ (E2). E2 in the blood binds to an estrogen receptor (ER) inside liver cells, causing the ER to change shape and dimerize. The promoter region of the vitellogenin gene contains estrogen response elements (EREs) or incomplete ERE (ERE-like) sequences to which dimerized ER/E2 complexes bind to initiate gene expression and produce vitellogenin (Nelson and Habibi 2013). The number of ERE or ERE-like sequences inside each vitellogenin gene's promoter region and the quantity and type of additional transcription factor-binding sites, such as GATA, are critical factors in determining the transcriptional potential of individual vitellogenin genes. Discrete vitellogenin gene subtypes within a species and across fish species have different promoter components and architectures (Mushirobira et al. 2018). Other factors may also affect the regulation of vitellogenesis in some fish species. For instance, hagfish make vitellogenins after eating and may respond to estrogen less strongly than other fishes (Nishimiya et al. 2011, 2017).

The mature vitellogenins that the hepatocytes have made are subsequently released into the bloodstream, where they travel to the ovarian follicle's capillaries and come into touch with the lemma. Then, via endocytosis, vitellogenins are taken up by developing oocytes through particular membrane vitellogenin receptors (Reading et al. 2011, 2014, 2017). The very low-density lipoprotein receptor (Vldlr) and other vertebrate vitellogenin receptors share the characteristic of possessing

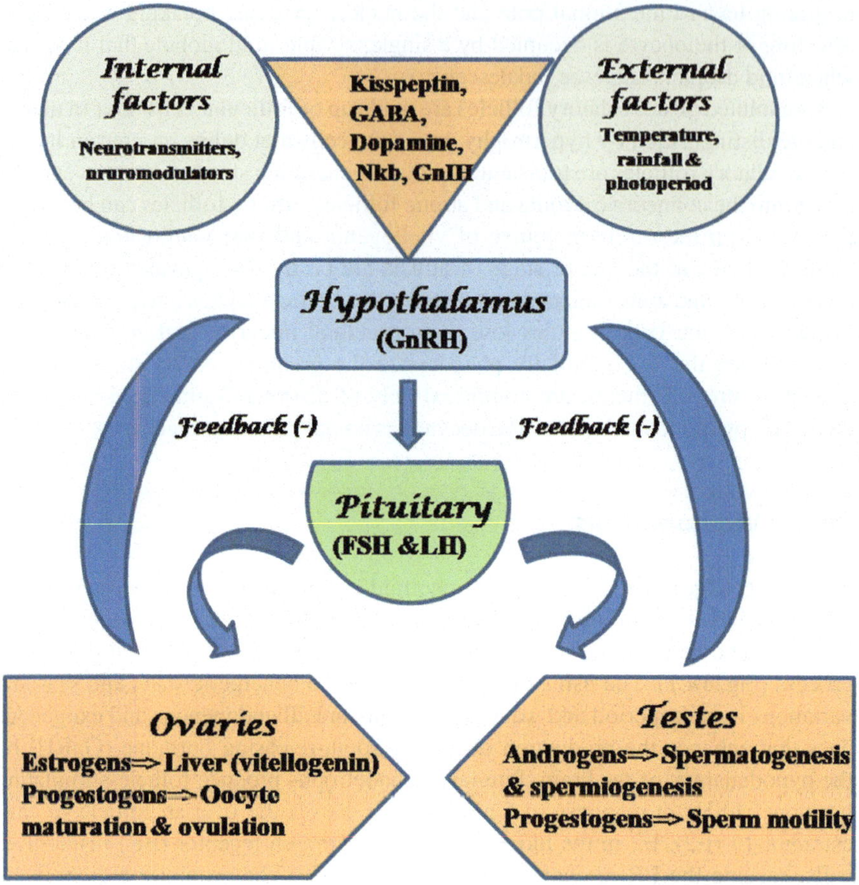

Fig. 18.1 The reproductive hypothalamic-pituitary-gonadal neuroendocrine axis controls

eight ligand-binding repeats (Lr8), which distinguishes the first reported fish vitellogenin receptor from them (Reading et al. 2017; Davail et al. 1998; Prat et al. 1998; Hiramatsu et al. 2004).

The Lr8 gene is not expressed during vitellogenesis, and the receptors must interact with vitellogenins before returning to the egg membrane and being used throughout the development process. In some species of fish, the Lrp13 vitellogenin receptor, the second type of vitellogenin receptor, binds a different vitellogenin ligand from the Lr8 receptor (Mushirobira et al. 2015). According to Mushirobira et al. (2015) and Reading et al. (2014), the expression of the Lrp13 gene and protein in salmonid and acanthomorph species is generally comparable to that of the Lr8 receptor. The inactive cathepsin D zymogens then colocalize with the imported vitellogenins. The subsequent drop in pH activates cathepsin D after vacuolar ATPases have acidified the lumen of the multivesicular bodies. Vitellogenin is then broken down by cathepsin D into proteins that are stored as yolk granules, globules, platelets, or liquid yolk in the ooplasm (Hiramatsu et al. 2002).

Numerous oviparous (egg-laying) fish species have been shown to induce Vtg, a yolk protein, in reaction to estrogens via an ESR-mediated pathway (Ryffel 1978). Male Vtg is generally recognized as a biomarker of exposure to environmental estrogens (Filby et al. 2006). At least 17 teleost species have at least two Vtg transcripts that have been identified so far (Hiramatsu et al. 2006). This study has cloned cDNAs that appear to be from the Vtg A group and encode one kind of Vtg (Finn and Kristoffersen 2007). The esr1 subtype is strongly linked with Vtg mRNA levels in the liver of largemouth bass (*Micropterus salmoides*) and Atlantic salmon (*Salmo salar*) (Sabo-Attwood et al. 2004; Meucci and Arukwe 2006). The function of esr subtypes in the in vivo and in vitro regulation of genes like Vtg in cinnamon clownfish, however, remains uncertain.

Several transforming growth factor (TGF) superfamily members in addition to E2 may potentially control vitellogenesis. The higher levels of mRNA during vitellogenesis may encourage follicle growth (Wang and Ge 2003a, b). Tgfb1 mRNA increases the expression of FSHR, suggesting that vitellogenesis may be modulated (Kohli et al. 2005). The ability of bone morphogenetic protein 15 (bmp15) to prevent precocious follicle maturation has also been demonstrated (Clelland et al. 2007). The condition of captive broodstock must be maintained, and when these cues are disrupted, abnormal oocyte and egg development results. Although severe cases of yolkless mutant fish have not been documented, disruption of photothermal parameters is excited vitellogenesis and receptor-mediated uptake of vitellogenins by the oocyte are necessary conditioning cues. As the days get shorter and the temperature drops in the fall, a temperate fish from the northern hemisphere known as the white perch begins vitellogenesis. It spawns in the spring (lengthening daylight and increasing temperature). The oocytes of these fish will collect neutral lipids (previtellogenic development) when they are raised at a consistent temperature and photoperiod but not vitellogenin-derived egg yolk. Constant photothermal temperatures suppressed vitellogenesis in Eurasian perch (*Perca fluiatilis*), according to a similar discovery (Milla et al. 2009).

In fish farming, the selection of reproducers is based on the presence of mature or vitellogenic oocytes. It is primarily based on using specific techniques to measure the oocyte diameter, the location of the germinal vesicle, and the frequency of various types of mature oocytes. In species like *A. gigas*, where there is not a duct that directly connects the ovary and urogenital papilla, endoscopic analysis can be used in place of cannulation to precisely choose the most appropriate breeders (Schorer et al. 2016). Following the completion of vitellogenesis, the oocytes are prepared for the next stage of oogenesis, which is the restart of meiosis (mature oocytes) and ovulation (Nagahama and Yamashita 2008). The oocyte maturation phase, which is the restart of meiosis, is distinguished by significant morphological changes in mature oocytes following vitellogenesis (Mylonas et al. 2010). Oocyte maturation was observed in the ovaries of several species of neotropical migratory fish during the hormonal induction phase. Matrinxa (Hainfellner et al. 2012), pacu (Criscuolo-Urbinati et al. 2012), piauc u (Pereira et al. 2017), and piapara have all demonstrated this process (Leonardo et al. 2004). They were also distinguished by the presence of oocytes with a diameter slightly larger than Vtg (Pereira et al. 2018).

Because oocyte maturation and ovulation occur at the end of the migration process in migratory species, it is expected that there are few reports or descriptions of these processes in their natural habitats. Interventions at this stage, however, would almost certainly result in the reproductive process being interrupted or altered. Ovulation occurs when the follicular layer, basal membrane, and theca layer divide synchronously, resulting in an opening through which the oocyte exits the follicle and enters the ovarian lumen. The postovulatory complex is composed of the cell layers that remain after the oocyte has been released. As a result, general information from periodic collections and descriptions of the macroscopic and/or histological characteristics of the ovaries are frequently the only sources of information on the spawning times of neotropical species (migratory or nonmigratory) in nature (Brandao et al. 2017). The key spawning signal is the presence of postovulatory complexes in the postspawning ovaries, which can be seen histologically (Ganias 2012). However, in natural collections, the length of the postovulatory complex reabsorption process is frequently short and difficult to detect.

According to this approach, descriptions of the existence, as well as the morphological and temporal aspects of the postovulatory complexes' resorption process, are more prevalent in captive hormone-induced spawning. According to extensive research, the basal membrane disintegrates and the postovulatory complexes vanish in common snook within 48 h of spawning (Grier et al. 2017). Few neotropical migratory species have determined the time of postovulatory complex reabsorption; in *P. fasciatum*, the process was monitored for 24 h and an extensive follicular reabsorption process was observed, but it did not completely remove the complex (Romagosa et al. 2005). For 4 days, the reabsorption process in lambari (*Astyanax bimaculatuslacustris*) was observed, with a reduction in complexes observed, accompanied by an intense apoptotic process but without complete reabsorption (Drummond et al. 2000).

18.3 The Abundance of Fish Vitellogenins

Multiple types of fish vitellogenins have developed through whole genome duplications in the vertebrate lineage, and their evolutionary history is complex. The structure and function of these several vitellogenins varied significantly across fish taxa. The yolk sac larvae and developing embryos with protein, carbohydrate, and lipid nourishment according to a predetermined schedule at various developmental time points. Many fish species have been identified, including gray mullet (*Mugil cephalus*) (Amano et al. 2007), barfin flounder (*Verasper moseri*) (Matsubara et al. 1999; Sawaguchi et al. 2008), striped bass (*Morone saxatilis)* (Williams et al. 2014), and white perch (Schilling et al. 2015; Yilmaz et al. 2018).

According to recent studies, fish Vg (s) may perform immune functions, act as a carrier molecule for proteins and steroid hormones, ions such as calcium, magnesium, and iron, and have been widely used as biomarkers for EDC exposure, aside from its traditional role as a nutrient reserve during embryonic development (Hara et al. 2016; Specker and Sullivan 1994; Mommsen and Walsh 1988). According to Reis-Henriques et al. (2000), ovarian Vg inside the oocyte can influence Vg synthesis in the liver. Furthermore, different Vg types expressed at different stages of oogenesis may play different roles during oocyte maturation and embryonic development (Pousis et al. 2011). We previously demonstrated that semi-purified conspecific Vg could induce full vitellogenesis (the production of Vg and its absorption into oocytes) in *C. batrachus* (Nath et al. 1997). Exogenous administration of mrigal, *Cirrhinus mrigala*, Vg into female catfish, *C. batrachus*, affects not only Vg production but also Vg integration into developing oocytes for growth and conversion into yolky oocytes, according to Nath and Maitra (2001). Furthermore, until the yolk sac is digested, yolk protein is responsible for early fish embryonic development. Furthermore, we recently discovered an immune-reactive protein that resembled vitellogenin in young catfish plasma (unpublished data). Based on the foregoing, it was concluded that Vg has the potential to promote growth.

18.4 Composition of Egg Yolks and Diet of Broodstock

Vitellogenins also transport minerals and vitamins including retinoids and carotenoids, as well as essential ions like calcium, magnesium, iron, zinc, and copper (Finn 2007; Specker and Sullivan 1994). In freshwater settings where metal ions are barely available for ingestion, ion transport via vitellogenins is crucial for embryo survival in species such as the masu salmon (*Oncorhynchus masou*) and mosquito fish.

The Vtg-derived yolk can make up as much as 80–90% of the dry mass of an ovulated egg in some species (Reading et al. 2017). The main lipid transfer protein superfamily, which also includes serum lipoproteins including low-density lipoprotein (LDL) and very low-density lipoprotein (VLDL), includes vitellogenins. They include phosphate, fat (20% of their weight), carbohydrates, and proteins (Smolenaars et al. 2007). About 11–12% of the residues in vitellogenin polypeptides are made up of the amino acid alanine. There are polyalanine regions in the vitellogenins of several species, including eels (Reading et al. 2009). During embryonic gluconeogenesis, alanine in particular may play a crucial role in the metabolic process that breaks down carbohydrates.

In contrast, less calcium and magnesium ions, which are abundant in seawater, are discovered in the egg yolks of marine fishes such as red seabream, barfin flounder, and Pacific herring (*Clupea pallasii*), suggesting that vitellogenin is not as important in delivering these ions to marine fish embryos (Reading et al. 2017). The vast majority of these lipids (approximately 80%) are phospholipids, which are carried into developing fish oocytes to generate demersal (sinking) eggs with no apparent oil droplets. Phospholipids, triacylglycerides, and wax- or sterol-esters can account for around 70%, 8% to 12%, and 4% of total egg lipids in marine fishes that lay pelagic eggs in the absence of significant oil droplets, respectively.

Other fish may produce enormous oil droplets covering more than 50% of the ooplasm in their pelagic eggs, which are often comprises triacylglycerides and wax- or sterol-esters, neutral lipids. These lipids assist pelagic eggs in floating, particularly the eggs of the golden perch (*Macquaria ambigua*) (Anderson et al. 1990). A meal with up to 40% protein enhances broodstock growth and larval survival, according to Al-Hefedh et al. However, broodstock lipid content may decrease (Al-Hafdeh 1999). However, it is crucial to note that in this study, fertility was substantially greater in fish given a low-protein diet (25–35% vs. 40–45%) (Al-Hafdeh 1999).

Along with proteins, which make up the majority of an egg yolk's weight, lipids play important functions in embryo development. They function as structural elements of cell membranes, as well as energy storage and signaling channels, among other critical roles and activities in the developing embryo. The essential fatty acids, which include homolipids and heterolipids, are required for embryo and larval growth and survival; however, the suitable supply and concentration of fatty acids in broodstock diets vary depending on the species (Leray et al. 1985; Izquierdo et al. 2001). Furthermore, it is thought that the first maternal contribution to the egg dictates the amount of *n*-3 highly unsaturated fatty acids (HUFA) required (Tuncer and Harrell 1992; Harrell et al. 1995). Polyunsaturated fatty acids (PUFAs) must be consumed because fish cannot manufacture them. Freshwater animals, according to research, require linoleic acid 18:2 (*n*-6) and alpha-linolenic acid 18:3 (*n*-3), but marine species require longer and more unsaturated eicosapentaenoic and docosahexaenoic acids (Takeuchi and Watanabe 1982). Docosahexaenoic acid 22:6 (*n*-3) and eicosapentaenoic acid 20:5 (*n*-3) have been demonstrated to influence egg quality and embryo survival (Carrillo and Zanuy 1995). However, white bass (*Morone chrysops*) ova from broodstock fed an *n*-6/*n*-3 diet had lower levels of eicosapentaenoic acid 20:5, and docosahexaenoic acid 22:6 (*n*-3) was shown to be preferentially integrated within ova over other *n*-3 PUFAs (Fuller et al. 2017). Because phosphatidylcholine is a component of the lipid cargo delivered by vitellogenins, it may be necessary for embryonic development. It has been demonstrated that the amount of vitamins A, E (including -tocopherol), C, and ascorbic acid in the broodstock diet affects fish eggs. Increasing dietary-tocopherol concentrations, a kind of vitamin E, has been shown to increase egg quality (Watanabe et al. 1991).

18.5 Role of Vitellogenins in Ornamental Fish

According to Yilmaz et al. 2017, good and poor zebrafish (*Danio rerio*) eggs have distinct proteomic profiles. Proteins involved in protein synthesis, energy, and lipid metabolisms, as well as several vitellogenin products and lectins, were found to be deficient in low-quality zebrafish eggs. Similar conclusions were reached regarding the importance of vitellogenins in Eurasian perch (Castets et al. 2012). Furthermore, upregulated proteins linked with endo-lysosomal activities, autophagy, apoptosis, and some oncogene products were shown to be more often present in low-quality zebra fish eggs (Yilmaz et al. 2017). According to the upregulation of endosome and lysosome activity-related proteins, the transition of oocytes to final maturation may be hampered, and this may include proteolysis of vitellogenin-derived yolk (Yilmaz et al. 2017). Similarly, efforts to combat apoptosis, which may occur during ovarian aresia associated with broodstock stress, may be reflected in increased upregulation of proteins with oncogene-related characteristics (Yilmaz et al. 2017). Rpl36–001 and Rpl36–002 are ribosomal proteins found only in high-quality zebrafish eggs. Four additional proteins are only found in low-quality eggs (tubulin zgc: 55461–001, carbonyl reductase cbr1–001, casein kinase zgc:86598–001, and 20, 30 -cyclic nucleotide 30 phosphodiesterase cnp-201). Furthermore, Prx2, Prx5, Prx6, and transferrin were shown to be down regulated in low quality Eurasian perch eggs, indicating that oxidative stress may be a factor influencing egg quality. As a result, these protein-related characteristics may be utilized to assess the quality of an egg. Zebrafish treated with *L. rhamnosus* had more vitellogenic follicles and a higher gonadosomatic-index (GSI) (Gioacchini et al. 2010, 2011a). The expression of the cyp19a gene in the ovary, hepatic Vtg, and er was shown to be similar (Gioacchini et al. 2011a). *L. rhamnosus* CICC 6141 and *L. casei* BL23 were recently employed to corroborate these findings (Qin et al. 2014).

Recently, zebrafish ovarian tissue was examined using a unique approach known as Fourier transform infrared (FT-IR) microspectroscopy (Giorgini et al. 2010). Through the examination of oocytes at various developmental stages, certain IR-vibrational patterns that are well connected with the selective absorption of Vtg and the maturation process were revealed. Analysis of multiple vibrational bands from pre-vitellogenic to mature oocytes revealed increased lipidic and glucidic components, a larger effect of hydration and phosphorylation activities, and a change in protein secondary structures (Carnevali et al. 2009). The injection of *L. rhamnosus* IMC 501 altered the chemical makeup of the oocytes, enhancing the vitellogenic process. Surprisingly, the ovaries of treated females displayed biochemical changes associated with protein secondary structure as well as phospholipidic and glucidic patterns. Concurrently, the levels of mRNA and the enzyme cathepsin L in vitellogenic oocytes extracted from probiotic-treated fish were examined and found to be higher, equivalent to mature oocytes obtained from controls (Giorgini et al. 2010). Because E2 was a regulator of these genes in cinnamon clownfish, higher expression of esr and Vtg in the gonads and livers is probable. In summary, we discovered that: (1) E2 increases esr and Vtg mRNA expression in cinnamon clownfish and (2) immature cinnamon clownfish exposed to E2 have greater gene mRNA expression. We concluded that E2 regulates esr and Vtg in cinnamon clownfish.

During vitellogenesis, the pituitary gland secretes growth hormone (GH), but the fish spend their efforts on reproduction rather than growth. Mina Moussavi, Erik R. Nelson, and colleagues investigated the role of GH in the regulation of Vg in goldfish. The goldfish Vg cDNA was largely cloned (1348 amino acids) and featured a highly conserved lipoprotein N-terminal region as a probe for Northern blot investigation. In vitro treatment of female liver with recombinant goldfish demonstrated an increase in Vg mRNA early in their recrudescence (in September), but GH had no effect later in vitellogenesis and throughout the post-spawning phase

(February–June). There was a link between GH activity and baseline growth hormone receptor expression levels (GHR). GH treatment, like Vg, boosted the amount of GHR transcript in September. In conclusion, the current study adds to our understanding of the role of GH in the season-dependent regulation of Vg and GHR gene expression. The gene expression of the white sturgeon, *Acipenser transmontanus*, was studied using the vitellogenin cDNA sequence. The livers of both males and females, as well as the undifferentiated gonads of fish treated with estrogen, were discovered to contain estrogen-induced vitellogenin mRNA. In both control and estrogen-treated guys, vitellogenin mRNA levels were low in the testicles. Only six to seven amino acids were missing from the amino terminus of a 186-kDa protein produced by the cDNA. According to comparisons with silver lamprey, Xenopus, and chicken vitellogenin sequences, the overall structure of the domains of the yolk protein was highly conserved. The guppy (*Poecilia reticulata*) is an ideal model for environmental estrogen studies because of its large caudal fin's remarkable regeneration potential. Vitellogenin (Vtg), the most frequently used biomarker of environmental estrogens, was investigated in this study to determine if caudal fins might be utilized to detect them.

Lipovitellin and phosvitin are yolk proteins found in goldfish oocytes, and each of these proteins has a variety of molecular weight variants. In prepared goldfish yolks, we were unable to detect an electrophoretic band that would have indicated undegraded vitellogenin. So, contrary to Hori et al. (1979) hypothesis that the entire vitellogenin is stored in goldfish oocytes, it looks likely that vitellogenin is stored in the oocytes and proteolytic processing results in proteins in the yolk.

On sodium dodecyl sulfate-gel electrophoresis, the native goldfish yolk protein is divided into two fractions: lipovitellin and phosvitin; the former produces a class of at least two large polypeptides (mol. wt = 105,000 and 110,000) and a class of up to four small polypeptides (mol. wt = 19,000–25,000), whereas the latter produces a class of at least two polypeptides. When compared to vitellogenins from other oviparous vertebrate groups, the goldfish's multiple molecular weight variants have a lower molecular weight, higher lipid content, and a lower phosphorus concentration. Similarly, some yolk proteins (lipovitellins and phosvitins) may be present in goldfish oocytes. These proteins differ from those of other oviparous vertebrate groups in that they have smaller molecular weights, higher lipid content, and lower phosphorus content. A lack of additional data suggests that the properties of vitellogenin and yolk proteins found in goldfish can be transferred to other teleost species; more research is needed to substantiate this assumption.

18.6 Oocyte Competence Acquisition

The consequences of *L. rhamnosus* on maturational competence were established in a recent study: probiotic administration improved the responsiveness of incompetent follicles (stage IIIa) to MIH and they are in vitro maturation (Gioacchini et al. 2012). Changes in the expression of the genes LHR, MPR, activin-A1, and tgf1 in the same study confirmed the competence acquisition of IIIa follicles. During follicle development, rhamnosus treated zebrafish affected some significant and putatively regulated molecular activities as well as biological processes.

18.7 Oocyte Maturation

Probiotics have also been shown to improve follicular maturation. GVBD rates were higher in oocytes from females fed a probiotic-supplemented diet, according to in vitro maturation experiments (Gioacchini et al. 2010, 2011b). Higher levels of genes coding for signals driving oocyte maturation (lhcgr, cbr11, paqr8) were found in the ovaries of zebrafish treated with *L. rhamnosus* (Gioacchini et al. 2012); an opposing trend was seen in the transcription of local factors involved in the inhibition of oocyte maturation (bmp15, gdf9, and tgf). The use of numerous probiotic strains, including *Lactobacillus casei* BL23 and *Lactobacillus rhamnosus* CICC 6141, validated these findings Qin et al. (2014). Giorgini et al. 2012 found that rhamnosus showed changes in ooplasma components, including significant changes in the electrophoretic pattern during maturation and, to a lesser extent, at yolk protein levels

18.8 Stress and Ovarian Atresia

The physiological responses of fish to stress as well as the mechanisms underlying how stress affects fish growth, reproduction, and survival are being studied by Suarez-Bregua et al. 2018; Faught and Vijayan 2018). The degree of corticosteroid response to a particular stressor has been shown to vary significantly within a species, depending on elements such as gender, genetic background, temperature, and others (McBryan et al. 2013). Poor water quality, osmoregulatory challenges, crowding, malnutrition, illness, handling, and tank confinement are all known to harm fish reproductive capacity.

The major stress hormones, cortisol and adrenaline, in particular, block the reproductive hormone axis. Finally, stress affects reproductive output by reducing reproductive hormones (GnRH, FSH, LH, and others) (Wendelaar Bonga 2011). Because cortisol's impact on reproductive hormones differs between species, it is impossible to say which aspects of reproductive performance are directly influenced by it or any other stress-coping mechanism. Because the stress response releases energy reserves required to offset the detrimental stressor or injury, bioenergetic resources may be shifted away from gonad development during stages such as previtellogenesis (oocyte lipidation) and vitellogenesis (yolk deposition) (Baltzegar et al. 2013).

To time captive spawning effectively and avoid ovarian atresia, husbandry methods that lessen stress and understanding the reproductive cycle are essential. The role of probiotics in the interaction between autophagic and apoptotic processes was reported during the development of zebrafish follicles. *L. rhamnosus* therapy decreased apoptosis and increased follicular survival (Gioacchini et al. 2013).

Examining preovulatory follicles under electron microscopy revealed that these fish have more autophagosomes (stages III and IV). The results were supported by higher autophagic protein and gene expression in preovulatory follicles. However, the TUNEL assay provided additional evidence that the same cells' expression of apoptotic genes was decreased (Gioacchini et al. 2013).

18.9 Experimental Evaluation of Vitellogenin as a Predictive Biomarker for Reproductive Disruption

Vitellogenin (Vtg) is produced by male oviparous vertebrates in response to environmental estrogens, but it is unknown how elevated Vtg levels relate to how environmental estrogens affect reproductive success. Environmental research is currently concentrating on the significance of endocrine disruption for population and ecological health. An organism's fitness and ability to reproduce may be significantly affected by endocrine disruption, which can change the organizational and activational effects of reproductive hormones. Environmental estrogens have been the subject of the most research, even though other groups of endocrine-disrupting substances, such as antiestrogens, androgens, antiandrogens, progestins, and retinoid mimics, have been identified.

Fish have become a particularly useful model for studying estrogenic endocrine disruption because both males and females of fish produce vitellogenin (Vtg), the precursor to the proteins found in egg yolks, in response to estrogen or estrogen mimics. VTG production in males and immature females varies from nonexistent to low when compared to adult female fish, which exhibit seasonal cycles of serum VTG levels with peak values reaching tens of milligrams per milliliter. All vertebrates' main endogenous estrogen, estradiol, regulates vitellogenesis hormonally in mature females. The liver produces VTG in response to rising estradiol levels, which is then circulated to the ovary. VTG enters the oocyte through specific receptor-mediated endocytosis.

VTG is split up into smaller yolk proteins (phosvitin, lipovitellin, and beta-component) once it is inside the oocyte; these proteins then assemble form yolk globules or granules. VTG is a highly specific biomarker for estrogen exposure in fish due to the particular relationship between VTG synthesis and estrogen stimulation as well as the minimal background production of this protein in all except mature females. The steroidal estrogen ethinylestradiol, the insecticides methoxychlor and o, p'-DDT and their metabolites, phytoestrogens, and sewage effluent are among the groups of chemicals and chemical mixtures that have been shown to modify fish vitellogenesis in vivo. The effects of environmental estrogens on sex differentiation in fish have only lately been studied, although changes in vitellogenesis have been well reported both in vivo and in vitro. Adult males were exposed to natural and ambient estrogens in lab settings, and the effects on reproductive success (the number of children produced) have recently been studied. Adult males exposed to estradiol experienced lower medaka and fathead minnow hatching success. In medaka mating experiments, octylphenol and bisphenol A also decreased male fertility (percent fertilized) and embryo survival. Few research has looked at how to fish's embryonic exposure to ambient estrogens affects their ability to reproduce later in life, and none have looked at the relationship between vitellogenin induction, sex differentiation, and reproductive success.

A guideline for a Tier 2 fish test protocol has been published by the Endocrine Disruptor Screening Program (EDSP) of the U.S. Environmental Protection Agency (USEPA) to assess and define the dose–response of potential endocrine-disrupting chemicals (EDCs) on fish reproduction and reproductive development (USEPA 2015). A similar test guideline was also published by the Organization for Economic Co-operation and Development (OECD) Test Guidelines Program as OECD TG 240 (OECD 2015).

The Medaka Extended One-Generation Reproduction Test (MEOGRT), as its name suggests, aims to shed light on whether a test substance has the potential to negatively impact fish, possibly through endocrine disruption. The MEOGRT is specifically intended to identify impacts on reproduction, secondary effects on growth, development, and survival, and additional parameters that may be sensitive to the endocrine-disrupting effects of the test chemical (Flynn et al. 2017). The MEOGRT spans three generations, spanning impacts from hatch to the transition from juvenile life stages to sexual maturity because many of these effects could become apparent later in life, notably around the period of reproduction.

Despite its economic significance, little is known about its reproductive physiology. Numerous fish species' reproductive biology has recently been investigated by measuring the quantities of vitellogenin (Vtg) in the blood plasma (Susca et al. 2001). An egg yolk precursor protein called vitellogenin is created in the liver under the guidance of estrogen and released into the bloodstream (Prakash et al. 2007). Depending on the species of fish, this big protein has a high molecular weight of between 250 and 600 kDa (Utarabhand and Bunlipatanon 1996). Through a process known as vitellogenesis, Vtg serves as food for developing oocytes and embryos in mature female teleost oviparous fish (Romano et al. 2004). Following oocyte development, Vtg is enzymatically divided into smaller yolk proteins such as phosphorylated phosvitin, lapidated lipovitellin, and components (Zhang et al. 2011). Calcium and iron, as other elements, including lipids, carbohydrates, and phosphorus are the primary elements that contribute to the circulation of the Vtg molecule, according to Mananos et al. (1994). Adult vitellogenic females had the Vtg, whereas males and immature females did not (Fenske et al. 2001). Male and immature female vertebrates express the Vtg gene, but insufficient levels of circulating estrogen are unable to increase this protein's production (Palumbo et al. 2007).

However, if these organisms are given synthetic estrogens, primarily 17-estradiol, they will produce the Vtg (Leonardi et al. 2010). Many fish species have been found to successfully produce Vtg when estradiol is used as an inducer (Mendoza et al. 2011). The levels of Vtg in fish represent the stage of maturation in a female individual under natural circumstances (Matsubara et al. 1994). To manage fish broodstock for reproduction in most farmed species, including fish, knowledge of reproductive physiology, particularly vitellogenesis, is crucial. According to earlier research, Vtg has been isolated and purified from a variety of fish species, using double chromatography and ion exchange followed by gel filtration chromatography. In *L. calcarifer*, vitellogenin has never been purified and described, necessitating research into its vitellogenesis and reproduction. Understanding vitellogenesis better can help with farm management and maturity assessment of this economically significant species (Utarabhand and Bunlipatanon 1996).

18.10 Future Research Directions

Understanding the reasons for good egg quality is essential for broodstock management, particularly as environmental changes occur and the demand for animal-based protein moves increasingly in favor of aquaculture. The focus should be placed on comprehending and reducing the impacts of handling and other stressors on broodstock, as well as developing informed spawning techniques for conditioning broodstock, including identifying postovulatory aging and atresia and addressing nutritional needs. The need to recognize these effects and potential countermeasures could have a significant impact on egg quality and the aquaculture sector's ability to continue to prosper. The findings of recent studies on egg quality, such as those on metabolomic pathways and indicators of poor egg quality, can easily be built upon when examining how to manage stress, especially in light of postovulatory aging. Understanding the factors that contribute to high- and low-quality eggs requires an understanding of the maternal RNA contribution during oogenesis and embryogenesis, as well as the times when genes are covered and revealed during development. The use of markers for egg quality can also have an impact on broodstock selection to find the best spawning fish.

References

Al-Hafdeh YS (1999) Effects of dietary protein on growth and body composition of Nile tilapia, Oreochromis niloticus L. Aquac Res 30:385–393

Amano H, Fujita T, Hiramatsu N, Shimizu M, Sawaguchi S, Matsubara T, Kagawa H, Nagae M, Sullivan CV, Hara A (2007) Egg yolk proteins in grey mullet (Mugil cephalus): purification and classification of multiple lipovitellins and other vitellogenin-derived yolk proteins and molecular cloning of the parent vitellogenin genes. J Exp Zool A Ecol Genet Physiol 307:324–341

Anderson AJ, Arthington AH, Anderson S (1990) Lipid classes and fatty acid composition of the eggs of some Australian fish. Comp Biochem Physiol 98:267–270

Baltzegar DA, Reading BJ, Douros JD, Borski RJ (2013) Role for leptin in promoting glucose mobilization during acute hyperosmotic stress in teleost fishes. J Endocrinol 220:61–72

Bobe J (2015) Egg quality in fish: present and future challenges. Anim Front 5:66–72

Brandao LED, Nascimento AV, Marcon L, Santos JE, Santiago KB, Rizzo E, Bazzoli N (2017) Comparative analyses of reproductive activity in Schizodonknerii (Steindachner, 1875) (Characiformes: Anostomidae) in three sections of the Sa˜o Francisco River basin. J Appl Ichthyol 33:1118–1124

Carnevali O, Conti C, Ferraris P, Garavaglia MG, Gioacchini G, Giorgini E, Rubini C, Sabbatini S, Tosi G (2009) FT-IR microspectroscopy on molecular building of zebrafish oocytes. J Mol Struct 938:207–213

Carrillo M, Zanuy S (1995) Manipulación de la reproducción de losteleósteos y calidad de las puestas. In: Proceedings of the Actas del V Congreso Nacional de Acuicultura, Sant Carles de la Rápita, Spain, vol 10–13, pp 1–9

Castets MD, Schaerlinger B, Silvestre F, Gardeur JN, Dieu M, Corbier C, Kestemont P, Fontaine P (2012) Combined analysis of percafluviatilis reproductive performance and oocyte proteomic profile. Theriogenology 78:432–442

Chapman RW, Reading BJ, Sullivan CV (2014) Ovary transcriptome profiling via artificial intelligence reveals a transcriptomic fingerprint predicting egg quality in striped bass, Morone saxatilis. PLoS One 9:e96818

Clelland E, Tan Q, Balofsky A, Lacivita R, Peng C (2007) Inhibition of premature oocyte maturation: a role for bone morphogenetic protein 15 in zebrafish ovarian follicles. Endocrinol 148:5451–5458. https://doi.org/10.1210/en.2007-0674

Criscuolo-Urbinati E, Urbinati EC, Kuradomi RY, Batlouni SR (2012) The administration of exogenous prostaglandin may improve ovulation in pacu (Piaractusmesopotamicus). Theriogenology 78:2087–2094

Davail B, Pakdel F, Bujo H, Perazzolo LM, Waclawek M, Schneider WJ, Le Menn F (1998) Evolution of oogenesis: the receptor for vitellogenin from the rainbow trout. J Lipid Res 39:1929–1937

Drummond CD, Bazzoli N, Rizzo E, Sato Y (2000) Postovulatory follicle: a model for experimental studies of programmed cell death or apoptosis in Teleosts. J Exp Zool 287:176–182

Enberg K, Dunlop ES, Jorgensen C (2008) Fish growth. In: Jorgensen SE, Fath B (eds) Encyclopedia of ecology. Elsevier, Amsterdam, pp 1564–1572

Faught E, Vijayan MM (2018) Maternal stress and fish reproduction: the role of cortisol revisited. Fish 19:1016–1030

Fenske M, Aerle RV, Brack S, Tyler CR, Segner H (2001) Development and validation of a homologous zebrafish (Danio rerio Hamilton-Buchanan) vitellogenin enzymelinked immunosorbent assay (ELISA) and its application for studies of estrogenic chemicals. Comp Biochem Physiol C Toxicol Pharmacol 129:217–232

Filby AL, Thorpe KL, Tyler CR (2006) Multiple molecular effect pathways of an environmental oestrogen in fish. J Mol Endocrinol 37:121–134

Finn RN (2007) Vertebrate yolk complexes and the functional implications of phosvitins and other subdomains in vitellogenins. Biol Reprod 76:926–935

Finn RN, Kristoffersen BA (2007) Vertebrate vitellogenin gene duplication in relation to the "3R hypothesis": correlation to the pelagic egg and the oceanic radiation of teleosts. PLoS One 169:1–12

Flynn K, Lothenbach D, Whiteman F, Hammermeister D, Touart LW, Swintek J, Tatarazako N, Onishi Y, Iguchi T, Johnson R (2017) Summary of the development of the US environmental protection agency's medaka extended one generation reproduction test (MEOGRT) using data from 9 multigenerational medaka tests. Environ Toxicol Chem 36:3387–3403

Fuller AS, Rawles SD, McEntire ME, Bader TJ, Riche M, Beck BH, Webster CD (2017) White bass (Morone chrysops) preferentially retain n-3 PUFA in ova when fed prepared diets with varying FA content. Lipids 52:823–836

Ganias K (2012) Thirty years of using the postovulatory follicles method: overview, problems and alternatives. Fish Res 117–118:63–74

Gioacchini G, Maradonna F, Lombardo F, Bizzaro D, Olivotto I, Carnevali O (2010) Increase of fecundity by probiotic administration in zebrafish (Danio rerio). Reproduction 140:953–959

Gioacchini G, Carnevali O, Giorgini E, Vaccari L, Bianchi V, Borini A (2011a) Evaluation of human oocytes ageing by focal plane array (FPA) fourier transform infrared (FT-IR) imaging spectroscopy. Fertil Steril 96(Supplement):S238–S239

Gioacchini G, Lombardo F, Merrifield DL, Silvi S, Cresci A, Avella MA, Carnevali O (2011b) Effects of probiotic on zebrafish reproduction. J Aquac Res Dev S1:002. https://doi.org/10.4172/2155-9546.S1-002

Gioacchini G, Giorgini E, Merrifield DL, Hardiman G, Borini A, Vaccari L, Carnevali O (2012) Probiotics can induce follicle maturational competence: the Danio rerio case. Biol Reprod 86(3):65

Gioacchini G, Dalla Valle L, Benato F, Fimia GM, Nardacci R, Ciccosanti F, Piacentini M, Borini A, Carnevali O (2013) Interplay between autophagy and apoptosis in the development of Danio rerio follicles and the effects of a probiotic. Reprod Fertil Dev 25:1115–1125

Giorgini E, Conti C, Ferraris P, Sabbatini S, Tosi G, Rubini C, Vaccari L, Gioacchini G, Carnevali O (2010) Effects of lactobacillus rhamnosus on zebrafish oocyte maturation: an FTIR imaging and biochemical analysis. Anal Bioanal Chem 398(7–8):3063–3072

Giorgini E, Gioacchini G, Maradonna F, Ferraris P, Carnevali O (2012) Melatonin effects on Fundulus heteroclitus reproduction. Reprod Fertil Dev, 24(6):794–803

Grier HJ, Neidig CL, Quagio-Grassiotto I (2017) Development and fate of the postovulatory follicle complex, postovulatory follicle, and observations on folliculogenesis and oocyte atresia in ovulated common snook, Centropomus undecimalis (Bloch, 1792). J Morphol 278:547

Hainfellner P, Munoz ME, Freitas GA, de Souza TG, Batlouni SR (2012) Spawning failure in Brycon amazonicus may be associated with ovulation and not with final oocyte maturation. Arq Bras Med Vet Zootec 64:515–517

Hara A, Hiramatsu N, Fujita T (2016) Vitellogenesis and choriogenesis in fishes. Fish Sci 82:187–202

Harrell RM, Woods LC (1995) Comparative fatty acid composition of eggs from domesticated and wild striped bass (Morone saxatilis). Aquaculture 133:225–233

Hiramatsu N, Ichikawa N, Fukada H, Fujita T, Sullivan CV, Hara A (2002) Identification and characterization of proteases involved in specific proteolysis of vitellogenin and yolk proteins in salmonids. J Exp Zool 292:11–25

Hiramatsu N, Chapman RW, Lindzey JK, Haynes MR, Sullivan CV (2004) Molecular characterization and expression of vitellogenin receptor from white perch (Morone americana). Biol Reprod 70:1720–1730

Hiramatsu N, Matsubara T, Fujita T, Sullivan CV, Hara A (2006) Multiple piscine vitellogenins: biomarkers of fish exposure to estrogenic endocrine disruptors in aquatic environments. Mar Biol 149:35–47

Hori SH, Kodama T, Tanahashi K (1979) Induction of vitellogenin synthesis in goldfish by massive doses of androgens. Gen Comp Endocrinol 37(3):306–320

Izquierdo MS, Fernández-Palacios H, Tacon AGJ (2001) Effect of broodstock nutrition on reproductive performance of fish. Aquaculture 197:25–42

Kohli G, Clelland E, Peng C (2005) Potential targets of transforming growth factor-b1 during inhibition of oocyte maturation in zebrafish. Reprod Biol Endocrinol 3:1

Leonardi M, Vera J, Terifeno E, Puchi M, Marin V (2010) Vitellogenin of the Chilean flounder (Paralichthysadpersus) as a biomarker of the south pacific. Part 1: induction, isolation and identification. Fish Physiol Biochem 36(3):757–765

Leonardo AFG, Romagosa E, Borella MI, Batlouni SR (2004) Induced spawning of hatchery-raised Brazilian catfish, cachara Pseudoplatystomafasciatum (Linnaeus, 1766). Aquaculture 240:451–461

Leray C, Nonnote G, Roubaud D, Leger C (1985) Incidence of (n-3) essential fatty acid deficiency on trout reproductive processes. Reprod Nutr Dev 25:567–581

Mananos E, Zanuy S, Menn FL, Carillo M, Nunez J (1994) Sea bass (Dicentrarchuslabrax L.) vitellogenin. I- induction, purification and partial characterization. Comp Biochem Physiol B 107:205–216

Matsubara T, Wada T, Hara A (1994) Purification and establishment of ELISA for vitellogenin of Japanese sardine (Sardinoposmelanostictus). Comp Biochem Physiol B 109(4):545–555

Matsubara T, Ohkubo N, Andoh T, Sullivan CV, Hara A (1999) Two forms of vitellogenin, yielding two distinct lipovitellins, play different roles during oocyte maturation and early development of barfin flounder, Veraspermoseri, a marine teleost that spawns pelagic eggs. Dev Biol 213:18–32

McBryan TL, Anttila K, Healy TM, Schulte PM (2013) Responses to temperature and hypoxia as interacting stressors in fish: implications for adaptation to environmental change. Integr Comp Biol 53:648–659

Mendoza R, Santilla O, Revol A, Aguilera C, Cruz J (2011) Alligator gar (Atractosteus spatula, Lacepede 1803) vitellogenin: purification, characterization and establishment of an enzyme-linked immunosorbent assay. Aquac Res 43:649–661

Meucci V, Arukwe A (2006) Transcriptional modulation of brain and hepatic estrogen receptor and P450arom isotypes in juvenile Atlantic salmon (Salmo salar) after waterborne exposure to the xenoestrogen, 4-nonylphenol. Aquat Toxicol 77:167–177

Milla S, Mandiki SNM, Hubermont P, Rougeot C, Mélard C, Kestemont P (2009) Ovarian steroidogenesis inhibition by constant photothermal conditions is caused by a lack of gonadotropin stimulation in Eurasian perch. Gen Comp Endocrinol 163:242–250

Mommsen TP, Walsh PJ (1988) Vitellogenesis and oocyte assembly. In: Hoar WS, Randall DJ (eds) Fish physiology, vol XI a. Academic Press, New York, NY, pp 347–406

Mushirobira Y, Mizuta H, Luo W, Todo T, Hara A, Reading BJ, Sullivan CV, Hiramatsu N (2015) Molecular cloning and partial characterization of a low-density lipoprotein receptor-related protein 13 (Lrp13) involved in vitellogenin uptake in the cutthroat trout (Oncorhynchus clarkii). Mol Reprod Dev 82:986–1000

Mushirobira Y, Nishimiya O, Nagata J, Todo T, Hara A, Reading BJ, Hiramatsu N (2018) Molecular cloning of vitellogenin gene promoters and in vitro and in vivo transcription profiles following estradiol-17β administration in the cutthroat trout. Gen Comp Endocrinol 267:157–166

Mylonas CC, Fostier A, Zanuy F (2010) Broodstock management and hormonal manipulations of fish reproduction. Gen Comp Endocrinol 165:516–534

Nagahama Y, Yamashita M (2008) Regulation of oocyte maturation in fish. Develop Growth Differ 50:195–219

Nath P, Maitra S (2001) Role of two plasma vitellogenins from Indian major carp (Cirrhinusmrigala) in catfish (Clarias batrachus) vitellogenesis. Gen Comp Endocrinol 124:30–44

Nath P, Bhakta M, Maitra S, Sarkar S (1997) Vitellogenin induces vitellogenesis in the catfish, Clarias batrachus. In: Kawashima S, Kikuyama S (eds) Proceedings of the XIIIth international congress of comparative endocrinology, Yokohama, pp 1475–1479

Nelson ER, Habibi HR (2013) Estrogen receptor function and regulation in fish and other vertebrates. Gen Comp Endocrinol 192:15–24

Nishimiya O, Kunihiro Y, Hiramatsu N, Inagawa H, Todo T, Matsubara T, Reading BJ, Sullivan CV, Hara A (2011) Molecular characterization and expression analysis of estrogen receptor and vitellogenins in inshore hagfish (Eptatretusburgeri). Indian J Sci Technol 4:194–195

Nishimiya O, Katsu Y, Inagawa H, Hiramatsu N, Todo T, Hara A (2017) Molecular cloning and characterization of hagfish estrogen receptors. J Steroid Biochem Mol Biol 165:190–201

Organisation for economic co-operation and development [OECD] (2015) Test No. 240: Medaka Extended One Generation Reproduction Test (MEOGRT). OECD Publishing, Paris

Palumbo AJ, Casenave JL, Jewel W, Doroshov SI, Tjeerdema RS (2007) Induction and partial characterization of California halibut (Paralichthyscalifornicus) vitellogenin. Comp Biochem Physiol A 146:200–207

Pereira TSB, Boscolo CNP, Moreira RG, Batlouni SR (2017) The use of mGnRHa provokes ovulation but not viable embryos in Leporinus macrocephalus. Aquac Int 25:515–529

Pereira TSB, Boscolo CNP, Moreira RG, Batlouni SR (2018) Leporinus elongatus induced spawning using carp pituitary extract or mammalian GnRH analogue combined with dopamine receptor antagonists. Anim Reprod 15(1):64–70

Pousis C, De Giorgi C, Mylonas CC, Bridges CR, Zupa R, Vassallo-Agius R et al (2011) Comparative study of liver vitellogenin gene expression and oocyte yolk accumulation in wild and captive Atlantic bluefin tuna (Thunnus thynnus L.). Anim Reprod Sci 123:98105

Prakash O, Goswomi SV, Sehgal N (2007) Establishment of ELISA for murrel vitellogenin and chriogenin, as biomarkers of potential endocrine disruption. Comp Biochem Physiol C 146:540–551

Prat F, Coward K, Sumpter JP, Tyler CR (1998) Molecular characterization and expression of two ovarian lipoprotein receptors in the rainbow trout, Oncorhynchus mykiss. Biol Reprod 58:1146–1153

Qin C, Xu L, Yang Y, He S, Dai Y, Zhao H, Zhou Z (2014) Comparison of fecundity and offspring immunity in zebrafish fed lactobacillus rhamnosus CICC 6141 and Lactobacillus casei BL23. Reproduction 147:53–64

Reading BJ, Hiramatsu N, Sawaguchi S, Matsubara T, Hara A, Lively MO, Sullivan CV (2009) Conserved and variant molecular and functional features of multiple egg yolk precursor proteins (vitellogenins) in white perch (Morone americana) and other teleosts. Mar Biotechnol 11:169–187

Reading BJ, Hiramatsu N, Sullivan CV (2011) Disparate binding of three types of vitellogenin to multiple forms of vitellogenin receptor in white perch. Biol Reprod 84:392–399

Reading BJ, Hiramatsu N, Schilling J, Molloy KT, Glassbrook N, Mizuta H, Luo W (2014) Lrp13 is a novel vertebrate lipoprotein receptor that binds vitellogenins in teleost fishes. J Lipid Res 55:2287–2295

Reading BJ, Sullivan CV, Schilling J (2017) Vitellogenesis in fishes. In: Reference module in life sciences. Elsevier, Amsterdam

Reis-Henriques MA, Ferreira M, Silva L, Dias A (2000) Evidence for an involvement of vitellogenin in the steroidogenic activity of rainbow trout (Oncorhynchus mykiss) vitellogenic oocytes. Gen Comp Endocrinol 117:260–267

Romagosa E, Batlouni SR, Borella MI, Leonardo AFG (2005) Involuc¸a˜o dos folı'culos po's-ovulato'riosem Pseudoplatystomafasciatum (Pisces, Teleostei). Bol Inst Pesca 31(2):129–135

Romano M, Rosanova P, Anteo C, Limatola E (2004) Vertebrate yolk protein: a review. Mol Reprod Dev 69:109

Ryffel GU (1978) Synthesis of vitellogenin, an attractive model for investigating hormoneinduced gene activation. Mol Cell Endocrinol 12:213–221

Sabo-Attwood T, Kroll KJ, Denslow ND (2004) Differential expression of largemouth bass (Micropterus salmoides) estrogen receptor isotypes alpha, beta, and gamma by estradiol. Mol Cell Endocrinol 218:107–118

Sawaguchi S, Ohkubo N, Amano H, Hiramatsu N, Hara A, Sullivan CV, Matsubara T (2008) Controlled accumulation of multiple vitellogenins into oocytes during vitellogenesis in the barfin flounder, Varasper moseri. Cybium Int J Ichthyol 32:262

Schilling J, Loziuk PL, Muddiman DC, Daniels HV, Reading BJ (2015) Mechanisms of egg yolk formation and implications on early life history of white perch (Morone americana). PLoS One 10:e0143225

Schorer M, Moreira RG, Batlouni SR (2016) Selection of pacu females to hormonal induction: effect of age and of evaluation methods. Bol Inst Pesca 42:901–913

Smolenaars MM, Madsen O, Rodenburg KW, Van der Horst DJ (2007) Molecular diversity and evolution of the large lipid transfer protein superfamily. J Lipid Res 48:489–502

Specker JL, Sullivan CV (1994) Vitellogenesis in fishes: status and perspectives. In: Davey KG, Peter RE, Tobe SS (eds) Perspectives in comparative endocrinology. Ottawa, ON, National Research Council of Canada, pp 304–315

Suarez-Bregua P, Guerreiro PM, Rotllant J (2018) Stress, glucocorticoids and bone: a review from mammals and fish. Front Endocrinol 526:1–8

Sullivan CV, Chapman RW, Reading BJ, Anderson PE (2015) Transcriptomics of mRNA and egg quality in farmed fish: some recent developments and future directions. Gen Comp Endocrinol 221:23–30

Susca V, Corriero A, Bridges CR, De Metrio G (2001) Study of the sexual maturity of female bluefi n tuna: purifi cation and partial characterization of vitellogenin and its use in an enzyme-linked immunosorbent assay. J Fish Biol 58(3):815

Takeuchi T, Watanabe T (1982) Effects of various polyunsaturated fatty acids on growth and fatty acid compositions of rainbow trout Salmo gairdneri, coho salmon Onchorhynchus kisutch, and chum salmon Onchorhynchus Keta. Bull Jpn Soc Sci Fish 48:1745–1752

Tuncer H, Harrell RM (1992) Essential fatty acid nutrition of larval striped bass (Morone saxatilis) and palmetto bass (M. saxatilis x M. chrysops). Aquaculture 101:105–121

US Environmental Protection Agency [USEPA], Office of Chemical Safety and Pollution Prevention (2015) OCSPP 890.2200: Medaka extended one generation reproduction test (MEOGRT), Endocrine Disruptor Screening Program

Utarabhand P, Bunlipatanon P (1996) Plasma vitellogenin of grouper (Ephinephelusmalabaricus): isolation, and properties. Comp Biochem Physiol C 115(2):101–110

Wang Y, Ge W (2003a) Involvement of cyclic adenosine 30,50- monophosphate in the differential regulation of activin betaA and betaB expression by gonadotropin in the zebrafish ovarian follicle cells. Endocrinology 144:491–499

Wang Y, Ge W (2003b) Spatial expression patterns of activin and its signaling system in the zebrafish ovarian follicle: evidence for paracrine action of activin on the oocytes. Biol Reprod 69:1998–2006

Watanabe T, Lee M, Mizutani J, Yamada T, Satoh S, Takeuchi T, Yoshida N, Kitada T, Arakawa T (1991) Effective components in cuttlefish meal and raw krill for improvement of quality of red sea bream (Pagrus major) eggs. Nippon Suisan Gakkaishi 57:681–694

Wendelaar Bonga SE (2011) Hormonal responses to stress: hormone response to stress. In: Encyclopedia of fish physiology. Academic Press, San Diego, CA, pp 1515–1523

Williams VN, Reading BJ, Amano H, Hiramatsu N, Schilling J, Salger SA, Williams TI, Gross K, Sullivan CV (2014) Proportional accumulation of yolk proteins derived from multiple vitellogenins is precisely regulated during vitellogenesis in striped bass (Morone saxatilis). J Exp Zool A Ecol Genet Physiol 321:301–315

Yilmaz O, Patinote A, Nguyen TV, Com E, Lavigne R, Pineau C, Sullivan CV, Bobe J (2017) Scrambled eggs: proteomic portraits and novel biomarkers of egg quality in zebrafish (Danio rerio). PLoS One 12:e0188084

Yilmaz O, Patinote A, Nguyen T, Bobe J (2018) Multiple vitellogenins in zebrafish (Danio rerio): quantitative inventory of genes, transcripts and proteins, and relation to egg quality. Fish Physiol Biochem 44:1–17

Zhang Y, Qu Q, Sun D, Liu X, Suo L, Zhang Y (2011) Vitellogenin in Amur sturgeon (Acipenser schrenckii): induction, purification and changes during the reproductive cycle. J Appl Ichthyol 27:660–665

Future Prospective of Vitellogenin Research

19

M. Chellapackialakshmi and C. Ravi

Abstract

Vitellogenin (Vtg) is an egg yolk precursor protein that harbors the nutrients used for embryo development. Vtg is present in the plasma of female oviparous vertebrates that carry out vitellogenesis through endogenous estrogen regulation mechanism. Types of Vtg and its various functional aspects have been explained. Endocrine-disrupting chemicals (EDC) are a group of xenobiotics that adversely impact the environment and disrupts the normal secretion of hormones. Exposure to EDCs continuously would affect reproduction and population of fishes. The previous findings regarding Vtg are summarized such as new Vtg model generation, binding affinity of Vtg receptor, biomarker studies, and advance technologies for egg development. Based on these implications of various studies, future research prospectives on vitellogenin have been discussed.

Keywords
Vitellogenin · Multiplicity of Vtg · Biomarker · Vtg model

Abbreviations

CRISPR	Clustered regularly interspaced short palindromic repeats
DDT	Dichloro diphenyl trichloroethane
EDC	Endocrine-disrupting chemicals
FSH	Follicle-stimulating hormone
LH	Luteinizing hormone
LLTP	Large lipid transfer protein

M. Chellapackialakshmi · C. Ravi (✉)
Department of Zoology, Thiagarajar College, Madurai, Tamil Nadu, India

V. Baskaralingam, R. Vanichviriyakit (eds.), *Vitellogenin in Fishes-Diversification, Biological Properties, and Future Perspectives*,
https://doi.org/10.1007/978-981-99-5340-0_19

LvH	Lipovitellin heavy chains
LvL	Lipovitellin light chains
PGC	Primordial germ cells
PRR	Pattern recognition receptor
RNA	Ribonucleicacids
TTX	Tetraodotoxin
VLDL	Very low-density lipoprotein
Vtg	Vitellogenin
Yp	Yolk protein

19.1 Introduction

Several fishes are oviparous, that is, their eggs are fertilized externally and develop into healthy embryo. Oocyte development occurs in a series of stages, with the production of primordial germ cells (PGCs), their transition into oogonia, and finally their development into oocytes. Vitellogenesis process involves maternal information and required molecules for early embryo development, such as RNAs, proteins, lipids, vitamins, and hormones are deposited in the emerging oocytes. Vitellogenin (Vtg) is a element of the large lipid transfer protein (LLTP) superfamily and is one of the most significant proteins deposited in oocytes (Sun and Zhang 2015).

In evolutionary aspects, Vtg is homologous among a large variety of animals from insects to chickens (Lucey 2014). Vtg is the precursor of egg yolk proteins predominantly seen in the females of all oviparous species such as fish, amphibians, reptiles, birds, major invertebrates and platypus. It consists of sugar, lipid, phosphorus, and other elements such as calcium, iron, and zinc (Hara et al. 2016).

19.2 Chronicles of Vitellogenin

In 1900, Chicken oogenesis has been studied, in that oocyte growth was predicted by utilizing amino acids, which were thought to be responsible for formation of egg yolk proteins. These proteins are initially developed in the maternal liver and then transported by blood to embryonic oocytes. Based on the biochemical analysis on the African clawed frog *Xenopus laevis*, Vtg was determined to be the precursor of egg yolk proteins (Bergink and Wallace 1974). Total number of vitellogenin (Vtg) genes present in various vertebrates has been depicted in Table 19.1.

19.3 Vitellogenin in Fishes

The vitellogenesis process starts with the release of glycoprotein gonadotropins from the pituitary gland. There are two types of gonadotropins such as follicle-stimulating hormone (FSH) and luteinizing hormone (LH). The two gonadotropins bind to membrane receptors in the ovarian follicle and activate steroid synthesizing

Table 19.1 Vitellogenin gene (Vtg) in vertebrates (Sun and Zhang 2015)

S. No	Common name	Scientific name	Number of Vtg genes
1.	Chicken	*Gallus gallus*	3
2.	Africa frog	*Xenopus laevis*	4
3.	Nematode	*Caenorhabditis elegans*	6
4.	Zebra fish	*Danio rerio*	7
5.	Carp	*Cyprinus carpio*	2
6.	Medaka	*Oryzias latipes*	4
7.	Striped bass	*Morone saxatilis*	3
8.	White perch	*Morone Americana*	3
9.	Teleost	–	Multiple

enzymes which produce 17β-estradiol, the primary hormone responsible for vitellogenesis (Lucey 2014). Accumulation of estradiol stimulates the liver to produce Vtg and then transported through the bloodstream to the ovary. It enters into the oocyte through specific receptor-mediated endocytosis (Cheek et al. 2001).

Vtg plays a vital role in the growth of oocytes. During the artificial reproduction, Vtg is synthesized in the liver and transported to egg cells and transformed into yolk protein (Yp), which is an essential nutrient for the development of embryo. Yp contains phosvitin, lipovitellin, and beta-component which accumulate in yolk globules (Hara et al. 2016; Reading et al. 2018).

Vtg is present in the plasma of female fish which carry out vitellogenesis process by hormonal regulation mechanism by estradiol which is the foremost endogenous estrogen in all vertebrates. Though male fish also has Vtg genes, they cannot synthesize automatically because of lack of estrogen. Conversely, Vtg may be synthesized by stimulation of specific, sensitive exogenous estrogen, and it has high expression level in many organisms (Guo et al. 2019). During vitellogenesis process, it reacts with antibodies that rose against egg extracts and particularly expressed in female blood serum. When estrogen is injected to fish, it induces the production of Vtg (Hara et al. 2016).

Exposure of chemical mixtures alters fish vitellogenesis in vivo, including the alkylphenols, nonylphenol, and octylphenol, the steroidal estrogenethinylestradiol, the pesticides methoxychlor, and *o*, *p*'-DDT and their metabolites and phytoestrogens (Cheek et al. 2001).

19.4 Types of Vitellogenin

Fish Vtgs can be categorized as complete and incomplete Vtg. LvH, Pv, LvL, β′-c, C-terminal coding domain are the five egg yolk protein regions, which is further subdivided into type A (VtgA) and type B (VtgB) that constitute complete Vtg. Lipovitellin is a dimer consisting of a heavy (LvH) and a light (LvL) chains, and it is rich in amino acids and lipids essential for embryonic development. Phosvitin has high phosphorus content and serine residues, which in turns bind calcium useful for osteogenesis. The incomplete Vtg mainly consists of LvH, LvL, and VtgC; moreover, it is a major lipoprotein in fish eggs (Hara et al. 2016).

19.5 Functions of Vtg

Vtg plays a major role in antibacterial activity and enhances phagocytosis of microbes (Liu et al. 2022). It is a multivalent pattern recognition receptor (PRR) capable to bind conserved components of bacteria and virus. It may act either as effector destabilizing cell walls or as a bridging molecule to improve phagocytosis via opsonization (Zhang et al. 2011; Carducci et al. 2019). Apart from immune functions, Vtg and yolk proteins exhibited antioxidant activity, essential for protection against oxidative damage (Sun and Zhang 2015). Vitellogenin subdomain is a binding protein which transfers tetraodotoxin (TTX) from liver to ovary in *Takifugu pardalis*. Accumulation of toxins in eggs used as a repellent against predators and as pheromone to attract males (Yin et al. 2017).

19.6 Vtg as Biomarker

A biomarker is a tool used to measure the level of exposure and the hazardous consequences of one or more chemical contaminants that have an impact on biochemical, cellular, physiological, and behavioral characteristics of organisms. Due to the link between endocrine alteration and biological research, several biomarkers can forecast the effects of reproductive or other endocrine disruption (Carducci et al. 2019).

19.7 Endocrine-Disrupting Chemicals (EDC)

Endocrine disruption is the perturbation of endogenous hormone function by chemicals. It can alter both the organizational and activational effects of reproductive hormones, probably having an intense effect on an organism's ability to reproduce. EDCs are a group of xenobiotics adversely impact the environment and disrupts the normal secretion of hormones in bodies when entering into animal. It can hinder with later phase of the ovarian cycle such as oocyte maturation and spawning. In this manner, physiological disorders occur, impact the reproductive, nervous, and endocrine systems of animals and humans or even causing carcinogenicity (Hirmatsu et al. 2017; Guo et al. 2019), and due to anthropogenic activity cause severe effects on the aquatic environment. Particularly, xenoestrogens is an endocrine-disrupting chemical that mimic natural estrogens. The major sources of these xenoestrogens are sewage effluents and agricultural and livestock wastes which undergo bioaccumulation and biomagnification processes (Murphy et al. 2004).

In general, Vtg synthesis can be induced by exposure to estrogens and endocrine-disrupting chemicals (EDCs) often found in polluted environments. Some chemical

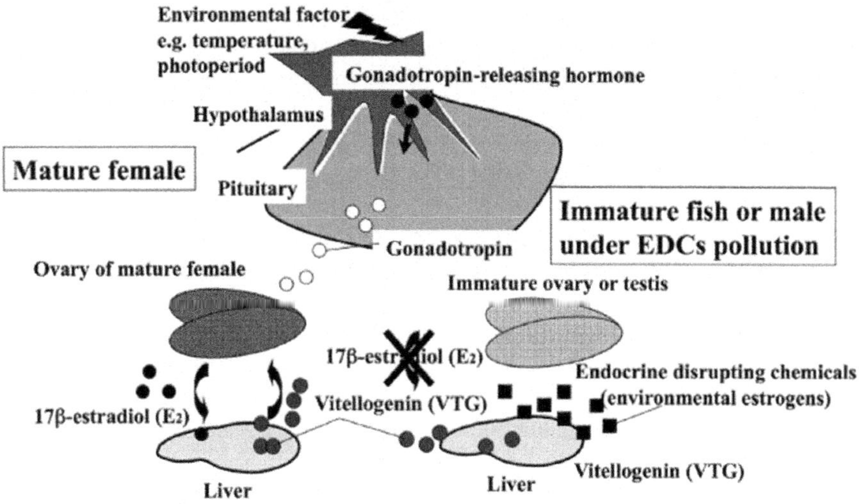

Fig. 19.1 Illustration of mechanism of vitellogenesis and environmental estrogens in fish (Soyano et al. 2010). Environmental factors or Endocrine disrupting chemicals affects gonadotropin relasing hormone to synthesis 17β estradiol as a result immature ovary or testis is produced. The routine mechanism of formation of mature ovary or testis is disturbed by EDC

compounds shows estrogen-like activity is associated with anthropgenic activities and is mostly present in aquatic environments (Hara et al. 2016). Fish is one of the accepted models for analyzing estrogenic endocrine disruption because endocrine disruptors include environmental estrogens, antiestrogens, androgens, antiandrogens, antiprogestins, and retinoid mimics may hinders hormones involved in fish reproduction. During the sexual differentiation phase of this vertebrate, these hormones may disrupt sexual differentiation or its sex may be even changed. Exposure to EDCs continuously would affect reproduction and population of fishes (Cheek et al. 2001; Guo et al. 2019). Mechanism of vitellogenin (Vtg) related to endocrine-disrupting chemicals (EDC) has been depicted in Fig. 19.1.

19.8 Implications of the Study

19.8.1 Immune Role of Vtg

Vtgs also play immune-relevant roles. Vtg showed hemagglutinating activity against the erythrocytes of chicken, toad, and grass carp. Vtg isolated from the ovaries of protochordate amphioxus (*Branchiostomajaponicum*) exerted antibacterial activity against the Gram-negative *E. coli* (Zhang et al. 2005). Growth of Gram-positive bacteria such as *Staphylococcus aureus*, *Bacillus subtilis,* and *Streptococcus pyogenes* and Gram negative bacteria including *E. coli*, *E. aerogenes,* and *Pseudomonas putida* have been inhibited by Vtg of pink barb *Puntius conchonius* (Shi et al. 2006). Both Gram-positive and Gram-negative bacteria were susceptible to Vtg of scallops

(*Patinopecten yessoensis*) and exerted resistance to pathogens. When *Caenorhabditis elegans* was infected with pathogens, the survival rate decreased because of Vtg-knockdown (Fischer et al. 2013).

19.9 Impact of EDC on Reproduction

The exposure of fishes to EDCs is an extensive phenomenon in aquatic environments around globe based on the field survey. EDCs may have a direct effect on fertilization or survival rates if they alter the ratios of various Vtgs present in the blood or deposited in the egg. Of late, only maleor immature fish have been targeted in Vtg-based EDC studies (Hara et al. 2016). For example, Kepone and *o*, *p*-DDD bind to the maturation-inducing steroid (4-pregnen-17,20,21-triol-3-one), receptor sites, and inhibit final oocyte maturation (Murphy et al. 2004).

19.10 Multiplicity of Vtg

Multiplicity of Vtg constitutes the major source of egg yolk nutrients sustaining development of oviparous animals. Preliminarily, two types of Vtg in Salmonidae were analyzed for its structure, domains, motifs, and expression profile during ovarian development (Hara et al. 2016). Spiny-rayed teleosts (*Acanthomorpha*sp) consists of multiple Vtg such as two paralogous complete forms of VtgA (VtgAa and VtgAb) as well as an incomplete form of Vtg and VtgC (Yilmaz et al. 2021).

19.11 Knock Out of Multiple Vtg

Vitellogenins are essential for the action at different stages during reproduction and embryonic development. The selective knockout of multiple Vtg forms carried out various experiments in zebrafish. From the research findings revealed that Vtg is not only responsible for embryo development but also has novel regulatory effects on fecundity and fertility. Using a multiple CRISPR/Cas9 genome editing technology showed increased fecundity (Yilmaz et al. 2021; Carducci et al. 2019).

19.12 Binding Affinity of Vtg and Its Receptor

In the ovary, Vtg receptor protein is primarily found near the oocyte membrane. In white perch, a novel lipoprotein receptor (low-density lipoprotein receptor-related protein 13, or Lrp13) may be involved in Vtg binding in addition to the Lr8-type Vtg receptor (Hara et al. 2016).

In acanthomorph fish spawning pelagic eggs, the heavy chain of VtgAa lipovitellin is highly degraded during oocyte maturation and producing a pool of free amino acids that generates an osmotic gradient. Controversially, the LvH derived from

VtgAb is involved in growth and maturation of oocyte and is used in late larval stages (Reading and Sullivan 2011; Carducci et al. 2019).

19.13 Role of Vtg as Biomarker

During the developmental stages of fishes, it is exposed to environmental estrogens as a result it will have an organizational or permanent effect on gonad morphology and reproductive function. However, it will have an activational or transitory effect on Vtg production (i.e., exposed animals will produce Vtg only while the stimulus is present). In this regard, Vtg acts as a tremendous biomarker of existing estrogenic exposure, but it may not specify organizational effects such as altered sex differentiation and impaired reproductive function (Cheek et al. 2001). The harmful effects of environmental estrogens have lead to vitellogenin to serve as a biomarker in assessing the EDC effects in teleosts (Carducci et al. 2019). In mature females, Vtg is a highly specific biomarker for estrogen exposure in fish.

19.14 Advance Technologies Used to Improve Egg Quality

Fish oogenesis primarily involves the production of the yolk globules, egg envelope, and oil globules and requires very specialized regulatory systems that are important for seed integrity and egg quality. In this process, primary oocyte grows by several orders of magnitude while synthesizing essential needs for fertilization and support complete development of a new life (Hara et al. 2016).

19.15 Oil Globules in Embryo Development

Numerous fish eggs contains neutral lipids composed of triglycerides and phospholipids produced from Vtg are together referred as "oil globules". These globules originate from a group of serum proteins. Accumulation of oil globule plays a significant role in the development of embryos and juvenile fish.

There are two models put forth:

1. A mechanism whereby free fatty acids from VLDL are liberated by blood lipase and then absorbed into the ovum to form an oil globule.
2. A process where VLDL enters the body directly through a receptor and causes the ovum to produce oil globules and release fatty acids.

The protein was released in the granulosa cell layer of the ovaries of cutthroat trout. Its expression reached peak level during the oil droplet stage and declined during vitellogenesis process (Hara et al. 2016).

19.16 Future Prospective

19.16.1 Vitellogenesis Process

In fishes, Vtgs undergo a second proteolysis that can vary on the basis of producing pelagic or demersal eggs and based on embryonic development. In salmonids, this second proteolysis has not been proved, due to the spawning of their eggs in freshwater (Hirmatsu et al. 2017). The subsequent proteolysis occurs during embryogenesis, but there is no adequate information regarding this.

19.16.2 EDC-Based Studies

Several Vtgs may form in the blood of mature female fish, and those that aggregate in eggs are predicted to serve as novel biomarkers of endocrine dysfunction brought by EDCs (Hara et al. 2016). In fish, there are only a few studies on the developmental exposure to environmental estrogens on subsequent reproductive success and studies not have yet focused on to understand the link between vitellogenin induction, sex differentiation of exposed fish, and reproductive success (Cheek et al. 2001).

Vtg and yolk proteins are also used in the detection of EDC contamination which develops new Vtg-based bioassays, which will be helpful to detect environmental pollution (Wang et al. 2017; Carducci et al. 2019). There is a knowledge gap in the assessment of the risks of EDC exposure to fishes, wildlife, and human, which requires critical acquaintance of reproductive and developmental physiology (Murphy et al. 2004).

19.16.3 Generation of Multiple Vtg Model

The primary use of Vtg protein in fishes has been to detect the gender of farmed fish or to measure the estrogenic activity of fish. Further, research work concentrated on "Multiple Vtg model," which may play a vital role to generate data regarding polymorphous Vtg from diverse fish species (Hara et al. 2016). Phylogenetically diverse species with various modes of reproduction should be included in Vtg multiplicity, which may improve the classification hierarchy of different types of Vtg molecules considering their primary structure and physiological functions. Since Vtg multiplicity in fish appears to be normal and different types of Vtg respond differently response kinetics and peak production levels, the magnitude of Vtg multiplicity in the target species. When beginning the development of Vtg assays for determining fish reproductive status or exposure to estrogenic EDCs, it is important to take into account the degree of Vtg multiplicity in a target species (Hirmatsu et al. 2017). Research is needed in these aspects to determine the specific role and molecular mechanisms of each Vtg in zebrafish (Yilmaz et al. 2021).

19.16.4 Binding Affinity to Vtg Receptors

There is a lacunae in this field regarding interaction of several Vtg subtypes with the Vtg receptor. Many unidentified lipoprotein receptors exist in fish, and there is a scope to discover the affinity of these receptors for Vtg. It has not been thoroughly investigated how receptor proteins interact with their Vtg ligands (Hara et al. 2016), and there is ample scope for exploration in this area.

19.16.5 Vtg Act as Biomarker

Vtg can also used as a biomarker to detect estrogenic effect at an early phase (Guo et al. 2019). Production of vitellogeninin male and juvenile fish is due to the exposure to these chemicals. This brings out Vtg a useful biomarker (Lucey 2014). In future, investigations of the biological importance of multiple vitellogenins are required, and it will provide the next generation of biomarkers, which could clarify potential mechanisms of these chemicals impair fish reproductive function (Hirmatsu et al. 2017).

19.17 Advance Technologies for Egg Quality Improvement

Inorder to protect the reproductive processes of wild fish and to enhance the quality of eggs and seeds in farmed fish, advanced technologies must be developed. Each oil globule molecule expresses in a different position within the follicles, and the dynamics of that expression are still unclear during oogenesis (Hara et al. 2016).

19.18 Coupling Vtg Model to Bioenergetic Model

Monte Carlo methods involve repetitive model simulations with random input values produced from probability distributions. Model predictions are outcome of probability distributions and correlation analysis that isused to identify the inputs which contributes for prediction variability.

Bioenergetics models are generated based on dynamic energy budgets, and it is used to explain the rates at which individuals allocate energy for maintenance, reproduction, growth, and development. The impact of endocrine disruption on energy allocation and the subsequent ecological impact on reproduction and growth might be simulated by coupling the vitellogenesis model to a bioenergetics model. This model is focused toward a computational biology framework to understand endocrine disruption in fish and for easy to link reproductive endocrine biomarkers of exposure to reproductive endpoints with ecological importance (Murphy et al. 2004).

19.19 Conclusion

Vtg is a glycoprotein essential for oocyte development during oogenesis. Continous exposure of EDC affects fish reproduction and reduces population. In order to retain Vtg and its function, synthesizing new Vtg biomarkers or developing multiplicity of Vtg is recommended in laboratory aspects. Hence, these techniques will be concerned in future to improves egg quality of fish and reduce mortality rate.

Acknowledgments The authors are thankful to the authorities of Thiagarajar College, Madurai, Tamil Nadu for the institutional facilities.

Conflict of Interest The authors have no conflict of interest to declare.

References

Bergink EW, Wallace RW (1974) Precursor product relationship between amphibian vitellogenin and yolk proteins, lipovitellin and phosvitin. J Biol Chem 249:2897–2903; https://pubmed.ncbi.nlm.nih.gov/4828327

Carducci F, Biscotti MA, Canapa A (2019) Vitellogenin gene family in vertebrates: evolution and functions. Eur Zool J 86(1):233–240

Cheek AO, Brouwer TH, Carroll S, Manning S, McLachlan JA, Brouwer M (2001) Experimental evaluation of vitellogenin as a predictive biomarker for reproductive disruption. Environ Health Perspect 109(7):681–690

Fischer M, Regitz C, Kull R, Boll M, Wenzel U (2013) Vitellogenins increase stress resistance of *Caenorhabditis elegans* after *Photorhabduslumінescens* infection depending on the steroid signaling pathway. Microbes Infect 15:569–578

Guo D, Qiu J, Li Y, Yang G, Qian Y (2019) Application of molecular biological biomarkers to endocrine disruption studies. J Sci Res 22(5):17037–17041

Hara A, Hirmatsu N, Fujitha T (2016) Vitellogenesis and choriogenesis in fishes. Fish Sci 82:187–202

Hirmatsu N, Matsubara T, Weber GM, Sullivan CV, Hara A (2017) Vitellogenesis in aquatic animals, pp 694–697

Liu R, Khang L, Wang G, Jiang Z, Ba X, Liu L (2022) Effect of simming on the induction of vitellogenin in conger eel (*Conger myriaster*). Front Mar Sci 9:887074

Lucey SM (2014) Characteristics of fish yolk proteins and a method for inducing vitellogenin

Murphy CA, Rose KA, Peter T (2004) Modelling vitellogenesis in female fish exposed to environmental stressors: predicting the effects of endocrine disturbance due to exposure to PCB mixture and cadmium. Reprod Toxicol 19:395–409

Reading BJ, Sullivan CV (2011) The reproductive organs and processes-vitellogenesis in fishes. Fish Physiol 1:635–646

Reading BJ, Andersen LK, Ryu YW, Mushirobira Y, Todo T, Hirmatsu N (2018) Oogenesis and egg quality in fin fish: yolk formation and other factors influencing female fertility. Fishes 3(45):1–28

Shi X, Zhang S, Pang Q (2006) Vitellogenin is a novel player in defense reactions. Fish Shellfish Immunol 20:769–772

Soyano K, Aoki J, Itashiki Y, Park CB, Nagae M, Takao Y, Lee YD, Yeo I, Zhong J (2010) Contaminations by endocrine disrupting chemicals in coastal waters of the East China Sea coastal environmental and ecosystem issues of the East China Sea, pp 215–226

Sun C, Zhang S (2015) Immune relevant and antioxidant activities of viteelogenin and yolk proteins in fish. Nutrients 7:8818–8829

Wang J, Ma S, Zhang Z, Zheng M, Dong Y, Ru S (2017) Vitellogenin in caudal fin of guppy (*Poecilia reticulata*) as a less invasive and sensitive biomarker for environmental estrogens. Sci Rep 7(7647):1–12

Yilmaz O, Patinote A, Com E, Pineau BJ (2021) Knock out of specific maternal vitellogenins in zebra fish (*Daniorerio*) evokes vital changes in egg proteomic profiles that resemble the phenotype of poor quality eggs. BMC Genomics 22(308):1–22

Yin X, Kiriake A, Ohta A, Kitani Y, Ishizaki S, Nagashima Y (2017) A novel function of vitellogenin subdomain, vWF type D, as a toxin-binding protein in the pufferfish *Takifugu pardalis* ovary. Toxicon 136:56–66

Zhang S, Sun Y, Pang Q, Shi X (2005) Hemagglutinating and antibacterial activities of vitellogenin. Fish Shellfish Immunol 19:93–95

Zhang S, Wang S, Li H, Li L (2011) Vitellogenin, a multivalent sensor and an antimicrobial effector. Int J Biochem Cell Biol 43(3):303–305

SPRINGER NATURE

GPSR Compliance

The European Union's (EU) General Product Safety Regulation (GPSR) is a set of rules that requires consumer products to be safe and our obligations to ensure this.

If you have any concerns about our products, you can contact us on ProductSafety@springernature.com

In case Publisher is established outside the EU, the EU authorized representative is:

Springer Nature Customer Service Center GmbH
Europaplatz 3
69115 Heidelberg, Germany

The manufacturer's authorised representative in the EU is Springer Nature Customer Service Centre GmbH, Europaplatz 3, 69115 Heidelberg, Germany. If you have any concerns regarding our products, please contact ProductSafety@springernature.com

Printed and bound by CPI Group (UK) Ltd, Croydon, CR0 4YY
01/10/2026
02225867-0009